In Extremis

Jürgen P. Kropp · Hans Joachim Schellnhuber
Editors

In Extremis

Disruptive Events and Trends in Climate and Hydrology

Springer

Editors
Dr. Jürgen P. Kropp
Potsdam Institut für
Klimafolgenforschung (PIK)
Telegraphenberg A31
14412 Potsdam
Germany
kropp@pik-potsdam.de

Prof. Dr. Hans Joachim Schellnhuber
Potsdam Institut für
Klimafolgenforschung (PIK)
Telegrafenberg 31 A
14473 Potsdam
Germany
john@pik-potsdam.de

ISBN 978-3-642-42260-7 ISBN 978-3-642-14863-7 (eBook)
DOI 10.1007/978-3-642-14863-7
Springer Heidelberg Dordrecht London New York

Cover design: eStudio Calamar S.L., Heidelberg

Printed on acid-free paper

Springer is part of Springer Science+Business Media (www.springer.com)

Preface

Each day on Earth is pervaded by natural disasters or extreme events which can threaten any location on the globe but which take place inhomogeneously in space and time. The scale of economic damage and loss of life has increased during recent decades, and yet adequate risk management methods are not yet in place. Due to the interdisciplinary nature of these problems satisfactorily answers from different fields of expertise are needed. Consequently, this book represents a summarizing overview on original work performed at the interface between science and practice. It provides a more unifying overview than would be possible for certain disciplines alone. Cooperation and mutual learning are key elements in the still emerging field of transdisciplinarity.

In this book, extreme events, trends, and correlations are the major fields of interest. Why is this so important? Humankind increasingly tends to initiate/unfold activities in risk-prone areas and has itself become a key player in environmental change. Both things in combination affect the quality of life and the viability of social as well as natural systems. To prevent and manage related risks novel approaches that deepen the understanding of extreme situations are required. Therefore, this book focuses on catastrophes, which are understood as extreme events. Incidents of this kind occur irregularly with differing severity and frequency. Typical examples include river and sea floods, cyclones, droughts, heat or cold spells, and related events such as storms, tornados, or landslides. Due to the tremendous damage such events can cause there is a particular interest in minimizing the risks. This calls for identification and application of appropriate measures based on accurate description and deeper understanding of the respective processes. One approach in this context focuses on the use and features of empirical data and in particular extreme values statistics, which has a long tradition in disciplines like hydrology or climate sciences.

From a mathematical point of view, extreme value theory has led to some remarkable achievements, in particular the generalized extreme value distribution. However, the application of most methods is restricted since they require identically and independently distributed data. Thus, normally, stationarity is needed, i.e., data have to have the same average and variance at any time, in order to apply standard extreme value statistics. These criteria are generally not fulfilled when we focus on real-life problems. There are many sources of non-stationarities, for example, global warming, or other slow changes of environmental conditions, like surface sealing, which imply modifications of river run-off in a catchment. The second limitation for many methods regards the absence of temporal correlations, i.e., mutual independence of the values. Regrettably, persistence is abundant in climate and hydrology. Originally described for hydrological records, a notable type of memory comprising an infinite correlation length - long-term persistance - has recently also been analysed in many other types of data, most importantly in temperature records, long-term persistence, has recently also been analysed in many other data, paramountly in temperature records. Against the background of climate change, the assessment of long-term phenomena has gained further importance. Typical questions of interest address whether the recent warming is compatible with natural fluctuations or whether it is a systemic change caused by anthropogenic influences – or a mixture of both. Therefore, appropriate techniques are needed which can distinguish between persistence and trends. Another common issue is how these two features affect the statistics of extreme events such as in the case for river runoff records, which not rarely comprise both long-term correlations and non-stationarities. Similar problems have been tackled and most of them have been advanced to a satisfying extent. Finally, empirical data series often have limited length and contain gaps.

Accordingly, there is a need to employ more sophisticated approaches and methods that can overcome these two statistical limitations. In order to evaluate extent to which these methods are appropriate in addressing the problems described above one must assess the following questions:

- Are these techniques suitable to open new ways in risk assessment?
- Which are adequate modeling strategies to cope with complexity and uncertainty?
- Are new concepts able to supply knowledge in practical applications?

These questions were one major motivation for the preparation of this book which hopefully can function at least as a precursor for the remaining work. This book represents the termination of the project "Scaling Analysis of Hydrometeorological Times Series," which was generously supported by the German Federal Ministry of Education and Research. The aim was to fructify novel statistical methods, which partly already have successfully been employed in modern physics, and in particular for relevant questions and problems of water management. The main goal was to extract an optimum of information relevant for decision making from extensive empirical databases available through the water management administration. The separation of random fluctuations and deterministic temporal structures has extensive consequences for the appraisement of the usual hydrological measures, in particular for those which describe the relevant long-term phenomena such as extreme discharges. In particular, this applies to problems that are related to climate change. After this short introduction, let us now briefly discuss how this book is organized. Structured in three parts, the book presents a broad spectrum of interdisciplinarity and describes advances of methods under different foci.

Extremes, Uncertainty, Reconstruction

In the first part the main issues are extremes, the involved uncertainties, and the reconstruction of extreme events from the theoretical as well as applied point of view. The lack of knowledge about quantities and the difficulty in qualifying epistemic uncertainty is common to many aspects of global environmental change. In situations of deep uncertainty it is often impossible to fulfill the rather demanding informational requirements for assessing probabilities. The outcome might be controversial policy advice if it is based on incomplete or poor data. Consequently, J.F. Eichner et al. investigate in Chap. 1 the statistics of return intervals as well of maxima under the influence of long-term correlations. In the second and third contributions in this part, bootstrap techniques are elaborated in order to achieve additional data, thus allowing assessment of uncertainties. In Chap. 2, for example, M. Mudelsee presents a non-parametric kernel estimation method and bootstrap is used to assess occurrence rates and associated confidence bands. H.W. Rust et al. elaborate in Chap. 3 bootstrap strategies for the generalized extreme value distribution in order to estimate confidence intervals assuming long-range dependence. Different approaches are thus combined, i.e., ensemble techniques, stochastic model techniques and the iterative amplitude-adjusted Fourier transform method in order to achieve sufficient data for confidence estimations. In Chap. 4, P. Dostal and his co-workers follow a different approach. They use regression analysis, kriging methods, and a run-off model to reconstruct historical events. A particular situation for developing countries is introduced by R. Kumar and C. Chatterjee (Chap. 5) who apply L-moments to estimate regional flood frequency analyses for gauged and ungauged basins.

Extremes, Trends, and Changes

The second part of the book is more application-oriented, because it provides case studies and discusses certain problems and fallacies in extreme value statistics, in particular with regard to trends. In Chap. 6, Z.W. Kundzewicz provides an overview on how intense precipitation, intense river flow, and climate change are interrelated and which methodological efforts are implied. This is extended by M. Radziejewski (Chap. 7) who analyses and discusses trend detection in river floods. For this purpose he applied certain trend tests in order to estimate the significance of trends. Finally, M. Kallache et al. (Chap. 8) apply an assessment framework considering auto-correlations in data and non-stationarity in extremes. The subsequent Chap. 9, by S. Trömel and C.-D. Schönwiese, assesses extreme values and trend analysis based on statistical modeling for precipitation time series. For this aim they apply a generalized time series decomposition technique. The last chapter (Chap. 10) of this part, by D. Rybski and J. Neumann, reviews the well-known Pettitt test. They show that this test - which is often used to assess break points - is sensitive to correlations.

Long-Term Phenomena and Nonlinear Properties

The last part of the book is devoted to long-term phenomena and nonlinear properties in data series. In Chap. 11 D. Rybski and his co-workers analyse auto-correlation and temporal multifractal behavior of daily river discharge time series. They found that daily run-off series can be characterized via a power-law decay of the autocorrelation function and established a cross-over time that usually is several weeks. The following chapter, Chap. 12, tries to extract long-term structures from data via linear and nonlinear dimensionality reduction techniques. Similar techniques as in the first chapter of this part were applied by V.N. Livina et al. (Chap. 13), whereas seasonality effects on nonlinear properties are emphasized. The outcome highlighted that their approach allows one to filter out more properties from data than by previously used techniques. Finally R. Donner (Chap. 14) focuses on spatial correlations in a river catchment.

Although each chapter tackles a different main focus, they share a common nexus, namely *extremes, trends, and correlations in hydrology and climate*. The specific problems are approached from various directions and therefore provide different emphases, giving either theoretical or practical, methodological or applied, and temporal or spatial perspective.

We wish to thank first and foremost the authors for their cooperation, and many of our colleagues in particular Carsten Walther and Diego Rybski. We also wish to thank Dr. Claus Ascheron and Adelheid Duhm from Springer-Verlag for their continuous help during the preparation of this book. The methodological approaches described in this book can be useful in the design and management of environmental systems, for policy development, environmental risk reduction, and prevention strategies. We hope that this book can also be used as a textbook for graduate students, for teachers at universities preparing courses or seminars, and for researchers in a variety of disciplines who are about to encounter the phenomena described here in their own work.

Potsdam, Germany — Jürgen P. Kropp
August 2010 — Hans Joachim Schellnhuber

Contents

Acronyms

ACF	autocorrelation function
AP1824	Area Precipitation 1824
ASA	autocorrelation spectrum analysis
CDF	cumulative distribution function
CMDS	classical multidimensional scaling
CWT	continuous wavelet transformation
d	wind direction
DEM	digital elevation model
DFA	detrended fluctuation analysis
DJF	December, January, February
DWD	Deutscher Wetterdienst (German Weather Service)
ECHAM	European Centre Hamburg Model
ENSO	El Niño/Southern Oscillation
EOF	empirical orthogonal function
EXP	exponential distribution
FA	(standard) fluctuation analysis
FAR	fractionally integrated auto-regressive process
FARIMA	fractionally integrated auto-regressive moving average process
FFM	Fourier filtering method
FFT	fast Fourier transform
FMA	February, March, April
GEV	generalized extreme value distribution
GLO	generalized logistic distribution
GNO	generalized normal distribution
GPD	generalized Pareto distribution
GRDC	Global Runoff Data Centre
HadRM	Hadley Centre regional model
i.i.d.	independent and identically distributed
ISOMAP	isometric feature mapping
JJA	June, July, August
LARSIM	large area runoff simulation model

LRD	long-range dependence
m a.s.l.	meter above sea level
MC	Monte Carlo simulations
MDS	multidimensional scaling
MICE	modelling impact of climate change
ML	maximum likelihood
MLE	maximum likelihood estimation
MRL	mean residual life plot
M-SSA	multichannel singular system analysis
$\mathcal{N}(\mu, \sigma^2)$	Gaussian/normal distribution with mean μ and variance σ^2
N	precipitation
NAO	North Atlantic oscillation
NLDR	nonlinear dimensionality reduction
OND	October, November, December
PC	principal component
PCA	principal component analysis
PDF	probability density function
PE3	Pearson type III distribution
POT	peak over threshold
PP	point process
PUB	prediction of ungauged basins
PWM	probability-weighted moments
R	environment for statistical computing `www.r-project.org`
RC	reconstructed component
ROI	region of influence
SLP	sea level pressure
SRD	short-range dependence
SSA	singular system analysis
t	air temperature
u	wind speed
UNF	uniform distribution
$W\mathcal{N}(\mu, \sigma^2)$	Gaussian/normal white noise
WAK	Wakeby distribution
WTMM	wavelet transform modulus maxima

Contributors

Yosef Ashkenazy The Jacob Blaustein Institute for Desert Research, Ben-Gurion University, 84990 Midreshet Ben-Gurion, Israel, ashkena@bgu.ac.il

Armin Bunde Institut für Theoretische Physik Justus-Liebig-Universität Giessen, 35392 Giessen, Germany, bunde@physik.uni-giessen.de

Katrin Bürger Dostal UmweltConsulting, 79102 Freiburg, Germany, kabu100@gmx.de

Chandranath Chatterjee Agricultural and Food Engineering Department, Indian Institute of Technology, Kharagpur-721302, West Bengal, India, cchatnih@yahoo.com

Reik Donner Potsdam Institut für Klimafolgenforschung, 14412 Potsdam, Germany, reik.donner@pik-potsdam.de

Paul Dostal Environmental Modelling Group University Mainz, 55099 Mainz, Germany, dostal@uni-mainz.de

Jan F. Eichner Justus-Liebig-Universität Giessen, 35392 Giessen, Germany jan.f.eichner@physik.uni-giessen.de

Shlomo Havlin Minerva Center and Department of Physics, Bar-Ilan University, 52900 Ramat-Gan, Israel havlin@ophir.ph.biu.ac.il

Florian Imbery Meteorologisches Institut Universität Freiburg, 79085 Freiburg, Germany, florian.imbery@meteo.uni-freiburg.de

Malaak Kallache Climpact 75009 Paris, France, mk@climpact.com

Jan W. Kantelhardt Fachbereich Physik und Zentrum für Computational Nanoscience Martin-Luther-Universität, 06099 Halle, Germany, kantelhardt@physik.uni-halle.de

Eva Koscielny-Bunde Institut für Theoretische Physik Justus-Liebig-Universität Giessen, 35392 Giessen, Germany, Eva.Koscielny-Bunde@physik.uni-giessen.de

Jürgen P. Kropp Potsdam Institut für Klimafolgenforschung, 14412 Potsdam, Germany, kropp@pik-potsdam.de

Rakesh Kumar National Institute of Hydrology Surface Water Hydrology Division Roorkee-247667, Uttaranchal, India, rakesh@nih.ernet.in

Zbigniew W. Kundzewicz Research Centre for Agricultural and Forest Environment, Polish Academy of Sciences 60809 Poznań, Poland, zkundze@man.poznan.pl

Holger Lange Norsk institutt for skog og landskap, 1431 Ås, Norway, holger.lange@skogforsk.no

Gunnar Lischeid Leibniz-Zentrum für Agrarlandschaftsforschung, 15374 Müncheberg, Germany, lischeid@zalf.de

Valerie N. Livina School of Environmental Sciences, University of East Anglia Norwich, NR4 7TJ, UK, v.livina@uea.ac.uk

Miguel D. Mahecha Biogeochemical Model-Data Integration Group Max-Planck Institut für Biogeochemie 07701 Jena, Germany, miguel.mahecha@bgc-jena.mpg.de

Manfred Mudelsee Meteorologisches Institut Universität Leipzig 04103 Leipzig, Germany; Climate Risk Analysis, 06114 Halle, Germany mudelsee@climate-risk-analysis.com

Jörg Neumann Bayerisches Landesamt für Umwelt, 95030 Hof, Germany, joerg.neumann@lfu.bayern.de

Maciej Radziejewski Faculty of Mathematics and Computer Science, Adam Mickiewicz University, 61614 Poznań, Poland, maciejr@amu.edu.pl

Henning W. Rust Laboratoire des Sciences du Climat et de l'Environnement, 91191 Gif-sur-Yvette, France, henning.rust@lsce.ipsl.fr

Diego Rybski Institut für Theoretische Physik Justus-Liebig-Universität Giessen, 35392 Giessen, Germany, ca-dr@rybski.de

Hans Joachim Schellnhuber Potsdam Institut für Klimafolgenforschung 14412 Potsdam, Germany, john@pik-potsdam.de

Christian-D. Schönwiese Institut für Meteorologie und Geophysik, Geothe-Universität Frankfurt, 60054 Frankfurt, Germany, schoenwiese@meteor.uni-frankfurt.de

Jochen Seidel Dept. of Hydrology and Geohydrology University of Stuttgart 70569 Stuttgart, Germany, jochen.seidel@iws.uni-stuttgart.de

Silke Trömel Institut für Meteorologie Universität Bonn, 53121 Bonn, Germany, silke.troemel@uni-bonn.de

Part I
Extremes, Uncertainty, and Reconstruction

Extreme flooding at the river Danube in the town Riedlingen in Baden-Württemberg in February 1990. Photo courtesy of Gewässerdirektion Riedlingen, Germany

Chapter 1
The Statistics of Return Intervals, Maxima, and Centennial Events Under the Influence of Long-Term Correlations

Jan F. Eichner, Jan W. Kantelhardt, Armin Bunde, and Shlomo Havlin

We review our studies of the statistics of return intervals and extreme events (block maxima) in long-term correlated data sets, characterized by a power-law decaying autocorrelation function with correlation exponent γ between 0 and 1, for different distributions (Gaussian, exponential, power-law, and log-normal). For the return intervals, the long-term memory leads (i) to a stretched exponential distribution (Weibull distribution), with an exponent equal to γ, (ii) to long-term correlations among the return intervals themselves, yielding clustering of both small and large return intervals, and (iii) to an anomalous behavior of the mean residual time to the next event that depends on the history and increases with the elapsed time in a counterintuitive way. We present an analytical scaling approach and demonstrate that all these features can be seen in long climate records. For the extreme events we studied how the long-term correlations in data sets with Gaussian and exponential distribution densities affect the extreme value statistics, i.e., the statistics of maxima values within time segments of fixed duration R. We found numerically that (i) the integrated distribution function of the maxima converges to a Gumbel distribution for large R similar to uncorrelated signals, (ii) the deviations for finite R depend on both the initial distribution of the records and on their correlation properties, (iii) the maxima series exhibit long-term correlations similar to those of the original data, and most notably (iv) the maxima distribution as well as the mean maxima significantly depend on the history, in particular on the previous maximum. Finally we evaluate the effect of long-term correlations on the estimation of centennial events, which is an important task in hydrological risk estimation. We show that most of the effects revealed in artificial data can also be found in real hydro- and climatological data series.

J.F. Eichner (✉)
Justus-Liebig-Universität Giessen, 35392, Giessen, Germany
e-mail: jan.f.eichner@physik.uni-giessen.de

J.P. Kropp, H.-J. Schellnhuber (eds.), *In Extremis*,
DOI 10.1007/978-3-642-14863-7_1, © Springer-Verlag Berlin Heidelberg 2011

1.1 Introduction

In recent years there is growing evidence that many natural records exhibit long-term persistence [1.1, 1.2]. Prominent examples include hydrological data [1.3–1.6], meteorological and climatological records [1.7–1.11], turbulence data [1.12, 1.13], as well as physiological records [1.14–1.16], and DNA sequences [1.17, 1.18]. Long-term correlations have also been found in the volatility of economic records [1.19, 1.20] and the Internet traffic [1.21].

In long-term persistent records $(x_i), i = 1, \ldots, N$ with mean $\bar{x}$ and standard deviation σ_x the autocorrelation function $C_x(s)$ decays as a power law,

$$C_x(s) = \frac{1}{\sigma_x^2}\langle (x_i - \bar{x})(x_{i+s} - \bar{x})\rangle \equiv \frac{1}{\sigma_x^2 (N-s)} \sum_{i=1}^{N-s} (x_i - \bar{x})(x_{i+s} - \bar{x}) \sim s^{-\gamma}, \quad (1.1)$$

where γ denotes the correlation exponent, $0 < \gamma < 1$. Such correlations are named "long term" since the mean correlation time $T = \int_0^\infty C_x(s)\, \mathrm{d}s$ – diverges for infinitely long series (in the limit $N \to \infty$). Power law long-term correlations according to Eq. (1.1) correspond to a power spectrum $S(\omega) \sim \omega^{-\beta}$ with $\beta = 1 - \gamma$ according to the Wiener–Khinchin theorem.

Five example records are shown in Fig. 1.1, together with their histograms of the data values. These records are (a) the reconstructed river runoff of the Sacramento river (USA) [1.22], (b) the reconstructed temperature record of Baffin Island (CAN) [1.23], (c) the reconstructed precipitation in New Mexico (USA) [1.24], (d) the historical water level minima of the Nile River (EGY) [1.25], and (e) the reconstructed northern hemisphere temperature after Mann [1.26]. The data and more information about the data can be retrieved in the Internet [1.27]. The correlations properties of these records are shown in Fig. 1.2. The figure shows the detrended fluctuation analysis (DFA) (for detailed information on DFA techniques, please see Chap. 11 in this book or references [1.11, 1.15, 1.28]) of the data sets. The measured γ values are (a) 0.8, (b) 0.55, (c) 0.4, (d) 0.3, and (e) 0.1. The observed phenomena should also occur in heartbeat records, Internet traffic, and stock market volatility and have to be taken into account for an efficient risk evaluation.

In our work we focussed on long-term correlated signals with different distributions of the values (Gaussian, exponential, power law, and log-normal distributions) [1.30]. In the numerical procedures, we generated sequences of random numbers (x_i) of length $N = 2^{21}$ with either Gaussian, exponential, power-law, or log-normal distribution. The corresponding distribution densities $P(x)$ are given by

$$P_{\mathrm{Gauss}}(x) = \frac{1}{\sqrt{2\pi}\,\sigma} \exp\left(-x^2/2\sigma\right), \quad (1.2)$$

$$P_{\exp}(x) = \frac{1}{x_0} \exp(-x/x_0), \quad (1.3)$$

$$P_{\mathrm{power}}(x) = (\delta - 1) x^{-\delta}, \quad (1.4)$$

$$P_{\text{log-norm}}(x) = \frac{1}{\sqrt{2\pi}\,x} \exp\left[-(\ln x + \mu)^2/2\right]. \quad (1.5)$$

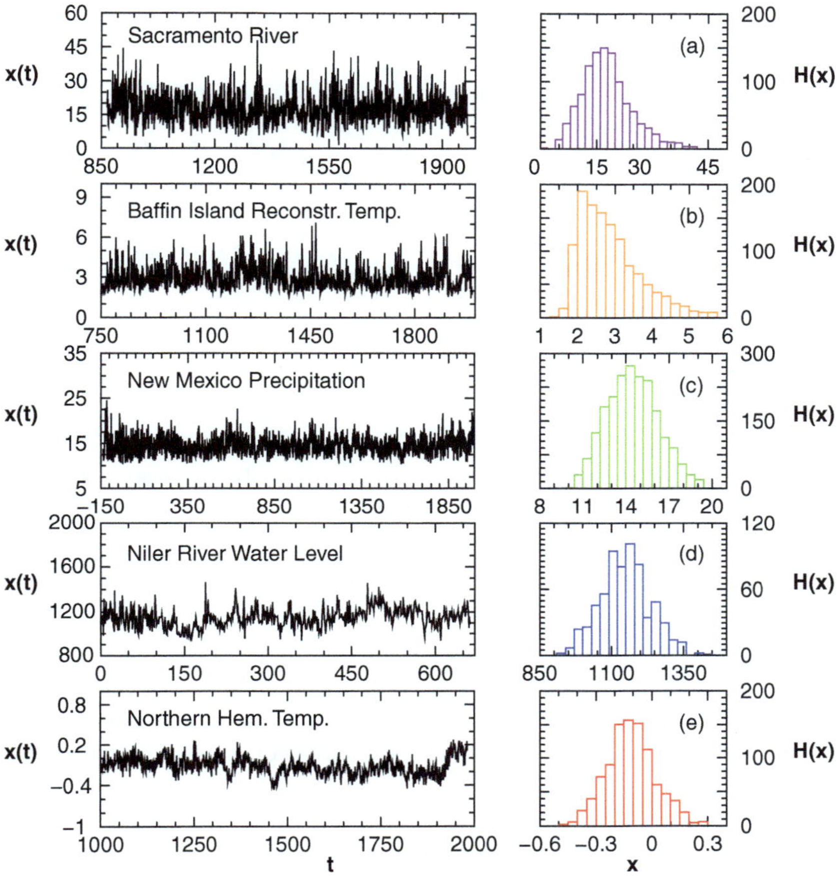

Fig. 1.1 Five annual data series $x(t)$ and their histograms $H(x)$; (**a**) reconstructed river run off of the Sacramento River (USA) [1.22], (**b**) reconstructed temperature record of Baffin Island (CAN) [1.23], (**c**) reconstructed precipitation in New Mexico (USA) [1.24], (**d**) historical water level minima of the Nile River (EGY) [1.25], and (**e**) the reconstructed northern hemisphere temperature after Mann [1.26]. The data can be retrieved in the Internet [1.27]

We chose, without loss of generality, $\sigma = 1$, $x_0 = 1$, and $\mu = 0.763$ in (1.2), (1.3), and (1.5) and select $\delta = 5.5$ in (1.4). The long-term correlations were introduced by the Fourier-filtering technique [1.31, 1.32]. To preserve the shape of the distribution after retransforming the data from Fourier space we applied an iterative method [1.30, 1.33, 1.34]. For each distribution density $P(x)$, we generated 150 data sets using 1 000 iterations, restoring the desired power spectrum by Fourier filtering and restoring the desired distribution by rank-ordered replacement of the values in each iteration until convergence is achieved. A full explanation of this iterative procedure is given in the appendix. For the Gaussian data one iteration is sufficient since Fourier filtering preserves the Gaussian distribution. We tested the quality of the long-term correlations of the data with detrended fluctuation analysis (DFA) and autocorrelation function analysis.

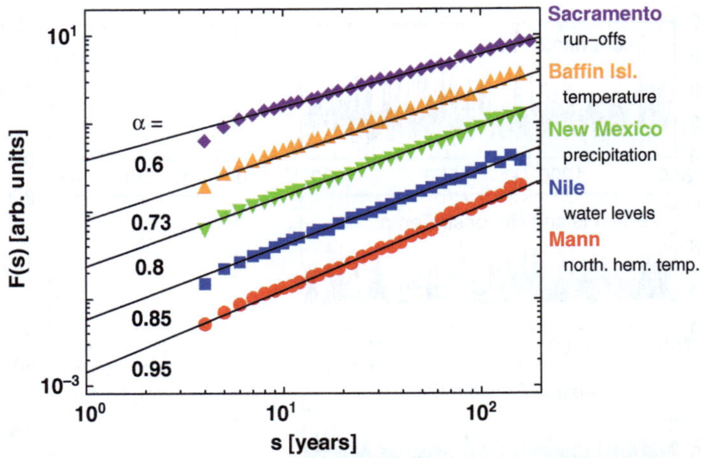

Fig. 1.2 DFA2-Analysis of the data sets introduced in Fig. 1.1. For explanation of the DFA-method, see [1.11, 1.15, 1.28]. The slopes α are related to the correlation exponents γ via $\alpha = 1 - \gamma/2$. The γ values are (from *top* to *bottom*) 0.8 for Sacramento river (*diamonds*), 0.55 for Baffin Island temperature (*triangle up*), 0.4 for New Mexico precipitation (*triangle down*), 0.3 for Nile river (square), and 0.1 for Mann temperature series (*circles*). Figure after [1.29]

In the following chapters we describe how the presence of such correlations affects the statistics of the return intervals r between events above a certain threshold value q and the statistics of block maxima m, i.e., maxima values taken from time segments of fixed duration R, as well as the estimation of centennial events Q_{100}.

1.2 Statistics of Return Intervals

The statistics of return intervals between well-defined extremal events is a powerful tool to characterize the temporal scaling properties of observed time series and to derive quantities for the estimation of the risk for hazardous events like floods, very high temperatures, or earthquakes. For Gaussian-distributed data it has been shown recently that the long-term correlations [1.1, 1.2], inherent, e.g. in river flows and temperature records, represent a natural mechanism for the clustering of the hazardous events [1.29]. The distribution density of the return intervals strongly depends on the history and can be well approximated by a stretched exponential [1.29, 1.35–1.37]. In addition, the mean residual time to the next extreme event increases with the elapsed time and depends strongly on the previous return intervals [1.29].

1.2.1 Mean Return Interval and Standard Deviation

For describing the recurrence of rare events exceeding a certain threshold q, defined in units of the standard deviations of the original distribution $P(x)$, we investigated

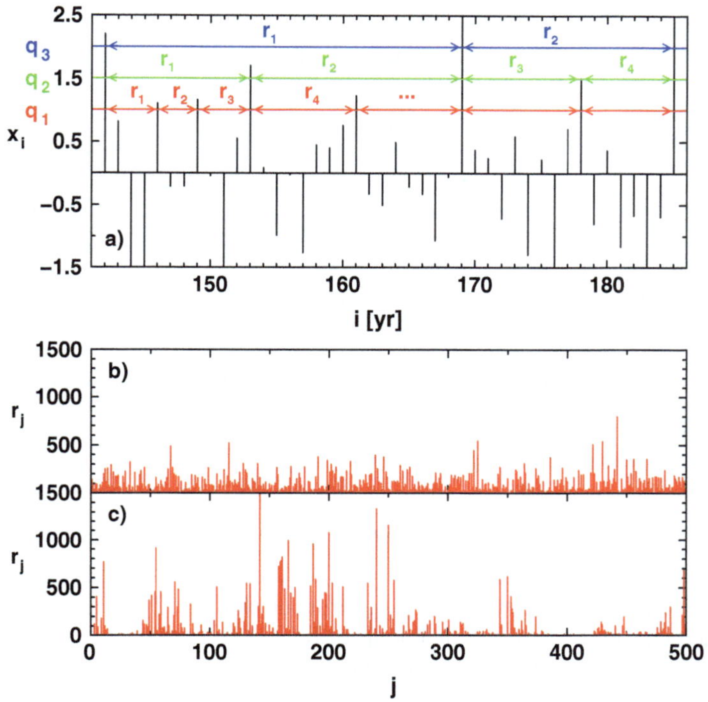

Fig. 1.3 Return intervals: (**a**) Return intervals r between events x_i above given thresholds q. (**b**) Sequence of return intervals r of long-term correlated data x_i with $\gamma = 0.4$ for a fixed threshold q so that the average r value $R_q = 100$ ($q = 2.327$). (**c**) Sequence of return intervals for uncorrelated data (shuffled x_i of (**b**)). In (**b**) more epochs with r values of small and large size appear compared to (**c**)

the statistics of the return intervals r between these events as illustrated in Fig. 1.3a. Figure 1.3b shows a section of the sequence of the return intervals for Gaussian long-term correlated data for a threshold q (quantile) chosen such that the mean return interval ("return period") R_q is approximately 100. Figure 1.3c shows the same section, but the original data were shuffled before destroying the correlations. One can see that there are more large r values and many more short r values in Fig. 1.3b compared to the uncorrelated case in Fig. 1.3c, although the mean return interval R_q is the same. The long and short return intervals in Fig. 1.3b appear in clusters [1.29], creating epochs of cumulated extreme events caused by the short r values, and also long epochs of few extreme events caused by the long r values. In the following we show, how R_q, the standard deviation σ_r, and the distribution density $P_q(r)$ of the return intervals are affected by the presence of long-term correlations as well as by different distribution densities $P(x)$ of the data.

First we considered the mean return interval R_q. For a given threshold q, there exist N_q return intervals r_j, $j = 1, 2, \ldots, N_q$, which satisfy the sum rule $\sum_{j=1}^{N_q} r_j = N$ for periodic boundary conditions. When the data are shuffled, the long-term correlations are destroyed, but the sum rule still applies with the same

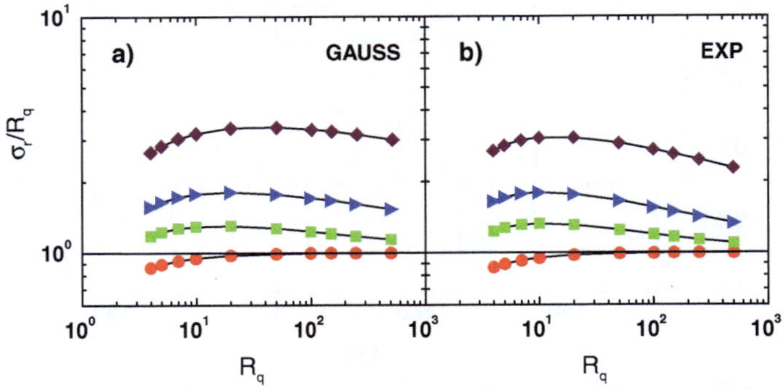

Fig. 1.4 Standard deviation of r values divided by R_q as a function of R_q for (**a**) Gaussian and (**b**) exponential uncorrelated data (*circles*) and long-term correlated data (*squares* $\gamma = 0.7$, *triangles* $\gamma = 0.4$, and *diamonds* $\gamma = 0.2$). The $\sigma_r(R_q)$ dependences are representative also for data with other distribution densities $P(x)$. Figure after [1.30]

value of N_q. Accordingly, for both long-term correlated and uncorrelated records, $R_q \equiv (1/N_q)\sum_{j=1}^{N_q} r_j$ is simply $R_q = N/N_q$, i.e., the return period (the mean return interval) is not affected by the long-term correlations. This statement can also be considered as the time series analogous of Kac's Lemma [1.37, 1.38]. Hence, R_q can be obtained directly from the tail of the (normalized) distribution density $P(x)$ of the values x_i via

$$\frac{1}{R_q} = \frac{N_q}{N} \cong \int_q^\infty P(x)\,dx. \tag{1.6}$$

The larger the q is, the larger the return period R_q will be. Accordingly, there is a one-by-one correspondence between q and R_q, which depends only on the distribution density $P(x)$ but not on the correlations.

However, there is no one-by-one correspondence between q and the variance $\sigma_r^2 \equiv (1/N_q)\sum_{j=1}^{N_q}[r_j - R_q]^2$ of the return intervals. Figure 1.4 shows (in units of R_q) the standard deviation σ_r for uncorrelated data and long-term correlated data with three correlation exponents. Due to the appearance of larger return intervals with increasing correlations (i.e., decreasing correlation exponent γ), the standard deviation also increases. The decrease of σ_r for small and large R_q is due to discretization effects and finite-size effects, respectively, which will be discussed in the following sections.

1.2.2 *Stretched Exponential and Finite-Size Effects for Large Return Intervals*

It is known that for uncorrelated records ("white noise"), the return intervals are also uncorrelated and (according to the Poisson statistics) exponentially distributed with $P_q(r) = \frac{1}{R_q}\exp(-r/R_q)$, where R_q is the mean return interval for the given threshold q (see, e.g., [1.39]). When introducing long-term correlations in Gaussian data the shape of $P_q(r)$ for large values of r, $r > R_q$, is changed to a stretched exponential [1.29, 1.30, 1.35–1.37] (see also [1.40])

$$P_q(r) \cong \frac{a_\gamma}{R_q}\exp[-b_\gamma (r/R_q)^\gamma], \tag{1.7}$$

where the exponent γ is the correlation exponent, and the parameters a_γ and b_γ are independent of q. Their dependence upon γ can be determined from the two normalization conditions that must hold for the (discrete) distribution density $P_q(r)$:

$$\sum_{r=1}^{\infty} P_q(r) = 1, \quad \sum_{r=1}^{\infty} r P_q(r) = R_q. \tag{1.8}$$

If we replace the sums for the discrete values of the return intervals r by integrals and disregard the deviations from (1.7) for small values of r, we can solve these two equations for a_γ and b_γ, obtaining

$$a_\gamma = \frac{\gamma\Gamma(2/\gamma)}{\Gamma^2(1/\gamma)} \quad \text{and} \quad b_\gamma = \frac{\Gamma(2/\gamma)}{\Gamma(1/\gamma)}, \tag{1.9}$$

where $\Gamma(x)$ is the gamma function. Hence, a_γ and b_γ are not free parameters but rather determined only by the correlation exponent γ (see also [1.37]). Hence, if γ is determined independently by correlation analysis, we can obtain a data collapse of all curves for different values of q by plotting $R_q P_q(r)$ versus r/R_q according to (1.7) [1.29], since the dependence of R_q on q is given by (1.6). However, we have to note that (1.7) does not hold for small values of r, causing some deviations in the values of parameters a_γ and b_γ and the data collapse.

Figure 1.5a displays the distribution densities $P_q(r)$ of the return intervals for long-term correlated Gaussian data for three different quantiles q ($q = 1.5, 2.0$, and 2.5), corresponding to three different return periods R_q ($R_q \simeq 15$, 44, and 161). When scaling the axis with R_q the three curves collapse to a single stretched exponential curve, shown in Fig. 1.5b. When shuffling the data first, i.e., destroying the correlations, the distribution density follows an exponential. In the correlated case many more small and large return intervals appear than in the uncorrelated case. Figure 1.5c indicates the stretched exponential shape. The fitted slopes of the lines are identical to the correlation exponents γ.

Figure 1.6 shows the rescaled distribution density function of the return intervals for the four different types of distributions of the original data [Gaussian, exponential, power-law, and log-normal according to (1.2), (1.3), (1.4), and (1.5)] with correlation exponents $\gamma = 0.2$ and 0.4 in a double logarithmic scale. The shapes of the corresponding stretched exponentials (1.7) are also plotted. The agreement is best

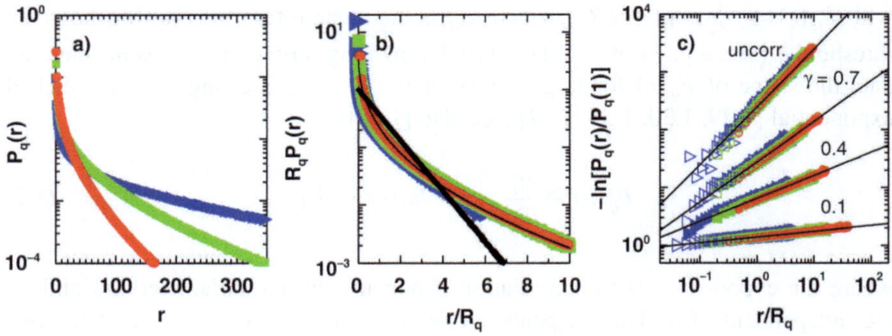

Fig. 1.5 (**a**) Distribution density $P_q(r)$ of the return intervals r for three thresholds $q = 1.5$ ($R_q \simeq 15$, *circles*), 2.0 (44, *square*), and 2.5 (161, *triangle*) for long-term correlated data ($\gamma = 0.4$). (b) When plotting $P_q(r)$ multiplied with R_q versus r/R_q, the three curves collapse to a single stretched exponential curve (*black curve*). When shuffling the data (i.e., destroying the correlations) the distribution density follows the Poisson statistics, i.e., an exponential curve (*straight black line*). (c) Double logarithmic plot of $-\ln(P_q(r)/P_q(1))$ as a function of r/R_q for $\gamma = 0.1, 0.4, 0.7$ as well as for the shuffled data (from *bottom* to *top*). The symbols correspond to the quantiles q from (**a**) and (**c**). The *straight lines* are shown for comparison and have the slope γ for the long-term correlated data and one for the uncorrelated shuffled data

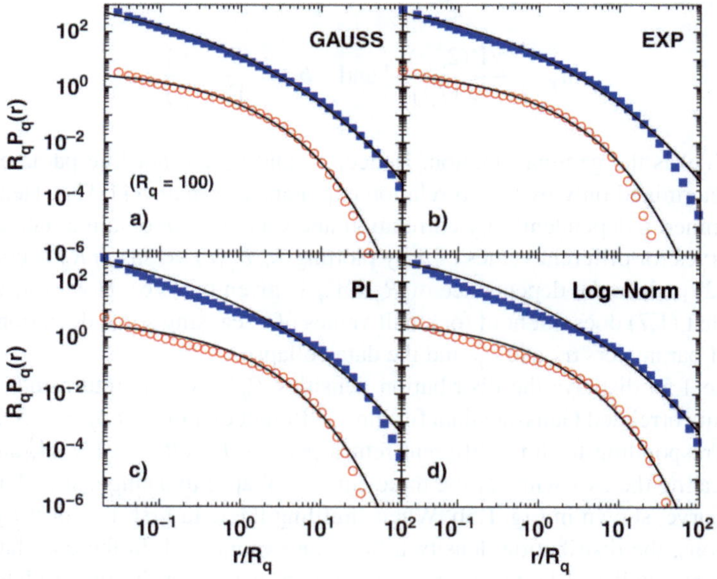

Fig. 1.6 Normalized distribution density function $R_q P_q(r)$ of return intervals r with $R_q = 100$ as a function of r/R_q for long-term correlated data with $\gamma = 0.4$ (*open symbols*) and $\gamma = 0.2$ (*filled symbols*; we multiplied the data for the *filled symbols* by a factor 100 to avoid overlapping curves). In (**a**) the original data were Gaussian distributed, in (**b**) exponentially distributed, in (**c**) power-law distributed with power -5.5, and in (**d**) log-normally distributed. All four figures follow quite well-stretched exponential curves (*solid lines*) over several decades. For small r/R_q values a power-law regime seems to dominate, while on large scales deviations from the stretched exponential behavior are due to finite-size effects. Figure after [1.30]

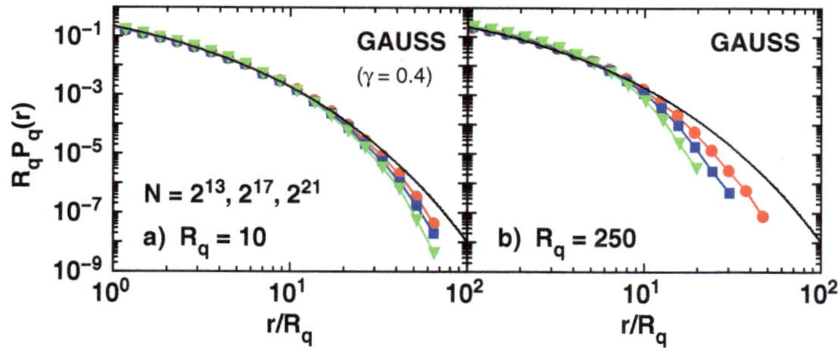

Fig. 1.7 Finite-size effects on $P_q(r)$. Deviations from the stretched exponential curve for large r/R_q and (**a**) small $R_q = 10$ as well as (**b**) large $R_q = 250$ in data with $\gamma = 0.4$ for three lengths $N = 2^{13}$ (*triangles*), 2^{17} (*squares*), and 2^{21} (*circles*) indicate finite-size effects. For large values of q and small lengths N the occurrence of very large r ($r \leq N - N/R_q$) is actually impossible. The finite-size effects for Gaussian-distributed data are representative for all four distribution densities $P(x)$ we considered. Figure after [1.30]

for the Gaussian data, but also for the other distributions the stretched exponential fit is a good approximation over several decades.

The deviations from the stretched exponential at the right tail of the curves in Fig. 1.6 (i.e., for $r \gg R_q$) are due to finite-size effects. Figure 1.7a, b shows that finite-size effects seem to cause a decrease of $P_q(r)$ for $r \gg R_q$ compared with the stretched exponential (1.7), unless we consider very long series. The series analyzed here have the lengths $N = 2^{13}$ (triangles), 2^{17} (squares), and 2^{21} (circles). The plots suggest a convergence in the asymptotic regime (large r) toward the stretched exponential curve (solid line). For small values of R_q the deviations are weaker since the number of return intervals, $N_q = N/R_q$, is much larger and the statistics is better.

Figure 1.8 compares $P_q(r)$ for simulated Gaussian-distributed data and three different correlation exponents with the five historical and reconstructed data sets introduced in Figs. 1.1 and 1.2. The solid lines, representing the theoretical curves with the measured γ values, match with $P_q(r)$ of the data (filled symbols). The dotted lines and the open symbols show the results for the shuffled data, when all correlations are destroyed. The shape of $P_q(r)$ becomes a simple exponential.

1.2.3 Power-Law Regime and Discretization Effects for Small Return Intervals

The curves in Fig. 1.6 exhibit significant deviations from the stretched exponential form for small values of r, $r < R_q$, which we have studied in detail. Figure 1.9 shows the left parts of $P_q(r)$ for three return periods R_q and again the four distribution densities $P(x)$. It reveals the occurrence of an additional intermediate scaling regime for $0.01 < r/R_q < 1$ for Gaussian and exponentially distributed data, see (a) and (b), and for $0.1 < r/R_q < 1$ for power-law and log-normally distributed

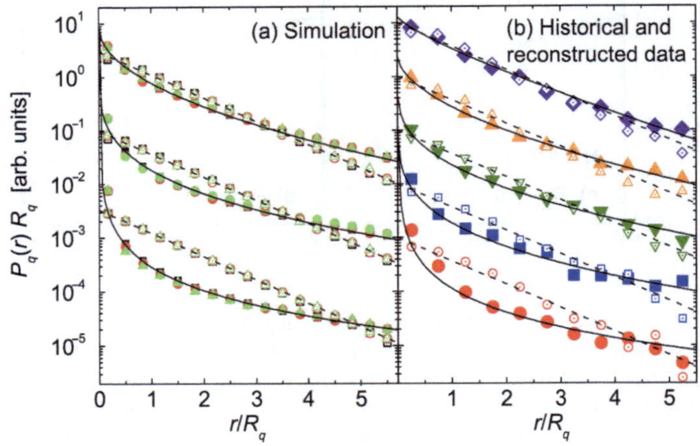

Fig. 1.8 (**a**) Distributions $P_q(r)$ of the return intervals r for the thresholds $q = 1.5$ ($R_q \approx 15$, *squares*), 2.0 ($R_q \approx 44$, *circles*), and 2.5 ($R_q \approx 161$, *triangles*) for simulated long-term correlated records with $\gamma = 0.5$ (*top*), 0.3 (*middle*), and 0.1 (*bottom*) (*filled symbols*) and for the corresponding shuffled data (*open symbols*). For the simulations, we used 1 000 records of length $N = 2 \times 10^6$ for each value of γ. (**b**) Distributions $P_q(r)$ of the return intervals r for the five climate records considered in Fig. 1.1 with same symbols, for both original data (*filled symbols*) and shuffled data (*open symbols*). The data have been averaged over all quantiles q with $R_q > 3$y and more than 50 return intervals. The *lines* are the theoretical curves following Eq. (1.7). Figure taken from [1.29]

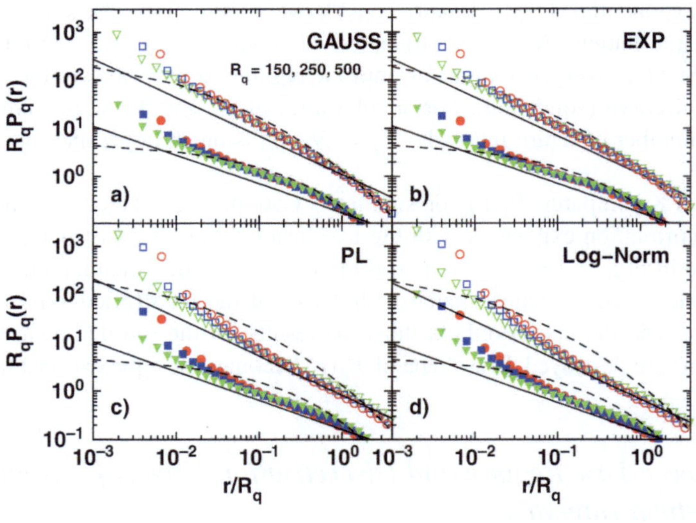

Fig. 1.9 Power-law regime of $P_q(r)$ for $r < R_q$ for $R_q = 150$ (*circles*), 250 (*squares*), and 500 (*triangles*) for (**a**) Gaussian, (**b**) exponentially, (**c**) power-law, and (**d**) log-normally distributed data with $\gamma = 0.4$ (*solid symbols*) and $\gamma = 0.2$ (*open symbols* shifted upward by a factor 10 to avoid overlapping curves). The *dashed line* is the stretched exponential (1.7), the *solid line* is a power law with slope $\gamma - 1$ as a guide to the eye. The power-law effect in (a) and (b) appears in the scale range $0.01 < r/R_q < 1$ and in (c) and (d) in $0.1 < r/R_q < 1$. For even smaller r/R_q values ($r = 1, 2, 3, \ldots$) all four figures show an upward trend which is due to the discreteness of the r values. Figure after [1.30]

data, see (c) and (d). The scaling behavior with R_q still holds in this new regime, such that data for different values of q collapse onto each other as long as $r \gg 1$. The scaling behavior in this regime might be characterized by a power law, giving rise to the two-branched ansatz [1.30]

$$R_q P_q(r) \sim \begin{cases} (r/R_q)^{\gamma'-1} & \text{for} \quad 1 \ll r \le R_q, \\ \exp[-b_\gamma (\frac{r}{R_q})^\gamma] & \text{for} \quad R_q < r \ll N, \end{cases} \tag{1.10}$$

replacing (1.7). For all distribution densities $P(x)$, $\gamma' \approx \gamma$ seems to be consistent with the data. However, we cannot fully exclude that γ' might depend slightly on the quantile q, in particular for power-law and log-normal distribution densities $P(x)$. The normalization factors and the parameter b cannot be calculated exactly for Eq. (1.10), which is a drawback of the two-branched distribution density.

Figures 1.6 and 1.7 show that the behavior of $P_q(r)$ as described by (1.10) becomes visible only for very long data sets. If short and, e.g., exponentially distributed data are studied, the stretched exponential regime shrinks and an asymptotic pure exponential decay of $P_q(r)$ can be observed. In this case, $P_q(r)$ displays a combination of (i) discretization effects for very small r values, (ii) power-law behavior for intermediate r values, and (iii) an exponential decay for large r values, resembling the return interval distribution obtained for seismic activity [1.41–1.43]. We cannot exclude the possibility that also seismic data are described by (1.10), i.e., the integrated probability density is described by a Weibull distribution.

For very small r values (close to 1) the continuous distribution density (1.10) has to be replaced by a discrete distribution. Figure 1.9 clearly shows the deviations from the power-law regime because of the discreteness of the r values. The first few points move upward ($r = 1$ and 2, maybe up to $r = 5$), because they accumulate probability from (impossible) smaller non-integer intervals. Hence, the data points for r close to 1 do not obey the scaling of the distribution density with R_q, and no data collapse can be achieved for them. For the power-law distributed data and the log-normal distributed data, Fig. 1.9c, d, the discretization effects even seem to suppress the full development of the power-law scaling regime.

Return intervals $r = 1$ are of particular relevance when answering the question if two consecutive events x_i and x_{i+1} will surpass the threshold q. Figures 1.10a, b show the distribution of the $r = 1$ values for Gaussian and exponentially distributed data as a function of the return period R_q (i.e., of the threshold q). In uncorrelated data the probability to have another extreme event directly after any extreme event must be $1/R_q$ (stars in Fig. 1.10), because each data point x_i is an extreme event with probability $1/R_q$ and there are no correlations. In long-term correlated data the probability to find another extreme event directly after an extreme event will be higher due to the persistence of the data. The figure shows, how this probability changes with increasing correlations (i.e., decreasing γ). Except for small values of γ the shape of the curves seems to be a power law of the form $P_q(r) \approx R_q^{-\gamma}$, in

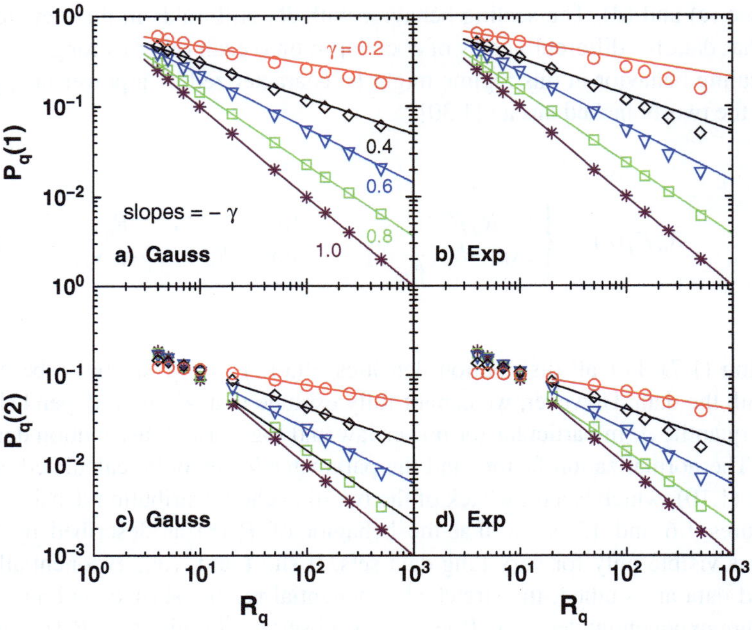

Fig. 1.10 Dependence of $P_q(1)$ and $P_q(2)$ on γ for various quantiles q. The distribution of the return intervals of size $r = 1$ is plotted versus R_q for (**a**) Gaussian and (**b**) exponentially distributed data with correlation exponents $\gamma = 0.2$ (*circles*), 0.4 (*diamonds*), 0.6 (*triangles*), 0.8 (*squares*) as well as uncorrelated data (*stars*). The stronger the correlations, i.e., the smaller the γ, the higher the probability that two consecutive events x_i and x_{i+1} exceed the given threshold q. The *straight lines* indicate power laws with exponents $-\gamma$, suggesting the discretization effects in Fig. 1.9 to be independent of q or R_q. Note that the larger the R_q, the stronger the correlation effect. (**c**) and (**d**) show the same for return intervals of size $r = 2$. Figure after [1.30]

agreement with (1.10) and the assumption $\gamma' = \gamma$.[1] The larger the R_q is chosen, the larger the difference between $P_q(1)$ in correlated and uncorrelated data. The distribution $P_q(2)$ of the $r = 2$ values (Fig. 1.10c, d) shows a similar power law for large values of R_q, but deviates for small R_q and large γ.

1.2.4 Long-Term Correlations of the Return Intervals

The form of the distribution density $P_q(r)$ of return intervals r between extreme events in long-term correlated data indicates that very short and very long return intervals are more frequent than for uncorrelated data. However, $P_q(r)$ does not quantify, if the return intervals themselves are arranged in a correlated fashion and if clustering of rare events may be induced by long-term correlations. In our previous work [1.29, 1.36] we reported that (i) long-term correlations in a Gaussian time

[1] The methods applied to create long-term correlated data [1.31, 1.33] become unprecise for very small values of γ and create data with an effective γ_{eff} slightly larger than γ.

series induce long-term correlations in the sequence of return intervals and (ii) that both correlation functions are characterized by the same correlation exponent γ. We showed that this leads to a clustering of extreme events, an effect that also can be seen in long climate records. We presented further support for our result (ii) showing that it is independent of the distribution density $D(x)$ of the data [1.30].

In order to determine the autocorrelation behavior of the return interval series (r_j), $j = 1, 2, \ldots, N_q$, for a given quantile q we calculate the autocorrelation function $C_r(s)$ of the return intervals

$$C_r(s) = \frac{1}{\sigma_r^2 (N_q - s)} \sum_{j=1}^{N_q - s} (r_j - R_q)(r_{j+s} - R_q). \tag{1.11}$$

Figure 1.11 shows $C_r(s)$ and $C_x(s)$ for data characterized by the four distribution densities $D(x)$ and four mean return periods R_q. One can see that for each distribution, the data approximately collapse to a single line, exhibiting the same slope as the original data. This scaling behavior shows that the return intervals are also long-term correlated, with the same value of γ as the original data. There is, however, one important difference in the correlation behavior: For the return intervals, the autocorrelation function $C_r(s)$ is significantly below the autocorrelation function $C_x(s)$ of the original data (also shown in Fig. 1.11) by a factor between 2 and 3, depending on the distribution. Accordingly, there is additional white noise in the return interval sequences that only weakly depends on the return period R_q.

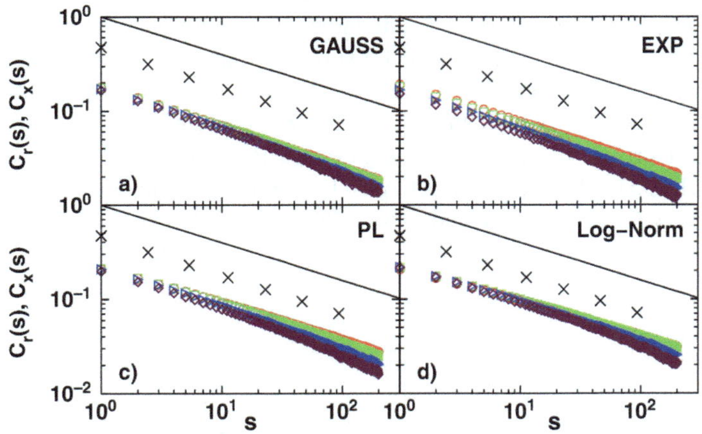

Fig. 1.11 Autocorrelation functions $C_x(s)$ of the original data ($\times$) and $C_r(s)$ of the return interval series for (**a**) Gaussian, (**b**) exponentially, (**c**) power-law, and (**d**) log-normally distributed data with $\gamma = 0.4$ for $R_q = 10$ (*circles*), 20 (*squares*), 50 (*triangles*), and 100 (*diamonds*). All curves show a power-law behavior indicating long-term correlations in the sequence of return intervals with the same correlation exponent γ (*straight lines*). Curves for large R_q show weaker correlations at large scales s due to finite-size effects. All results were averaged over 150 configurations of original data with length $N = 2^{21}$. Figure after [1.30]

We believe that this uncorrelated component is a consequence of the way the return intervals are constructed. Tiny changes in the threshold q can lead to large changes of several return intervals in the records. Thus, small white noise in the original data leads to larger noise in the return intervals, and this causes the significant additional random component. There is no crossover in the scaling behavior of the autocorrelation function except for finite-size effects.

1.2.5 Conditional Return Interval Distributions

In order to study the history effects on the distribution density $P_q(r)$ of the return intervals r we first consider the *conditional* distribution density $P_q(r|r_0)$, defined as the distribution density of all those return intervals that directly follow a given r_0 value in the sequence of return intervals. An explanation of conditional return intervals is given in Fig. 1.12. For uncorrelated data $P_q(r|r_0) = P_q(r)$. Figure 1.13 displays the ratio of the conditional distribution density $P_q(r|r_0)$ and the unconditional $P_q(r)$ for Gaussian and exponentially distributed data with six R_q values (20, 50, 100, 150, 250, and 500), for $r_0/R_q \simeq 1/4$, 1, and 4, as a function of r/R_q. For fixed r_0/R_q all data points collapse onto single curves, independent of q and independent of the distribution of the original data. We thus find scaling behavior of the conditional distribution density function

$$P_q(r|r_0) = \frac{1}{R_q} f_{r_0/R_q}(r/R_q) \tag{1.12}$$

like in (1.10). The long-term correlations in the sequence of r values cause a culmination of small r values for small r_0 and large r values for large r_0. The conditions $r_0 = R_q/4$ and $r_0 = R_q$ yield maxima for similar r values, $r \approx 0.2R_q$ and $r \approx 2R_q$, respectively, see Fig. 1.13a,b,d,e. The effect is weaker for small r/R_q but still important, because the $r = 1$ values are relevant for the clustering of extreme events. For $r_0 = 4R_q$ a strong enhancement of large return intervals $r \approx 10R_q$ is observed, see Fig. 1.13(c,f). The divergent behavior for large r_0/R_q indicates that the next extreme event is moving more and more to the future if the last return interval was huge. When the series of r values is shuffled, i.e., all correlations are destroyed,

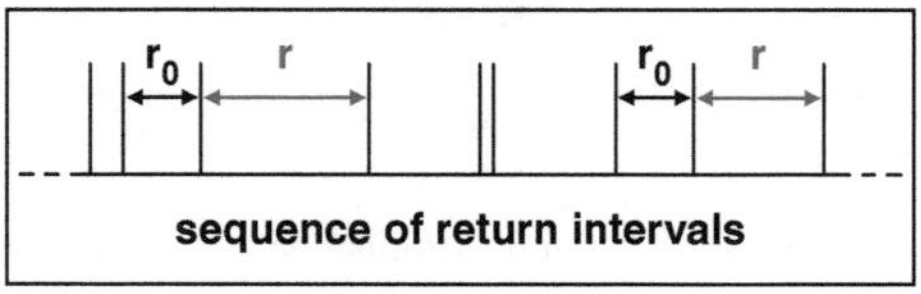

Fig. 1.12 Conditional return intervals are those r values that follow in the sequence of the return intervals directly after a r value of a distinct size r_0. For data with poor statistics, e.g. real data, it is convenient to use r_0 as a boundary value, i.e., considering all those r values that follow directly after a r value smaller or larger than, e.g., the median of all return intervals

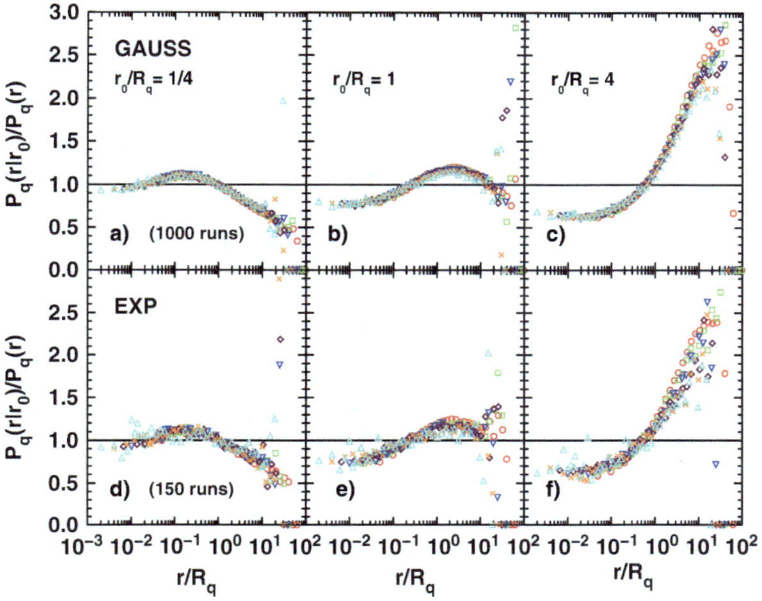

Fig. 1.13 Conditional distribution density $P_q(r|r_0)$ divided by $P_q(r)$ for three different conditions. (**a**)–(**c**) the ratio for long-term correlated ($\gamma = 0.4$) Gaussian-distributed original data for the conditions $r_0/R_q = 1/4$ (a), 1 (**b**), and 4 (**c**), averaged over 1000 runs. As a result of the long-term correlations small r_0 increases the probability to find small r values directly after r_0, while large r_0 increases the probability to find large r values directly after r_0. The maximum of the ratio moves from small r/R_q to large r/R_q with increasing condition r_0/R_q (the right cutoff is due to finite set length and therefore limited maximum r values). The data collapse consists of six different curves for quantiles with $R_q = 20$ (*circle*), 50 (*square*), 100 (*triangle down*), 150 (*diamond*), 250 ($\times$), and 500 (*triangle* up). (**d**)–(**f**) the same for exponentially distributed original data. The collapse seems noisier because the curves are averaged over 150 runs. For uncorrelated data, $P_q(r|r_0) = P_q(r)$ exactly, as indicated by the *straight line* at the ratio 1. Figure after [1.30]

$P_q(r|r_0)$ and $P_q(r)$ are identical, but still show the stretched exponential shape. For uncorrelated original data $P_q(r)$ and $P_q(r|r_0)$ both follow a simple exponential decay.

The conditional distribution densities for the five historical and reconstructed records and for artificial data with the same correlation exponents and similar set lengths are shown in Fig. 1.14 for just two conditions, r_0 smaller ($-$) or larger ($+$) the median r value. The splitting of the curves $P_q^+(r)$ and $P_q^-(r)$ for large r/R_q is a consequence of the long-term correlations.

1.2.6 Conditional Return Periods

In contrast to the mean return interval R_q, the conditional mean return interval $R_q(r_0) = \sum_{r=1}^{\infty} r P_q(r|r_0)$, i.e., the average return interval of those r values that follow directly an interval of size r_0, clearly exhibits correlation effects.

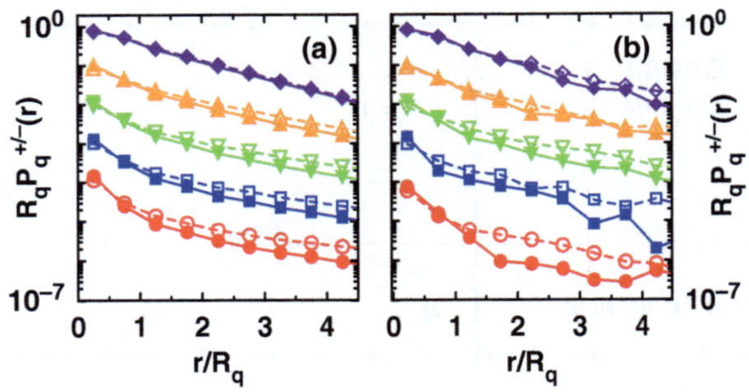

Fig. 1.14 Conditional distribution densities $R_q P_q^+(r)$ (*open symbols*) and $R_q P_q^-(r)$ (*full symbols*) averaged over all r_0 above and below the median return interval, respectively, versus r/R_q for (**a**) artificial records and (**b**) the five climatological records from Fig. 1.1 with the symbols from Fig. 1.2. The artificial data in (**b**) have the same γ values and mean record lengths as the climate records; we studied 1000 records of size $N = 1250$. Figure after [1.29]

Figure 1.15(a)–(d) shows $R_q(r_0)$ in units of R_q as a function of r_0/R_q for four values of R_q (5, 10, 50, and 250), $\gamma = 0.4$, and the four distribution densities $P(x)$ listed in (1.2), (1.3), (1.4), and (1.5). The correlation effect becomes apparent: after small r_0/R_q the next expected return interval $R_q(r_0)$ is smaller than R_q, and after large r_0/R_q it is much larger than R_q. Although the shapes of the (more or less) collapsing curves differ depending on the original distribution and on the chosen quantile q (i.e., on R_q), the tendency is the same for all four original distribution densities. We like to note again that finite-size effects, violating the scaling behavior, are most pronounced for the exponential distribution.

Figure 1.15e,f shows the conditional return periods $R_q(r_0)$ for two data sets, the observed annual temperatures of Prague [1.9] and the reconstructed annual temperatures of the northern hemisphere after Jones [1.44], and for the corresponding shuffled sets (open symbols). The original records exhibit long-term correlations with $\gamma \approx 0.7$ (a) and 0.4 (b). The correlation effect is shown by increasing $R_q(r_0)$ with increasing r_0 (filled symbols). The results for the shuffled data (open symbols) do not show a particular dependence on r_0.

Due to the tight conditions regarding r_0, the conditional mean return interval $R_q(r_0)$ requires very large statistics and is thus not suitable for studies of real recordings. The quantity can be improved by integrating over two ranges of r_0 values, e.g. r_0 larger or smaller than the return period R_q, resulting in only two conditions. Therefore, we define R_q^+ and R_q^- as the average return intervals that follow r_0 values either larger (+) or smaller (−) than R_q. Figure 1.16 shows R_q^+ and R_q^- (filled symbols) in units of R_q as a function of different correlation exponents γ for all four distributions. For uncorrelated data (results shown at $\gamma = 1$), R_q^+ and R_q^- coincide with R_q. The smaller the γ, the stronger the correlations, and the more pronounced is the difference between R_q^+ and R_q^-.

Figure 1.16 also shows the average conditional return intervals R_q^{++} and R_q^{--} of those r-values that directly follow two return intervals, r_0 and r_{-1}, both larger (++) or smaller (−−) than R_q. As expected, the correlation effect is even stronger

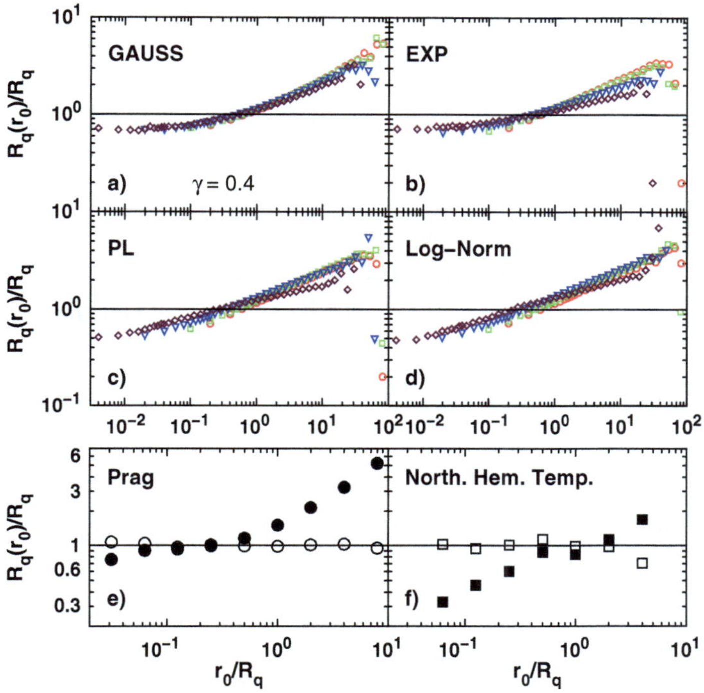

Fig. 1.15 Conditional return periods $R_q(r_0)$ in units of R_q versus the condition r_0/R_q for R_q = 5 (*circle*), 10 (*square*), 50 (*triangle*), and 250 (*diamond*) for all four original distributions with $\gamma = 0.4$. While Gaussian (**a**) and exponential (**b**) data show a data collapse with nearly no R_q dependence (except for stronger finite-size effects for exponential data at large r_0/R_q), Power-law (**c**) and log-normal (**d**) data show a slight R_q dependence resulting in different slopes for different R_q. All figures clearly display the memory effect in form of increasing $R_q(r_0)$ with increasing r_0, caused by the long-term correlations in the original data. In uncorrelated data $R_q(r_0) = R_q$, as indicated by the horizontal lines at ratio 1. Conditional return periods $R_q(r_0)$ for (**a**) the annual temperature of Prague (CZ, 1775–1992) [1.9] and (**b**) the reconstructed annual temperature data of the northern hemisphere (1000–1850) after Jones [1.44]. The original records exhibit long-term correlations with $\gamma \approx 0.7$ (**a**) and 0.4 (**b**). The *full circles* represent the observed data sets, while the *open triangles* represent data sets that are obtained by randomly shuffling the observed original records. Since $R_q(r_0)/R_q$ depends only on r_0/R_q, we averaged $R_q(r_0)/R_q$ over several q values in order to improve the statistics. For the reconstructed temperature record we limited ourselves to data up to 1850 AD in order to exclude possible clustering of rare events due to global warming

in R_q^{++} and R_q^{--} but the quantities require more statistics than R_q^{+} and R_q^{-}. All curves in Fig. 1.16 look very similar and suggest that an effect of the shape of the distribution of the original data is minor, when the data are long-term correlated.

By definition, $R_q(r_0)$ is the expected waiting time to the next event, when the two events before were separated by r_0. The more general quantity is the expected conditional residual waiting time $\tau_q(t|r_0)$ to the next event, when the time t has been elapsed (see Fig. 1.17). For $t = 0$, $\tau_q(0|r_0)$ is identical to $R_q(r_0)$. In general, $\tau_q(t|r_0)$ is related to $P_q(r|r_0)$ by

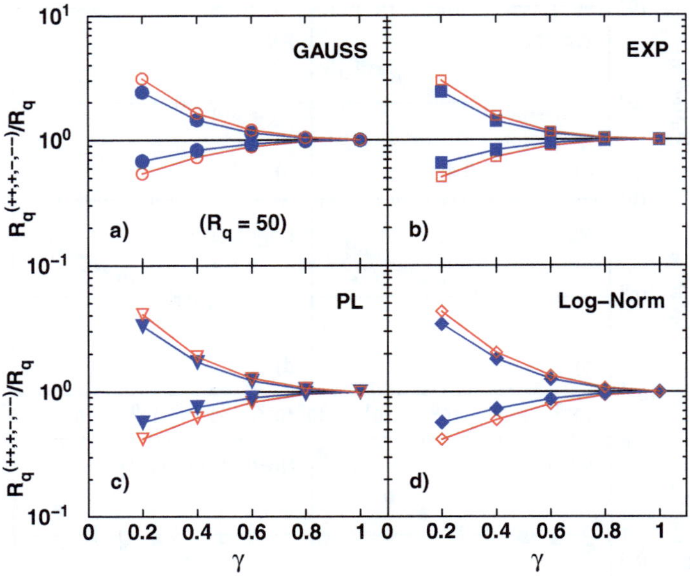

Fig. 1.16 Conditional return periods for long-term correlated data with five values of γ and fixed $R_q = 50$. (**a**) the single-conditional and the double-conditional return periods ($R_q^{+/-}$ and $R_q^{++/--}$) in units of R_q for Gaussian data. R_q^+ (*filled symbols, upper curve*) and R_q^- (*filled symbols, lower curve*) are the single-conditional return periods, where the predecessor r_0 is larger or smaller than R_q. R_q^{++} (*open symbols, upper curve*) and R_q^{--} (*open symbols, lower curve*) are the double-conditional return periods where the preceding and the pre-preceding return intervals are both larger or both smaller than R_q ($R_q^{++} = R_q(r|r_0 > R_q|r_{-1} > R_q)$, $R_q^{--} = R_q(r|r_0 < R_q|r_{-1} < R_q)$). The splitting of the curves is a consequence of the long-term correlations in the sequence of r values. In the uncorrelated case shown at $\gamma \geq 1$ all four conditional return periods degenerate to R_q. (**b**) The same for exponentially distributed data, (**c**) for power-law distributed data, and (**d**) for log-normally distributed data. All points were averaged over 20 runs with roughly 40000 r values. Figure after [1.30]

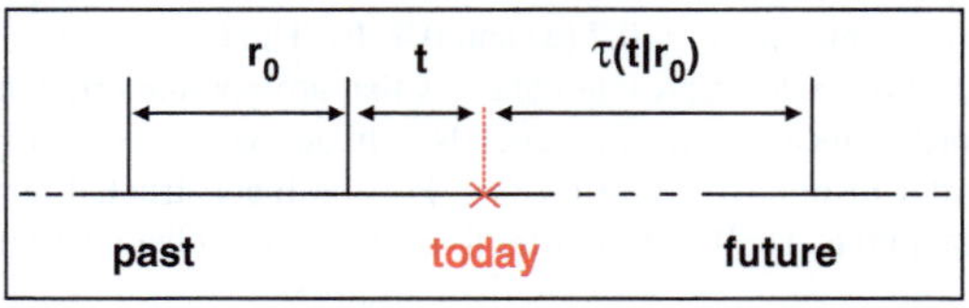

Fig. 1.17 The conditional residual waiting time $\tau_q(t|r_0)$ is defined as the average remaining time between *today* and the next extreme event, when the time between *today* and the last extreme event is of size t and the return interval between the two last extreme events is of the size r_0

$$\tau_q(t|r_0) = \int_t^\infty (r - t) P_q(r|r_0) dr \Big/ \int_t^\infty P_q(r|r_0) dr. \qquad (1.13)$$

For uncorrelated records, $\tau_q(t|r_0)/R_q = 1$ (except for discretization effects that lead to $\tau_q(t|r_0)/R_q > 1$ for $t > 0$, see [1.45]). Due to the scaling of $P_q(r|r_0)$,

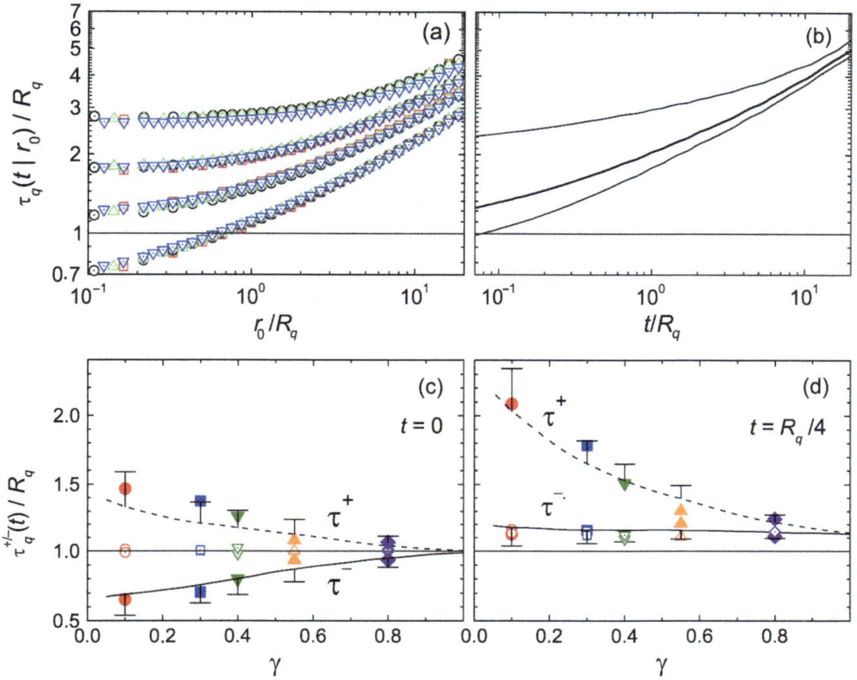

Fig. 1.18 (**a**) Mean conditional residual time to the next event $\tau_q(t|r_0)$ (in units of R_q) versus r_0/R_q for four q values ($q = 1.0, 1.25, 1.5$, and 1.75, *different symbols*) and four values of the elapsed time t since the last event ($t/R_q = 0, 0.25, 1$ and 4, from *bottom* to *top*). (**b**) Mean conditional residual time $\tau_q(t|r_0)$ as a function of t/R_q for $r_0/R_q = 1/8$ (*lower curve*) and $r_0/R_q = 8$ (*upper curve*). The *middle curve* represents the mean conditional residual time averaged over all r_0. (c,d) Mean conditional residual times $\tau_q^-(t)$ (*full line*) and $\tau_q^+(t)$ (*dashed line*)) averaged over all r_0 below and above the median return interval, respectively, for (c) $t = 0$ and (d) $t = R_q/4$. The symbols are for the five climatological records from Fig. 1.2, for both original data (*filled symbols*) and shuffled data (*open symbols*). For obtaining the curves in (**a, b**) we used the same statistics as in Fig. 1.8(**a**), and for the theoretical curves in (**c, d**) and the error bars we used the same statistics as in Fig. 1.14 (**a**). Figure taken from [1.29]

we expect that also $\tau_q(t|r_0)/R_q$ scales with r_0/R_q and t/R_q. Figure 1.18a shows that this is indeed the case. The data collapse for each value of t/R_q confirms the scaling property. The figure clearly displays the effect of the long-term memory: Small and large return intervals are more likely to be followed by small and large ones, respectively, and hence $\tau_q(0|r_0)/R_q \equiv R_q(r_0)/R_q$ is well below (above) 1 for r_0/R_q well below (above) 1. With increasing t, the expected residual time to the next event increases, as is also shown in Fig. 1.18b, for two values of r_0 (top and bottom curves). Note that only for an infinite long-term correlated record, the value of $\tau_q(t|r_0)$ will increase indefinitely with t and r_0. For real (finite) records, there exists a maximum return interval which limits the values of t, r_0, and $\tau_q(t|r_0)$. The middle curve shows the expected residual time averaged over all r_0, i.e., the unconditional residual time. In this case, the interval between the last two events is not taken explicitly into account, and the slower than Poisson decrease of the unconditional distribution function $P_q(r)$ leads to the anomalous increase of the mean residual time with the elapsed time [1.45]. Very recently, this approach (average over r_0) has been applied successfully to worldwide earthquake records [1.41]. For the case of long-term correlated records, however, like the hydroclimate records discussed here, the large differences between the three curves in Fig. 1.18b suggest that for an efficient risk estimation, also the previous return interval has to be taken into account and not only the distribution of the return intervals.

To reveal this intriguing behavior in the relatively short observed and reconstructed records, we improved the statistics (similar to Fig. 1.12) by studying the mean expected conditional residual waiting times $\tau_q^-(t)$ and $\tau_q^+(t)$ for r_0 below and above the median r_q^*, respectively. For uncorrelated data, both quantities are identical and coincide with R_q.

Figure 1.18c shows $\tau_q^+(0)/R_q$ and $\tau_q^-(0)/R_q$ versus γ for simulated records (lines) and the five representative climatological records (symbols). The difference between τ_q^+ and τ_q^- becomes more pronounced with decreasing value of γ, i.e., increasing long-term memory. The results for the climate records are in good agreement with the theoretical curves. The same comparison for $t/R_q = 1/4$ instead of $t = 0$ is shown in Fig. 1.18d. The behavior is qualitatively different: while $\tau_q^-(0)/R_q$ increases with increasing γ, $\tau_q^-(R_q/4)/R_q$ is rather constant. Again, the agreement between simulated and real records is quite satisfactory, revealing the strong effect of memory in the hydroclimate records that also results in the clustering of the extreme events. To show the significance of the results, we also analyzed the corresponding shuffled data. We obtained $\tau_q^+(0)/R_q \approx \tau_q^-(0)/R_q \approx 1$ and $\tau_q^+(R_q/4)/R_q \approx \tau_q^-(R_q/4)/R_q \approx 1.1$. In the second case, the shuffled data (following the Poisson distribution) show a slight increase of the residual time (1.1 instead of 1). This is a finite-size effect that already has been noticed in [1.45].

1.3 Statistics of Maxima

Extreme events are rare occurrences of extraordinary nature, such as floods, very high temperatures, or earthquakes. In studying the extreme value statistics of the corresponding time series one wants to learn about the distribution of the extreme events, i.e., the maximum values of the signal within time segments of fixed duration R, and the statistical properties of their sequences. In hydrological engineering, for example, extreme value statistics are commonly applied to decide what building projects are required to protect riverside areas against typical floods that occur once in 100 years. Many exact and empirical results on extreme value statistics have been obtained in the past years, for reviews, see e.g., [1.46–1.50]. Most of these results, however, hold only in the limit $R \to \infty$ and are based on statistically independent values of the time series. Both assumptions are not strictly valid in practice. Since observational data are always finite, predictions for finite time intervals R are required, and – most importantly – correlations cannot be disregarded.

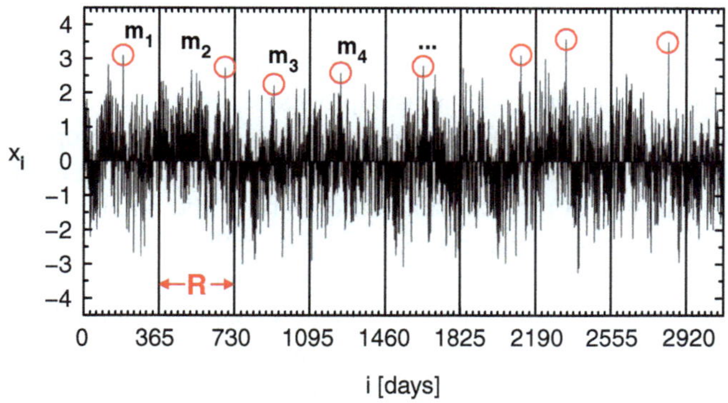

Fig. 1.19 Definition of maxima: A time series (x_i), $i = 1, \ldots, N$, of, e.g., daily data is separated into segments of length $R = 365$ days. The maximum values m_j (O) in each segment, e.g., annual maxima, define another time series (m_j), $j = 1, \ldots, N/R$. Figure taken from [1.60]

Figure 1.19 illustrates the definition of the series of maxima (m_j), $j = 1, \ldots, N/R$ of original data (x_i), $i = 1, \ldots, N$, within segments of size R for $R = 365$, i.e., for annual maxima if (x_i) represents daily data. According to traditional extreme value statistics the integrated distribution density of the maxima m_j converges to a Gumbel distribution for independently and identically distributed (i.i.d.) data (x_i) with Gaussian or exponential distribution density [1.46, 1.47, 1.49].

In long-term correlated records, the central assumption in the traditional extreme value statistics [1.46] is not fulfilled: extreme events cannot be viewed a priori as uncorrelated even when there is a long time span between them. Recently, there have been some approaches to include correlations in the study of extreme value statistics. For the special case of Gaussian $1/f$ correlations in voltage fluctuations in GaAs films extreme value statistics have been demonstrated to follow a Gumbel distribution [1.51]. A somewhat different asymptotic behavior was observed in experiments on turbulence and in the two-dimensional XY model [1.52, 1.53], see also [1.54]. Extreme value statistics have also been employed in studies of hierarchically correlated random variables representing the energies of directed polymers [1.55] and in studies of maximal heights of growing self-affine surfaces [1.56]. In the Edwards–Wilkinson model and the Kardar–Parisi–Zhang model for fluctuating, strongly correlated interfaces an Airy distribution function has been obtained as exact solution for the distribution of maximal heights very recently [1.57]. On the other hand, the statistics of extreme height fluctuations for Edwards–Wilkinson relaxation on small-world substrates are rather described by the classical Fisher–Tippet–Gumbel distribution [1.58]. Besides these recent results there is a theorem by S. M. Berman [1.59] (see also [1.47, 1.49]) stating that the maxima statistics of stationary Gaussian sequences with correlations converges to a Gumbel distribution asymptotically for $R \to \infty$ provided that $C_x(s)\log(s) \to 0$ for $s \to \infty$, which holds for long-term correlations.

In our recent work [1.60] we focussed on long-term correlated signals and showed numerically that (i) the asymptotic convergence of the integrated maxima distribution to the Gumbel formula occurs also for long-term correlated Gaussian or exponentially distributed signals (x_i), (ii) for finite R, the deviation of the integrated maxima distribution from the asymptotics depends significantly on the initial distribution of the data (x_i) and their long-term correlation properties, (iii) the maxima series (m_j) exhibit long-term correlations similar to those of the data (x_i), and – most notably – (iv) the distribution density of the maxima, the integrated maxima distribution, and the mean maxima significantly depend on the history, i.e., the previous maximum m_0. The last item implies that conditional mean maxima and conditional maxima distributions (with m_0 as condition) should be considered for improved extreme event prediction. We also showed that the conditional mean maxima for observational data [1.25, 1.61] have similar dependence on m_0 as for artificial long-term correlated data.

1.3.1 Extreme Value Statistics for i.i.d. Data

In classical extreme value statistics one assumes that records (x_i) consist of i.i.d. data, described by density distributions $P(x)$, which can be, e.g., a Gaussian or an exponential distribution. One is interested in the distribution density function $P_R(m)$ of the maxima (m_j) determined in segments of length R in the original series (x_i) (see Fig. 1.19). Note that all maxima are also elements of the original data. The corresponding integrated maxima distribution $G_R(m)$ is defined as

$$G_R(m) = 1 - E_R(m) = \int_{-\infty}^{m} P_R(m') - \mathrm{d}m'. \tag{1.14}$$

Since $G_R(m)$ is the probability of finding a maximum smaller than m, $E_R(m)$ denotes the probability of finding a maximum that exceeds m. One of the main results of traditional extreme value statistics states that for independently and identically distributed (i.i.d.) data (x_i) with Gaussian or exponential distribution density function $P(x)$ the integrated distribution $G_R(m)$ converges to a double exponential (Fisher–Tippet–Gumbel) distribution (often labeled as Type I) [1.46, 1.47, 1.49, 1.62], i.e.,

$$G_R(m) \to G\left(\frac{m-u}{\alpha}\right) = \exp\left[-\mathrm{e}^{\left(-\frac{m-u}{\alpha}\right)}\right] \tag{1.15}$$

for $R \to \infty$, where α is the scale parameter and u the location parameter. By the method of moments those parameters are given by $\alpha = \frac{\sqrt{6}}{\pi}\sigma_R$ and $u = m_R - n_e\alpha$ with the Euler constant $n_\mathrm{e} = 0.577216$ [1.47, 1.63–1.65]. Here m_R and σ_R denote the (R-dependent) mean maximum and the standard deviation, respectively. Note that different asymptotics will be reached for broader distributions of data (x_i) that belong to other domains of attraction [1.47]. For example, for data following a power-law distribution (or Pareto distribution), $P(x) = (x/x_0)^{-k}$, $G_R(m)$ converges to a Fréchet distribution, often labeled as Type II. For data following a distribution with finite upper endpoint, for example, the uniform distribution $P(x) = 1$ for $0 \leq x \leq 1$, $G_R(m)$ converges to a Weibull distribution, often labeled as Type III. These are the other two types of asymptotics, which, however, we do not consider here.

1.3.2 Effect of Long-Term Persistence on the Distribution of the Maxima

We began by studying how the convergence of the integrated maxima distribution $G_R(m)$ toward the Gumbel distribution (1.15) is affected by long-term correlations in the signal (x_i). Regarding the distribution density $P(x)$ of the signal, we compare results for a Gaussian distribution, $P(x) = 1/(\sqrt{2\pi})\exp(-x^2/2)$ $(-\infty < x < \infty)$ and an exponential distribution $P(x) = \exp(-x)$ $(0 < x < \infty)$. Artificial long-term correlated signals following these distributions can be generated by the Fourier-filtering method (see, e. g., [1.31]) and by the Schreiber–Schmitz iteration procedure [1.33, 1.34], respectively (see Appendix). In the Schreiber–Schmitz procedure we employed 1000 iterations for each record of length $N = 2^{21} \approx 2 \times 10^6$. We found that our results do not depend on the number of iterations if more than 100 iterations are used. We studied 150 configurations for most plots.

Figures 1.20 and 1.21 compare the maxima statistics for uncorrelated and long-term correlated data ($\gamma = 0.4$, see (1.1)), respectively. The results for Gaussian-distributed data are shown on the left and for exponential distributed data on the right. In panels (a,b) the unscaled distribution densities $P_R(m)$ of the maxima within segments of size R are shown for several values of R. Since (1.14) and (1.15) yield that for $R \to \infty$

$$P_R(m) \to \frac{1}{\alpha}\exp\left[-e^{\left(-\frac{m-u}{\alpha}\right)} - \frac{m-u}{\alpha}\right], \tag{1.16}$$

the distribution densities $P_R(m)$ can be scaled upon each other if $\alpha P_R(m)$ is plotted versus $(m-u)/\alpha$, see Figs. 1.20c,d and 1.21c,d. In Figs. 1.20e and 1.21e it is shown that the convergence toward (1.16) (continuous line) is rather slow in the case of a Gaussian distribution of the original data. In contrast, for an exponential distribution $P(x)$ the limiting Gumbel distribution is observed for both uncorrelated and long-term correlated data already for quite small segment sizes R. In Fig. 1.21d deviations occur only at very small R values ($R < 10$) where scaling breaks down due to the sharp cutoff of the exponential density distribution $P(x)$ at $x = 0$. In the long-term correlated case, where the correlation time T diverges, the fast convergence is particularly surprising, since the segment duration R can never exceed T. From a theoretical point of view, we expect a convergence toward the Gumbel limit only for very large R values. The reason for this fast convergence may be a rapid weakening of the correlations among the maxima with increasing values of R, as we will see in the next section (Fig. 1.23).

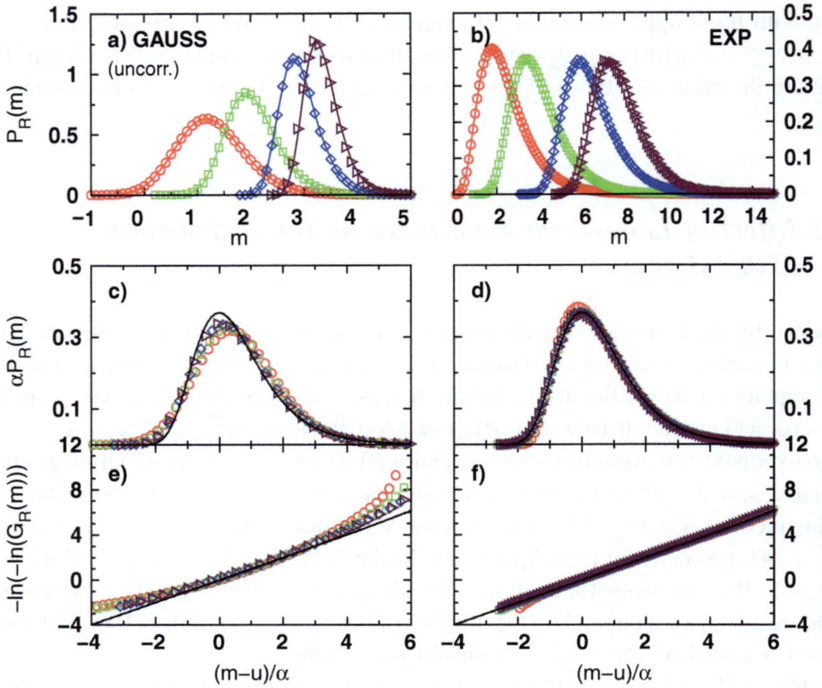

Fig. 1.20 Distributions of maxima in segments of length R for uncorrelated data with (**a, c, e**) Gaussian and (**b, d, f**) exponential distribution density $P(x)$. Panels (**a, b**) show the distribution density function $P_R(m)$ of the maximum values for four segment sizes $R = 6$ (*circles*), 30 (*squares*), 365 (*diamonds*), and 1500 (*triangles*). Panels (**c, d**) show that a collapse of all four curves to a single curve is achieved in both cases, when the m axis is replaced by $(m-u)/\alpha$ and $P_R(m)$ is multiplied by the scale parameter α. The *solid line* is the Gumbel distribution density, (1.16). Panels (e,f) show the corresponding integrated distribution $G_R(m)$ together with the Gumbel function (1.15). Note that exponential records converge to Gumbel much faster than Gaussian records. Figure taken from [1.60]

We concluded that the distribution $P(x)$ of the original data has a much stronger effect upon the convergence toward the Gumbel distribution than the long-term correlations in the data. Long-term correlations just slightly delay the convergence of $G_R(m)$ toward the Gumbel distribution (1.15). This can be observed very clearly in the plot of the integrated and scaled distribution $G_R(m)$ on logarithmic scale in the bottom panels of the Figs. 1.20 and 1.21.

Figure 1.22 shows a direct comparison of the distribution densities $P_R(m)$ for uncorrelated and correlated Gaussian and exponentially distributed data for $R = 365$ (corresponding to annual maxima). The distributions for the long-term correlated data exhibit a slight shift to the left and in particular a significant broadening of the left tail. The reason for this is that correlations cause some periods with many large values x_i and other periods with only relatively small values x_i. When picking the annual maxima from the correlated data the periods where small x_i values dominate will yield rather small annual maxima compared with uncorrelated

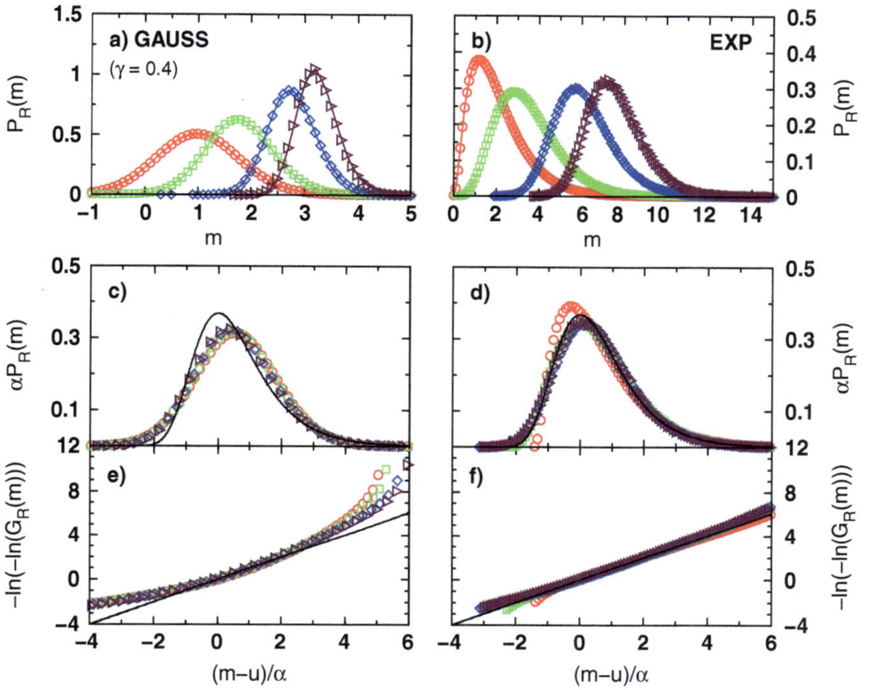

Fig. 1.21 Distributions of maxima in segments of length R for long-term correlated data with $\gamma = 0.4$; for explanations of the plots and the symbols see Fig. 1.20. The curves in (**a,b**) appear broader than in Fig. 1.20, because in correlated data more small m values are considered in $P_R(m)$ than in uncorrelated data. In (**d**) the curve for $R = 6$ (*circles*) differs most from the theoretical curve for uncorrelated data, an effect caused by the correlations together with a rather small R value: the left tail of $P_R(m)$ is strongly affected by the abrupt left end of the exponential. For larger R values this effect disappears and the Gumbel distribution is well approached. For the Gaussian data in (**c,e**) the Gumbel law (*solid line*) is again not well approached. Figure taken from [1.60]

data; this leads to the broadening of the left tail of $P_R(m)$. The largest events are still identified as annual maxima, and hence the right tail of the distribution density is hardly affected by correlations. Figure 1.22 clearly illustrates that the probability of a maxima exceeding an arbitrary but sufficiently large value m^*, $E_R(m^*)$ (see (1.14)), is not significantly different for correlated and uncorrelated data, both for $P(x)$ Gaussian and exponential.

1.3.3 Effect of Long-Term Persistence on the Correlations of the Maxima

The distributions of maxima considered in the previous section do not quantify, however, if the maxima values are arranged in a correlated or in an uncorrelated fashion and if clustering of maxima may be induced by long-term correlations in

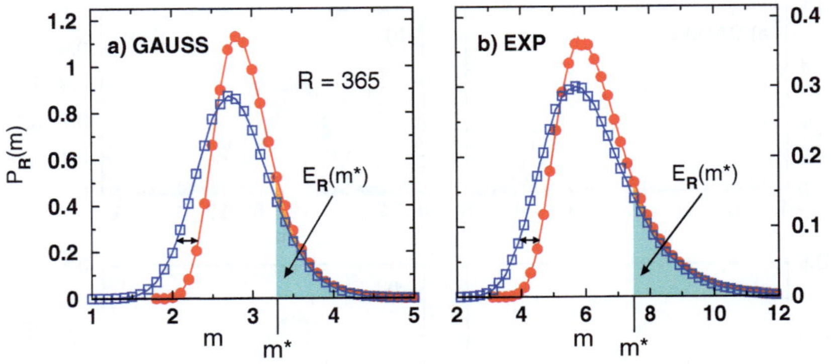

Fig. 1.22 Comparison of the distribution density $P_R(m)$ of maxima for uncorrelated (*circles*) and long-term correlated ($\gamma = 0.4$, squares) data for fixed $R = 365$. For both, (**a**) Gaussian and (**b**) exponentially distributed (x_i), the long-term correlations lead to a similar broadening of the left tail, while the right tail is hardly affected by correlations. In (**a**), the probability $E_R(m^*)$ of finding a maximum m larger than an (arbitrary but sufficiently large) $m^* = 3.3$ is 0.14 for correlated data (*gray area*) compared with 0.16 for uncorrelated data and in (**b**) 0.18 for correlated data compared with 0.2 for uncorrelated data for $m^* = 7.4$. The difference in the exceedance probabilities $E_R(m^*)$ for large m^* is marginal. Figure taken from [1.60]

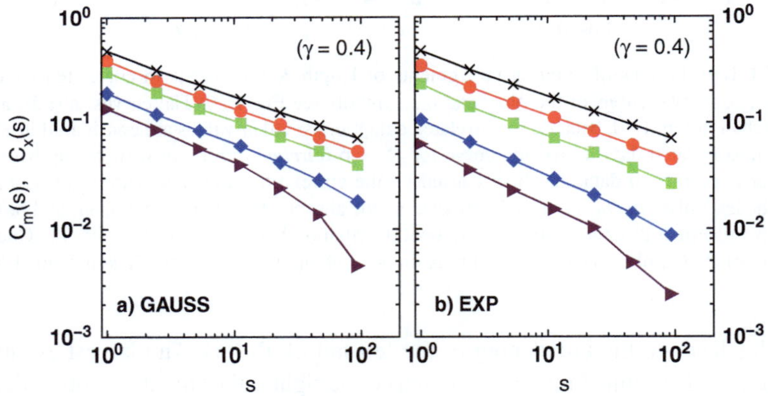

Fig. 1.23 Autocorrelation function $C_m(s)$ of the maxima (m_j) of (**a**) Gaussian and (**b**) exponentially distributed (x_i) for different R values, $R = 6$ (*circles*), $R = 30$ (*squares*), $R = 365$ (*diamonds*), and $R = 1500$ (*triangles*). The autocorrelation function $C_x(s)$ of the original data (x_i) (*crosses*) shows the slope $-\gamma = -0.4$. Figure taken from [1.60]

the data. To study this question, we have evaluated the correlation properties of the series of maxima (m_j), $j = 1, \ldots, N/R$, of long-term correlated data with Gaussian and exponential distribution. Figure 1.23 shows representative results for the maxima autocorrelation function

$$C_m(s) = \frac{\langle (m_j - m_R)(m_{j+s} - m_R) \rangle}{\langle (m_j - m_R)^2 \rangle}, \tag{1.17}$$

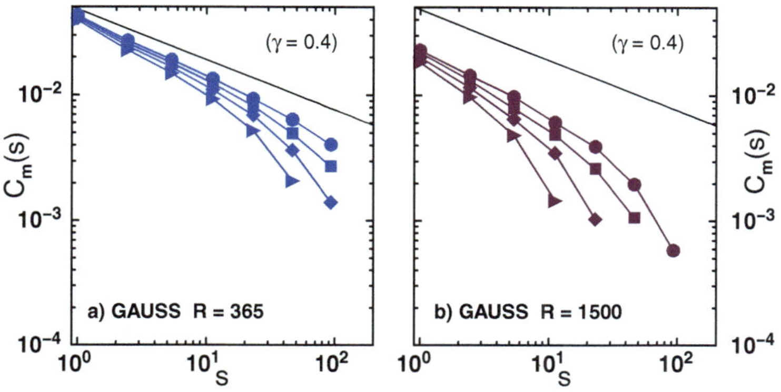

Fig. 1.24 Study of finite-size effects in the autocorrelation function $C_m(s)$ of sequences of maxima (m_j) for Gaussian distributed (x_i) with (**a**) $R = 365$ and (**b**) $R = 1\,500$. The set lengths are $N = 2^{21}$ (*circles*), 2^{20} (*squares*), 2^{19} (*diamonds*), and 2^{18} (*triangles*). The descent of the slopes of $C_m(s)$ from the slope of the straight line $(-\gamma = -0.4)$ with decreasing set length seems to be a finite-size effect. Figure taken from [1.60]

where m_R denotes the mean maximum value in the series and $\langle \ldots \rangle$ is the average over j similar to (1.1). The comparison with the scaling behavior of the autocorrelation function $C_x(s)$ of the original data (x_i) (see (1.1)) that follows a power-law decay, $C_x(s) \sim s^{-\gamma}$ with $\gamma = 0.4$, reveals the presence of long-term correlations with a correlation exponent $\gamma' \approx \gamma$ in the maxima series. Hence, large maxima m are more likely to be followed by large maxima and small maxima are rather followed by small maxima, leading to clusters of large and small maxima. We note that similar behavior has been observed for the series of return intervals between extreme events (see Sect. 1.2.4).

Figure 1.23 also shows that the series of maxima for the same R values appear less correlated for exponentially distributed data than for Gaussian-distributed data. Due to a wider distribution of the maxima in the exponential case (see Fig. 1.22) the autocorrelation function $C_m(s)$ is lower for maxima of exponential data compared to maxima of Gaussian data.

Figure 1.24 shows that the deviations of the autocorrelation function $C_m(s)$ from a power-law fit with slope $\gamma = 0.4$ for large values of R and s are presumably caused by finite-size effects. They become significantly smaller as the length N of the series is increased. In the case of uncorrelated data, the series of maxima is also uncorrelated, $C_m(s) = 0$ for $s > 0$ (not shown).

1.3.4 Conditional Mean Maxima

As a consequence of the long-term correlations in the series of maxima (m_j), the probability of finding a certain value m_j depends on the history, and in particular on the value of the immediately preceeding maximum m_{j-1}, which we will denote by m_0 in the following. This effect has to be taken into account in predictions and

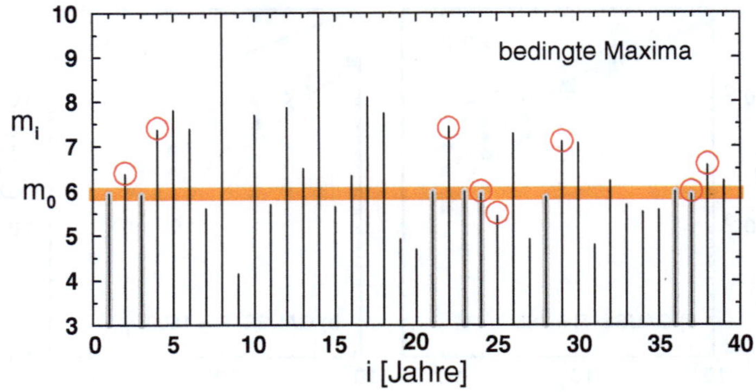

Fig. 1.25 Definition of conditional maxima. In the sequence of (*annual*) maxima only those m values (indicated by *circles*) are considered, which directly follow a maximum of approximate size $m_0 \approx 6$ (*gray band*). The new sequence of m values is the sequence of conditional maxima. Figure taken from [1.60]

risk estimations. For a quantitative analysis we considered conditional maxima as illustrated in Fig. 1.25, where all maxima following an $m_0 \approx 6$ (within the gray band), i.e., the subset of maxima which fulfill the condition of having a preceeding maximum close to m_0, are indicated by circles. The width Δm_0 sketched by the gray band around m_0 in Fig. 1.25 is set such that a sufficient number of approximately 700 conditional maxima is obtained for each record. The corresponding conditional mean maximum value $m_R(m_0)$ is defined as the average of all these conditional maxima. Note that $m_R(m_0)$ will be independent of m_0 for uncorrelated data.

Figure 1.26 shows the conditional mean maxima $m_R(m_0)$ versus m_0 for long-term correlated Gaussian and exponentially distributed data for four values of R. Of course, the mean maxima are larger for larger segment sizes R. This dependence is also observed for the unconditional mean maxima indicated by horizontal lines in Fig. 1.26. In addition to this trivial dependence, the conditional mean maxima significantly depend upon the condition, i.e., the previous maximum m_0, showing a clear memory effect. Evidently, this dependence is most pronounced for the small segment durations $R = 6$ and 30. However, it is still observable for the large $R = 365$ (most common for observational daily data) and even $R \approx 1500$ (beyond common observational limits). Note that the results for Gaussian and exponentially distributed data agree only qualitatively: while the m_0 dependence of $m_R(m_0)$ is quite close to linear for Gaussian data, there seems to be significant curvature for the exponentially distributed data, which is a remnant of the asymmetry of the exponential distribution.

Next we tested our predictions on real records which are known to exhibit long-term correlations. We have studied two data sets (i) the annual data of the Nile river water level minima [1.3, 1.25] and (ii) the reconstructed northern hemisphere annual temperatures by Moberg [1.61]. The Nile series is composed of 663 minimal water levels of the Nile river for the years 622–1284 AD (we use the last 660 data points), measured at Roda gauge near Cairo. Since the Nile data consist of annual

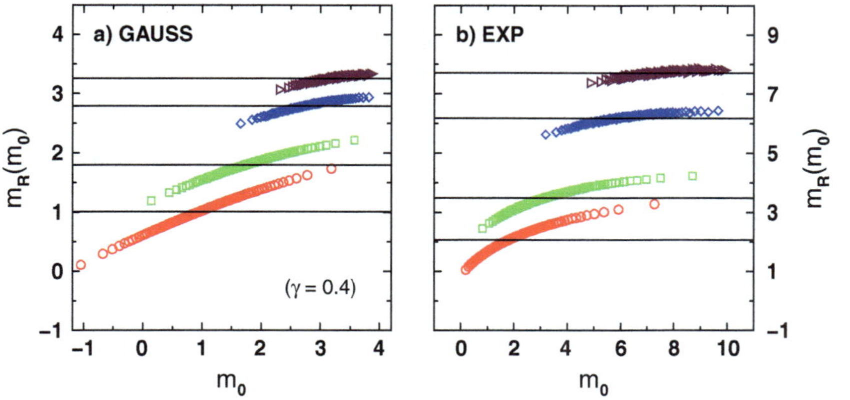

Fig. 1.26 Conditional mean maxima $m_R(m_0)$ for $\gamma = 0.4$ and $R = 6$ (*circles*), $R = 30$ (*boxes*), $R = 365$ (*diamonds*), and $R = 1500$ (*triangles*) versus m_0 for (**a**) Gaussian and (**b**) exponential data. The *straight lines* represent the unconditional means m_R for given R. The width Δm_0 for the condition m_0 was chosen such that approximately 700 m values were obtained for each m_0 in each of the 150 runs of $N = 2^{21}$ x values. Both figures show the memory effect in form of $m_R(m_0) > m_R$ for rather large m_0 (above m_R) and $m_R(m_0) < m_R$ for rather small m_0 (below m_R). Figure taken from [1.60]

minima, we studied extreme minima instead of maxima. The northern hemisphere temperature reconstruction in degree Celsius after Moberg covers the period from 1 to 1979 AD (we use the last 1968 data points) and was last updated in February 2005. The correlation properties of both records have been shown elsewhere [1.3, 1.29, 1.66] to be characterized by $C_x(s) \sim s^{-\gamma}$ with $\gamma \approx 0.3$ (see (1.1)).

In order to get sufficient statistics for the conditional means $m_R(m_0)$, we have considered six m_0 intervals for each value of R and have set the width Δm_0 of the band around m_0 such that there are no gaps between the bands. Figure 1.27 shows the results for three values of R, $R =$ 1y, 6y, and 12y. In all cases, the effect of the long-term correlations on the conditional mean minima and maxima $m_R(m_0)$ is clearly visible for both records: the conditional means are smaller for smaller condition value m_0 and larger for larger condition value.

To prove that the dependence upon m_0 is indeed due to the long-term persistence in the records, we have also studied randomly shuffled surrogate data, where all correlations are removed. As shown by the open symbols in Fig. 1.27 the m_0 dependence completely disappears, indicating that the dependence was due to the correlations in the data.

1.3.5 Conditional Maxima Distributions

The quantity $m_R(m_0)$ is the first moment of the conditional distribution density $P_R(m|m_0)$, which is defined as the distribution density of all maxima m_j that follow a given maximum value m_0 ($m_{j-1} \approx m_0$, see Fig. 1.25). Figure 1.28 shows

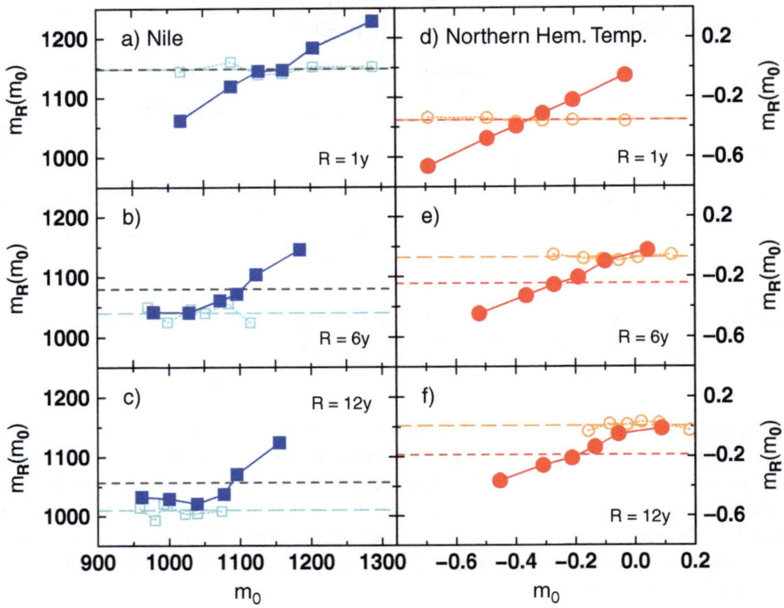

Fig. 1.27 (**a–c**) Conditional mean minima $m_R(m_0)$ for annual data of Nile river water level minima (*squares*) [1.25] and (**d–f**) conditional mean maxima for the reconstructed northern hemisphere annual temperatures after Moberg (*circles*) [1.61], for (**a, d**) $R = 1$y, (**b, e**) $R = 6$y, and (**c, f**) $R = 12$y. The *filled symbols* show the results for the real data and the *open symbols* correspond to surrogate data where all correlations have been destroyed by random shuffling. The shuffled data have a different m_0 range due to the broadening of the left tail of $P_R(m)$ in the correlated data (see Fig. 1.22). The unconditional mean minima (**a–c**) and maxima (**d–f**) are indicated by *dashed lines*; the *long-dashed lines* correspond to the shuffled data. Figure taken from [1.60]

$P_R(m|m_0)$ for two values of m_0 and again for Gaussian as well as for exponentially distributed long-term correlated data sets with $\gamma = 0.4$. When compared with the unconditional distribution density $P_R(m)$, the long-term correlations lead to a shift of $P_R(m|m_0)$ to smaller m values for small m_0 and to larger m values for large m_0, respectively. The conditional exceedance probability

$$E_R(m|m_0) = \int_m^\infty P_R(m'|m_0) - \mathrm{d}m'. \tag{1.18}$$

defines the probability of finding a maximum larger than m provided that the previous value was close to m_0. We found a strong dependence of $E_R(m|m_0)$ upon the condition m_0. Consequently, the difference between the unconditional probabilities $E_R(m)$ (see Fig. 1.22) and the corresponding conditional probabilities $E_R(m|m_0)$ depends strongly on m_0 in the presence of long-term correlations.

Next we quantified the effect of long-term correlations upon $E_R(m|m_0)$ for different conditions m_0 and different m values. Figure 1.29 shows the conditional exceedance probability $E_R(m|m_0)$ for six m values versus m_0. The m values were chosen such that the corresponding unconditional probabilities are $E_R(m) = 0.9, 0.5, 0.3, 0.1, 0.05$, and 0.01, respectively. For the Gaussian data and m corresponding to $E_R(m) = 0.5$ the curve $E_R(m|m_0)$ varies by factor of 2 depending on m_0, while the variation does not exceed a factor of 1.5 for the exponential data.

In general, the memory effect caused by the long-term correlations seems to be strongest for intermediate m values. For $E_R(m) \leq 0.5$, the larger the m value (i.e., the lower the curve in Fig. 1.29) the smaller is the apparent effect of the correlations on the difference between the conditional probabilities $E_R(m|m_0)$ (symbols) and the unconditional probabilities $E_R(m)$ (straight lines). Hence, Fig. 1.29 may suggest that the memory effect will disappear for very large m values. This, however, is not true. Figure 1.30 shows the ratios of the conditional exceedance probabilities $E_R(m|m_0)$ and the unconditional exceedance probabilities $E_R(m)$. The figure clearly shows an increase of the memory effect for larger m values, i.e., for more extreme events. This increase seems weaker for exponentially distributed data than for Gaussian-distributed data due to the less correlated maximum series of exponential data; however, the tendency is the same. As Fig. 1.30 shows $E_R(m|m_0)$ can differ up to a factor of 2 from $E_R(m)$ when considering the history m_0 in the presence of long-term correlations (with $\gamma = 0.4$). This effect has to be taken into account in predictions and risk estimations of large events.

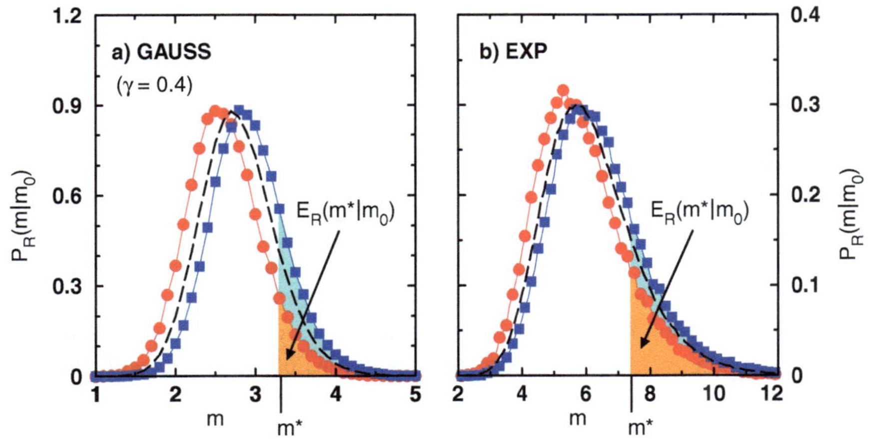

Fig. 1.28 (**a**) Conditional distribution density $P_R(m|m_0)$ of maximum values taken from correlated Gaussian data ($\gamma = 0.4$) with $R = 365$ and $m_0 = 2.06$ (*circles*) as well as $m_0 = 3.55$ (*squares*). (**b**) the same as (**a**) for exponentially distributed data with $m_0 = 4.10$ (*circles*) and $m_0 = 8.65$ (*squares*). The width Δm_0 around m_0 is set such that a sufficient number of approximately 700 conditional maxima is obtained for each of the 150 data sets considered here. The conditional exceedance probability $E_R(m^*|m_0)$, i.e., the probability to find a m value larger than an arbitrarily given m^* (see (1.14)), also depends on the history m_0. For example, in (**a**) $E_{365}(3.30|2.06) = 0.08$ (*gray area*) is significantly smaller than $E_{365}(3.30|3.55) = 0.20$ (*black area* plus *gray area*), and in (**b**) $E_{365}(7.35|4.10) = 0.14 < E_{365}(7.35|8.65) = 0.24$. Figure taken from [1.60]

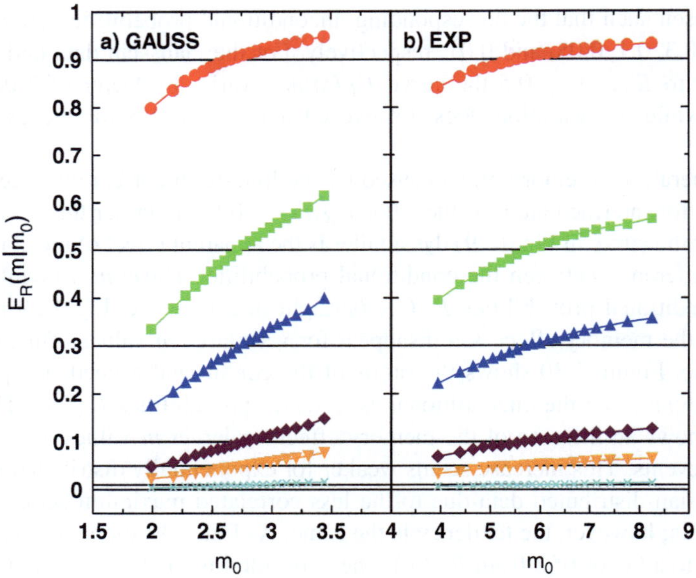

Fig. 1.29 Conditional exceedance probabilities $E_R(m|m_0)$ to find a maximum larger than a given m for (**a**) Gaussian and (**b**) exponentially distributed data with $\gamma = 0.4$ and $R = 365$ versus the condition m_0. The *straight lines* indicate the corresponding unconditional exceedance probabilities $E_R(m)$. The six m values were chosen such that $E_R(m) = 0.9$ (*circles*, $m = 2.15$ for Gaussian and 4.40 for exponential data), 0.5 (*squares*, $m = 2.75$ and 5.95), 0.3 (*triangles* up, $m = 2.95$ and 6.70), 0.1 (*diamonds*, $m = 3.35$ and 8.00), 0.05 (*triangles down*, $m = 3.55$ and 8.75), and 0.01 (*crosses*, $m = 3.95$ and 10.40). Each point in the graph is based on a statistics of 500 conditional m values and averaged over 150 runs of $N = 2^{21}$ x values. Figure taken from [1.60]

1.4 Centennial Events

In this section we discuss how the size of typical centennial events, i.e., typical maxima that occur once in 100 years, can be determined in practice. Such values are commonly used, e.g., in hydrological risk estimation for centennial floodings. If all distributions $P(x)$ and $P_R(m)$ of the considered time series were known exactly, i.e., for infinitely long records, two alternative definitions of typical centennial events would be possible as illustrated in Fig. 1.31a, b.

The *first* definition of a typical centennial event (see Fig. 1.31a) considers the distribution density $P(x)$ (here: a Gaussian distribution of x_i, representing daily data) and determines the quantile $q_{36\,500}$ (dashed line) that is exceeded by only 1/36 500 of all values of the distribution, i.e., on average $x_i > q_{36\,500}$ occurs once in 100 years = 36 500 days. For this definition the distribution $P(x)$ must be known for very rare events. Moreover, since $q_{36\,500}$ is based only on the distribution of the values x_i, it is unaffected by possible correlations and clustering of centennial events (see [1.29, 1.36]). This definition takes into account all events that exceed the quantile, regardless of whether they occur within the same 100-year period or not.

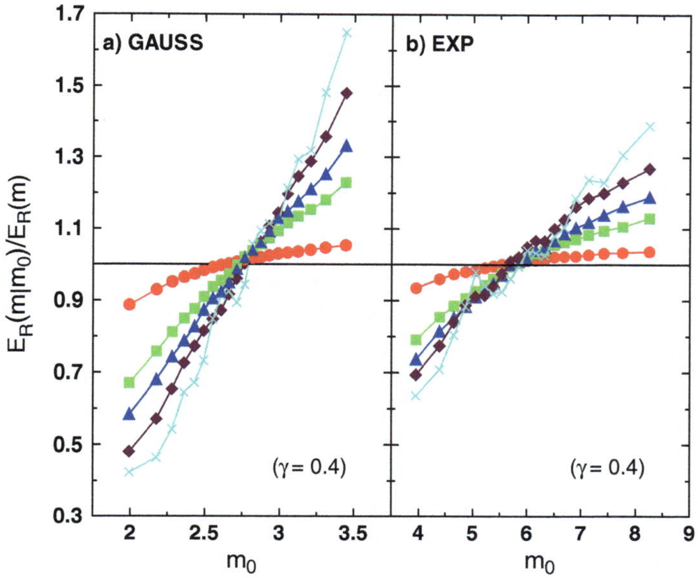

Fig. 1.30 Ratios of the conditional and unconditional exceedance probabilities $E_R(m|m_0)$ and $E_R(m)$ to find a maximum larger than m for (**a**) Gaussian and (**b**) exponentially distributed data with $\gamma = 0.4$ and $R = 365$. The symbols and the statistics are the same as in Fig. 1.29 except for the data corresponding to $E_R(m) = 0.05$ (*triangles down*), which are not shown here to avoid overlapping symbols. The effect of long-term correlations seems to be strongest for the largest m (*crosses*): depending on m_0 $E_R(m|m_0)$ varies from 0.4 up to 1.7 for Gaussian data, i. e., by a factor greater than four. For exponential data this factor is still greater than two. Figure taken from [1.60]

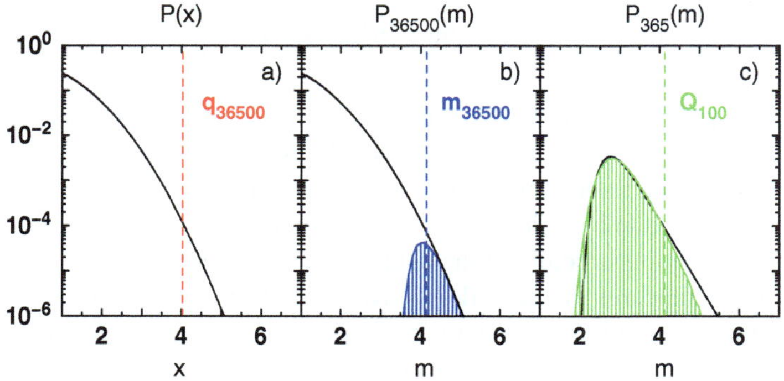

Fig. 1.31 Definition of centennial events: (**a**) the tail of a Gaussian distribution $P(x)$ (*solid black line*) with the $1/36\,500$ quantile (*dashed line*), (**b**) the Gaussian tail (*solid black line*) and the histogram $P_{36\,500}(m)$ of centennial maxima with the average $m_{36\,500}$ (*dashed line*), and (**c**) the histogram $P_{365}(m)$ of annual maxima and the corresponding Gumbel fit ((1.16), *solid line*) with its $1/100$ quantile Q_{100} (*dashed line*, (1.19)). The deviations of the fit from the histogram are caused by the underlying Gaussian distribution, for which $R = 365$ is not sufficiently large to reach the Gumbel limit. Figure taken from [1.60]

The *second* definition of a typical centennial event (see Fig. 1.31b) considers the distribution density $P_{36\,500}(m)$ of centennial maxima m within periods of 100 years (histogram in the figure) and determines the mean centennial maximum value $m_{36\,500}$ as the first moment of this distribution (*dashed line*). Clearly, this definition includes the effects of correlations, but multiple exceedances of the threshold within one period of 100 years are not regarded in $m_{36\,500}$ by definition. Still, many centennial events must have occurred in the record (x_i) to allow the determination of $P_{36\,500}(m)$.

In real data, however, the number of values obtained from records with typical durations of 30, 50, or 100 years is not sufficient to study the distribution densities $P(x)$ or $P_R(m)$ in order to determine directly the size of centennial events by calculation of $q_{36\,500}$ or $m_{36\,500}$. It is usually not known a priori if even one centennial event occurred within the observational period. This makes it very difficult to estimate the size of a typical centennial event. Therefore, a practical definition is needed. This *third* definition assumes that $P(x)$ is (most likely) in the domain of attraction of the Gumbel distribution. Then one applies the Gumbel fit formula (1.16) to the distribution density $P_R(m)$ of maxima m within (smaller) segments of size $R = 365$ days and approximates the typical centennial event from this Gumbel fit. This procedure is illustrated in Fig. 1.31c, where the histogram $P_{365}(m)$ of annual maxima and a fitted Gumbel curve (solid line) are shown. It is common practice in hydrology to estimate the size of a typical centennial event by calculating the threshold Q_{100} (dashed line in Fig. 1.31c), which is exceeded by only 1/100 of the fitted Gumbel distribution of annual maxima [1.63, 1.65]. In terms of the integrated Gumbel distribution $G_{365}(m)$ (see (1.14) and (1.15)) this definition corresponds to $G_{365}(Q_{100}) = 0.99$, which yields $Q_{100} = u - \alpha \ln(-\ln 0.99)$, i.e.,

$$Q_{100} = m_{365} - \left[\ln(-\ln 0.99) + n_e\right]\frac{\sqrt{6}}{\pi}\sigma_{365} \qquad (1.19)$$

or $Q_{100} \approx m_{365} + 3.14\sigma_{365}$ [1.63, 1.65]. Here, m_{365} and σ_{365} denote the annual average and its standard deviation, respectively, which are easily accessible also in short records. The three definitions for centennial maxima, $q_{36\,500}$, $m_{36\,500}$, and Q_{100} (most regarded in hydrology), are similar, but differ slightly depending on the underlying correlation structure of the data, as we will show now.

Figure 1.32 compares the quantile $q_{36\,500}$, the distribution density $P_{36\,500}(m)$ of centennial maxima, and the distribution of Q_{100} values for correlated and uncorrelated Gaussian-distributed data. To obtain a similar statistical basis for both, $P_{36\,500}(m)$ and the distribution of Q_{100}, segments of length 36 500 days (100 years) should be considered for each of the Q_{100} values. This was done for the data shown by filled circles in Fig. 1.32. However, since real observational records are often shorter, Fig. 1.32 also shows the distribution of the Q_{100} values based on segments of 10 950 days (30 years, squares). One can see that $P_{36\,500}(m)$ and also the corresponding (actual) mean centennial maximum $m_{36\,500}$ are less affected by the considered long-term correlations than the estimated centennial events Q_{100} based on Gumbel fits, while the quantile $q_{36\,500}$ is independent of correlations. Note also that

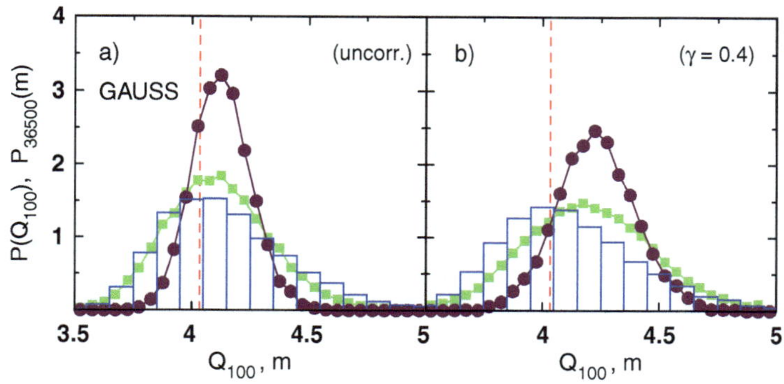

Fig. 1.32 Comparison of the quantile q_{36500} (*dashed vertical lines*), the distribution density $P_{36\,500}(m)$ of centennial maxima (*histograms*), and the distributions of the Q_{100} values based on 30 years (*squares*) and 100 years (*circles*) for Gaussian-distributed (**a**) uncorrelated and (**b**) long-term correlated ($\gamma = 0.4$) data; see Fig. 1.31 and Eq. (1.19) for the definitions. Each Q_{100} value was calculated from a different segment of length 36 500 days (= 100 years, *circles*) and 10 950 days (= 30 years, *squares*). The histograms were obtained from 150 artificial records with $N = 2^{21}$. Figure taken from [1.60]

the shift of the Q_{100} histogram due to the correlations is larger in contrast to the small shift of the $P_{36\,500}(m)$ histogram. For $P_{36\,500}(m)$ we observe mainly a broadening caused by the correlations (see Fig. 1.22). The shift of the Q_{100} histogram to the right is probably caused by the influence of the scale parameter in the fit formula, i.e., by the standard deviation σ_{365} that appears in Eq. (1.19). In addition, for the quantity Q_{100} a broadening of the histogram is observed for the correlated data, leading to a less accurate estimation of the typical centennial event. However, the distribution of the Q_{100} values still remains significantly narrower than $P_{36\,500}(m)$, which indicates that – for single records comprising just about 100 years – typical centennial events can be approximated somewhat more reliably using Q_{100} instead of the single maximum m picked from the record. The estimations of centennial events via Q_{100} based on just 30 years (Figs. 1.32c,d) are similarly reliable as those via the maximum picked from a 100-year series.

In order to compare the dependence of the actual and estimated sizes of centennial events and the accuracy of their estimation in long-term correlated records quantitatively, we have studied histograms like those in Fig. 1.32 for artificial data characterized by different correlation exponents γ. Figure 1.33 shows, as a function of γ, the constant value $q_{36\,500}$ (as defined in Fig. 1.31a), the (actual) mean centennial maximum $m_{36\,500}$ (diamonds) with the corresponding standard deviation $\sigma_{36\,500}$ and the mean estimated centennial maximum $\langle Q_{100}\rangle$ with its standard deviation. Again, each single value Q_{100} is calculated for a segment of length 10 950 days (30 years) in (a, squares) and 36 500 days (100 years) in (b, circles). The quantity $q_{36\,500}$ is (trivially) independent of the long-term correlations, because it is based only on the distribution density $P(x)$ and can hardly be calculated in practice, because $P(x)$ required for the calculation is usually not known. In contrast to $q_{36\,500}$, $m_{36\,500}$ and even more $\langle Q_{100}\rangle$ are significantly affected by strong long-term

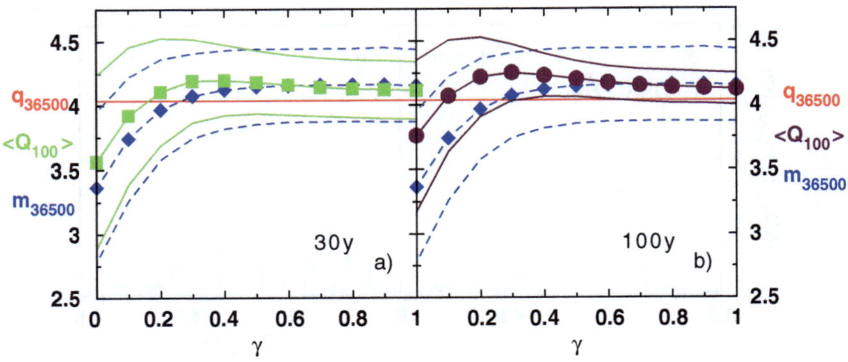

Fig. 1.33 Dependence of mean centennial maxima m_{36500} and estimated centennial events Q_{100} on the correlation exponent γ of the data; small values of γ indicating strong long-term correlations. The average (*actual*) centennial maxima (*diamonds*) and the corresponding upper fluctuations (range of one standard deviation indicated by *dashed lines*) is compared with the average Q_{100} value (*squares and circles*) and its lower fluctuations (*solid lines*). The Q_{100} values were obtained from segments of size (**a**) 10 950 days (30 years, *squares*) and (b) 36 500 days (100 years, *circles*). The *solid horizontal line* in both figures is the 1/36 500 quantile $q_{36\,500} = 4.034$. Figure taken from [1.60]

correlations ($\gamma < 0.3$). Both values decrease with increasing correlations (decreasing γ). Due to the strong long-term correlations large maxima tend to cluster, i.e., there are epochs where considerably more large maximum values occur than in weakly correlated and uncorrelated data [1.29]. As a consequence, there exist also epochs, where the maximum values are considerably lower than those in weakly or uncorrelated records. With increasing correlations ($\gamma < 0.3$) these periods of small maxima become more pronounced and more frequent, forcing the average centennial maximum $m_{36\,500}$ and also $\langle Q_{100} \rangle$ (which is based on annual maxima) to drop below the quantile $q_{36\,500}$. The corresponding standard deviations (dashed lines and solid lines), which characterize the widths of the histograms of $P_{36\,500}(m)$ and $P(Q_{100})$ and thus carry the information regarding the accuracy of the estimations for short data, increase with decreasing γ. For $\gamma > 0.4$ the mean centennial maximum $m_{36\,500}$ and $\langle Q_{100} \rangle$ are roughly constant.

The mean estimated centennial maximum $\langle Q_{100} \rangle$ tends to overestimate the size of the centennial events for $\gamma \leq 0.6$. However, the systematic deviation from $m_{36\,500}$ is even smaller for $\langle Q_{100} \rangle$ based on just 30 years of data (Fig. 1.33a) than for 100 years of data (Fig. 1.33b). In addition, for weakly correlated data, the corresponding standard deviations are lower than $\sigma_{36\,500}$, indicating more reliable estimations of centennial maxima with Q_{100} for short records. In conclusion, we found that the quantity HQ_{100} most regarded in hydrology is a very reliable predictor of typical centennial events in short records, but the values tend to be systematically larger than the $m_{36\,500}$ values for data with strong long-term correlations. However, in real hydrology data such as river runoff data the correlations are hardly stronger than $\gamma = 0.3$ [1.6, 1.67]. So the Q_{100} value is still a good estimator for centennial floods.

Finally, we studied the effect of long-term memory on Q_{100}. While there is hardly any memory to be expected in $m_{36\,500}$ because of the very large R value

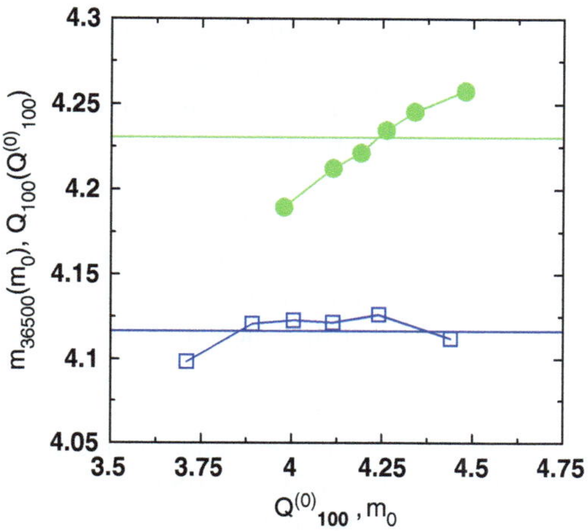

Fig. 1.34 Conditional average Q_{100} (*filled circles*) and conditional $m_{36\,500}$ (*squares*) for Gaussian long-term correlated ($\gamma = 0.4$) data. The $\langle Q_{100}(Q_{100}^{(0)})\rangle$ values show a correlation-based memory effect indicated by the upward trend, while the m_{36500} values seem to have no memory due to the large R value. The *horizontal lines* correspond to the unconditional average values 4.23 and 4.12 for $\langle Q_{100}\rangle$ and $m_{36\,500}$, respectively. The Q_{100} values were calculated from segments of length 36 500 days. Figure taken from [1.60]

(see Figs. 1.23 and 1.26), Q_{100} should still show some dependence on the history, because it is based on m_{365} and σ_{365}. Figure 1.34 shows, for long-term correlated Gaussian-distributed data, the conditional average $\langle Q_{100}(Q_{100}^{(0)})\rangle$ and the conditional $m_{36\,500}(m_0)$, i.e. the average Q_{100} value following a Q_{100} of size $Q_{100}^{(0)}$ and analogous for $m_{36\,500}(m_0)$. While the conditional average Q_{100} (filled circles) depends on the history, the conditional $m_{36\,500}$ (squares) fluctuate around the unconditional mean (solid line). Although the conditional effect on Q_{100} is relatively small, it remains measurable and can help to improve predictions of extreme events within given periods of time.

1.5 Conclusion

In summary we have studied the effect of long-term correlations in time series upon extreme value statistics and the return intervals between extreme events.

For the return intervals in long-term persistent data with Gaussian, exponential, power-law, and log-normal distribution densities, we have shown that mainly the correlations rather than the distributions affect the return interval statistics, in particular, the return interval distribution density, the conditional return interval distribution density, the conditional mean return intervals, and the correlation properties of the return interval series. The stretched exponential decay of the return interval distribution density for long return intervals is complemented by a power-law decay for small return intervals, which will dominate the observable behavior in most (rather short) observational data. Still, the scaling behavior with the mean return interval holds in both regimes. Besides, discreteness corrections as well as finite-size corrections have to be taken into account for very short and long return intervals. We have shown that the long-term persistence inherent in hydroclimate records represents a natural mechanism for the clustering of the hazardous events. As a consequence of the long-term memory, the mean residual time to the next event increases with the elapsed time and depends strongly on the previous return interval. We have demonstrated that this counterintuitive phenomenon can be seen in long climate records. We also showed that the calculation of double-conditional return periods might yield an improved risk estimation for the occurrence of hazardous events in practical applications.

Considering series of maxima within segments of size R of the original data, we have shown numerically that the maxima distribution functions still converge to the same type of Gumbel distributions as for uncorrelated data for increasing R. For finite values of R, however, some deviations occur especially for originally Gaussian-distributed data. Our extensive numerical simulations suggested that contrary to the common assumption in extreme value statistics, the maxima time series turn out to be *not* independently, identically distributed numbers. The series of maxima rather exhibit long-term correlations similar to those in the original data. Most notably we have found that the maxima distribution as well as the mean maxima significantly depend on the history, in particular on the value of the previous maximum. In addition, we have shown that long-term memory can lead to a slight systematic overestimation of centennial events if the approximation Q_{100} is considered. In general, Q_{100} is a surprisingly reliable approximation for centennial events, especially in short records.

Nevertheless, further work is needed to test if our findings are similar in other (non-Gaussian) initial distributions. In addition, we suggest that memory via the conditional mean maxima and conditional maxima distributions as well as conditional Q_{100} values should be considered for an improved risk estimation in long-term correlated data. It is also plausible that multiscaling, which occurs, e.g., in many hydrological time series, might have an even more significant impact on risk estimation and the prediction of extreme events like floods. Further work is definitely required to study the effects of multiscaling in time series upon extreme value statistics.

Acknowledgments This work has been supported by the Bundesministerium für Bildung und Forschung (under grant no. 0330271) and the Israel Science Foundation.

Appendix: Data Generation

The long-term correlations in the random numbers were introduced by the Fourier-filtering technique, see, e.g. [1.31]. However, for the non-Gaussian distributions, (1.3), (1.4), and (1.5), the shape of $D(x)$ in not preserved by Fourier filtering.

In these cases we applied an iterative algorithm introduced by Schreiber and Schmitz [1.33, 1.34]. The algorithm consists of the following steps: First one creates a Gaussian-distributed long-term correlated data set with the desired correlation exponent γ by standard Fourier filtering [1.31]. The power spectrum $S_{\mathrm{G}}(\omega) = F_{\mathrm{G}}(\omega)F_{\mathrm{G}}^{*}(\omega)$ of this data set is considered as reference spectrum (where ω denotes the frequency in Fourier space and the $F_{\mathrm{G}}(\omega)$ are the complex Fourier coefficients). Next one creates an uncorrelated sequence of random numbers (x_i^{ref}), following a desired distribution, e.g., exponential distribution (1.3). The (complex) Fourier transform $F(\omega)$ of the (x_i^{ref}) is now divided by its absolute value and multiplied by the square root of the reference spectrum

$$F_{\mathrm{new}}(\omega) = \frac{F(\omega)\sqrt{S_{\mathrm{G}}(\omega)}}{|F(\omega)|}. \tag{1.20}$$

After the Fourier back-transformation of $F_{\mathrm{new}}(\omega)$, the new sequence (x_i^{new}) has the desired correlations (i.e., the desired γ), but the shape of the distribution has changed toward a (more or less) Gaussian distribution. In order to enforce the desired distribution, we exchange the (x_i^{new}) by the (x_i^{ref}), such that the largest value of the new set is replaced by the largest value of the reference set, the second largest of the new set by the second largest of the reference set, and so on. After this the new sequence has the desired distribution and is clearly correlated. However, due to the exchange algorithm the perfect long-term correlations of the new data sequence were slightly altered again. So the procedure is repeated: the new sequence is Fourier transformed followed by spectrum adjustment, and the exchange algorithm is applied to the Fourier back-transformed data set. These steps are repeated several times, until the desired quality (or the best possible quality) of the spectrum of the new data series is achieved.

References

1.1. A. Bunde, J. Kropp, H.-J. Schellnhuber (eds.) *The Science of Disasters – Climate Disruptions, Heart Attacks, and Market Crashes*, edited (Springer, Berlin, 2002)
1.2. A. Bunde, S. Havlin (eds.), *Fractals in Science* (Springer, Berlin, 1994)
1.3. H.E. Hurst, R.P. Black, Y.M. Simaika, *Long-term Storage: An Experimental Study* (Constable & Co. Ltd., London, 1965)
1.4. B.B. Mandelbrot, J.R. Wallis, Wat. Resour. Res. **5**, 321 (1969)
1.5. I. Rodriguez-Iturbe et al., *Fractal River Basins - Change and Self-Organization* (Cambridge University Press, Cambridge, 1997)
1.6. J.W. Kantelhardt, E. Koscielny-Bunde, D. Rybski, P. Braun, A. Bunde, S. Havlin, J. Geophys. Res. (Atmosph.) **111**, D01106 (2006)
1.7. E. Koscielny-Bunde, A. Bunde, S. Havlin, Y. Goldreich, Physica A **231**, 393 (1996)
1.8. J.D. Pelletier, D.L. Turcotte, J. Hydrology **203**, 198 (1997)
1.9. E. Koscielny-Bunde, A. Bunde, S. Havlin, H.E. Roman, Y. Goldreich, H.-J. Schellnhuber, Phys. Rev. Lett. **81**, 729 (1998)
1.10. P. Talkner, R.O. Weber, Phys. Rev. E **62**, 150 (2000)

1.11. J.F. Eichner, E. Koscielny-Bunde, A. Bunde, S. Havlin, H.-J. Schellnhuber, Phys. Rev. E **68**, 046133 (2003)
1.12. M.F. Shlesinger, B.J. West, J. Klafter, Phys. Rev. Lett. **58**, 1100 (1987)
1.13. R.R. Prasad, C. Meneveau, K.R. Sreenivasan, Phys. Rev. Lett. **61**, 74 (1988)
1.14. C.-K. Peng, J. Mietus, J.M. Hausdorff, S. Havlin, H.E. Stanley, A.L. Goldberger, Phys. Rev. Lett. **70**, 1343 (1993)
1.15. A. Bunde, S. Havlin, J.W. Kantelhardt, T. Penzel, J.-H. Peter, K. Voigt, Phys. Rev. Lett. **85**, 3736 (2000)
1.16. J.W. Kantelhardt, T. Penzel, S. Rostig, H.F. Becker, S. Havlin, A. Bunde, Physica A **319**, 447 (2003)
1.17. C.-K. Peng, S.V. Buldyrev, A.L. Goldberger, S. Havlin, F. Sciortino, M. Simons, H.E. Stanley, Nature **356**, 168 (1992)
1.18. A. Arneodo, E. Bacry, P.V. Graves, J.F. Muzy, Phys. Rev. Lett. **74**, 3293 (1995)
1.19. Y.H. Liu, P. Cizeau, M. Meyer, C.-K. Peng, H.E. Stanley, Physica A **245**, 437 (1997)
1.20. Y. Liu, P. Gopikrishnan, P. Cizeau, M. Meyer, C.-K. Peng, H.E. Stanley, Phys. Rev. E **60**, 1390 (1999)
1.21. W.E. Leland, M.S. Taqqu, W. Willinger, D.V. Wilson, IEEE/ACM Transactions on networking **2**, 1 (1994); V. Paxson, S. Floyd, IEEE/ACM Transactions on networking **3**, 226 (1995)
1.22. D.M. Meko, M.D. Therrell, C.H. Baisan, M.K. Hughes, J. Amer. Wat. Resourc. Assoc. **37**, 4 (2001)
1.23. J.J. Moore, K.A. Hughen, G.H. Miller, J.T. Overpeck, J Paleolimnol **25**, 503 (2001)
1.24. H.D. Grissino-Mayer, *Tree Rings, Environment, and Humanity*, ed. by J.S. Dean, D.M. Meko, T.W. Swetnam (Radiocarbon 1996, Department of Geosciences, The University of Arizona, Tucson, AZ, (1996). pp. 191–204
1.25. B. Whitcher et al., Wat. Resour. Res. **38**, 1054 (2002), data obtained from: http://sunsite.univie.ac.at/statlib/S/beran (Accessed in Jan. 2003)
1.26. M.E. Mann, R.S. Bradley, M.K. Hughes, Geophys. Res. Lett. **26**, 759 (1999)
1.27. NOAA Paleoclimatology Program, http://www.ngdc.noaa.gov/paleo/recons.html.
1.28. C.-K. Peng, S.V. Buldyrev, S. Havlin, M. Simons, H.E. Stanley, A.L. Goldberger, Phys. Rev. E **49**, 1685 (1994)
1.29. A. Bunde, J.F. Eichner, J.W. Kantelhardt, S. Havlin, Phys. Rev. Lett. **94**, 48701 (2005)
1.30. J.F. Eichner, J.W. Kantelhardt, A. Bunde, S. Havlin, Phys. Rev. E **73**, 16130 (2006)
1.31. H.A. Makse, S. Havlin, M. Schwartz, H.E. Stanley, Phys. Rev. E **53**, 5445 (1996)
1.32. D.L. Turcotte, *Fractals and Chaos in Geology and Geophysics* (Cambridge University Press, Cambridge 1992)
1.33. T. Schreiber, A. Schmitz, Phys. Rev. Lett. **77**, 635 (1996)
1.34. T. Schreiber, A. Schmitz, Physica D **142**, 346 (2000)
1.35. A. Bunde, J.F. Eichner, J.W. Kantelhardt, S. Havlin, Physica A **330**, 1 (2003)
1.36. A. Bunde, J.F. Eichner, J.W. Kantelhardt, S. Havlin, Physica A **342**, 308 (2004)
1.37. E.G. Altmann, H. Kantz, Phys. Rev. E **71**, 056106 (2005)
1.38. M. Kac, Bull. Am. Math. Soc. **53**, 1002 (1947)
1.39. H.v. Storch and F. W. Zwiers, *Statistical Analysis in Climate Research* (Cambridge University Press, Cambridge 2001)
1.40. When considering the (different) problem of zero level crossing in long term correlated Gaussian data it has been proven by G. F. Newell and M. Rosenblatt in Ann. Math. Statist. **33**, 1306 (1962), that the probability of having no zero level crossing after t time steps is bounded from above by a stretched exponential. See also I. E. Blake and W. C. Lindsay, IEEE Trans. Inform. Theory **19**, 295 (1973) and S. N. Majumdar, Current Sci **77**, 370 (1999)
1.41. A. Corral, Phys. Rev. Lett. **92**, 108501 (2004)
1.42. V.N. Livina, S. Havlin, A. Bunde, Phys. Rev. Lett. **95**, 208501 (2005)
1.43. R. Shcherbakov, G. Yakovlev, D.L. Turcotte, J.B. Rundle, Phys. Rev. Lett. **95**, 218501 (2005)
1.44. P.D. Jones, K.R. Briffa, T.P. Barnett, S.F.B Tett, The Holocene **8**, 455 (1998); data obtained from: IGBP PAGES/World Data Center-A for Palaeoclimatology Data Contribution Series #1998-039. NOAA/NGDC Palaeoclimatology Program, Boulder CO, USA

1.45. D. Sornette, L. Knopoff, Bull. Seism. Soc. Am. **87**, 789 (1997)
1.46. E.J. Gumbel, *Statistics of Extremes* (Columbia University Press, New York, 1958)
1.47. M.R. Leadbetter, G. Lindgren, H. Rootzen, *Extremes and Related Properties of Random Sequences and Processes* (Springer, New York, NY, 1983)
1.48. J. Galambos, J. Lechner, E. Simin (eds.), *Extreme Value Theory and Applications* (Kluwer, Dordrecht, 1994).
1.49. J. Galambos *The Asymptotic Theory of Extreme Order Statistics*, (Wiley, New York, NY, 1978)
1.50. P. Embrechts, C. Klüppelberg, T. Mikosch, *Modelling Extremal Events*, ed. by I. Karatzas, M. Yor (Springer, Berlin, 1997)
1.51. T. Antal, M. Droz, G. Györgyi, Z. Racz, Phys. Rev. Lett. **87**, 240601 (2001)
1.52. S.T. Bramwell, P.C.W. Holdsworth, J.-F. Pinton, Nature **396**, 552 (1998)
1.53. S.T. Bramwell, K. Christensen, J.-Y. Fortin, P.C.W. Holdsworth, H.J. Jensen, S. Lise, J.M. Lopez, M. Nicodemi, J.-F. Pinton, M. Sellitto, Phys. Rev. Lett. **84**, 3744 (2000)
1.54. K. Dahlstedt, H. J. Jensen, J. Phys. A: Math. Gen. **34**, 11193 (2001)
1.55. D.S. Dean S.N. Majumdar, Phys. Rev. E **64**, 046121 (2001)
1.56. S. Raychaudhuri, M. Cranston, C. Przybyla, Y. Shapir, Phys. Rev. Lett. **87**, 136101 (2001)
1.57. S.N. Majumdar, A. Comtet, Phys. Rev. Lett. **92**, 225501 (2004); J. Stat. Phys. **119**, 777 (2005)
1.58. H. Guclu, G. Korniss, Phys. Rev. E **69**, 65104(R) (2004)
1.59. S.M. Berman, Ann. Math. Statist. **35**, 502 (1964)
1.60. J.F. Eichner, J.W. Kantelhardt, A. Bunde, S. Havlin, Phys. Rev. E **73**, 16130 (2006)
1.61. Data obtained from: http://www.ndcd.noaa.gov/paleo/recons.html. See also: A. Moberg, D. M. Sonechkin, K. Holmgren, N. M. Datsenko, W. Karln, Nature **433**, 613 (2005).
1.62. R.A. Fisher, L.H.C. Tippett, Proc. Camb. Phil. Soc. **24**, 180 (1928)
1.63. V. te Chow, *Handbook of Applied Hydrology* (McGraw-Hill Book Company, New York, NY, 1964)
1.64. A.J. Raudkivi, *Hydrology* (Pergamon Press, Oxford, 1979)
1.65. P.F. Rasmussen, N. Gautam, J. Hydrol. **280**, 265 (2003)
1.66. D. Rybski, private communication
1.67. E. Koscielny-Bunde, J.W. Kantelhardt, P. Braun, A. Bunde, S. Havlin, J. Hydrol. **322**, 120 (2006)

Karl Friedrich Hieronymus Baron von Münchhausen, born 11 May 1720 at Bodenwerder (Germany), died 22 February 1797 at the same place. In one of his stories, he claims to have escaped once from sinking into a swamp by pulling himself up by his own bootstraps.... In applied sciences, the bootstrap resampling approach helps the statistician to obtain meaningful results when the data distribution is unknown, namely by using the data as realizations of their own distribution (Painting by G. Bruckner, Rinteln (Germany), 1752. Reproduced with permission by Bibliographisches Institut & F. A. Brockhaus AG, Mannheim (Germany))

Chapter 2
The Bootstrap in Climate Risk Analysis

Manfred Mudelsee

Climate risk is the probability of adverse effects from extreme values of variables in the climate system. Because climate changes, so can the various types of climate risk (floods, storms, etc.) change. This field is of strong socioeconomic relevance. Estimates of climate risk variations come from instrumental, proxy and documentary records of past climate extremes and projections of future extremes. Kernel estimation is a powerful statistical technique for quantifying trends in climate risk. It is not parametrically restricted and allows realistic, non-monotonic trends. The bootstrap is a computing-intensive statistical resampling method used here to provide a confidence band around the estimated risk curve. Confidence bands, like error bars, are essential for a reliable assessment whether changes and trends are significant or came by chance into the data. This methodology is presented using reconstructed flood records of the central European rivers Elbe, Oder and Werra over the past five centuries. Trends in flood risk differ among rivers and also between hydrological seasons. The scientific conclusion is that flood risk analysis has to take into account the high spatial variability from orographic rainfall, as well as different hydrological regimes in winter and summer. In an ideal co-operation between experts, quantitative knowledge with uncertainty ranges (like the estimated trends in climate risk) should form the deliverable from scientists to policy makers and decision takers.

2.1 Introduction

Climate in its original definition refers to the mean state and variability of the atmosphere. Today a wider definition, including the hydrosphere, cryosphere and biosphere, is viewed as more appropriate to recognize the interdependences within that complex system. Climate changes affect many variables and many timescales; an upper limit is set by the age of the Earth ($\sim$4.6 Ga). Humans play a significant role

---

M. Mudelsee (✉)
Meteorologisches Institut Universität Leipzig, 04103 Leipzig, Germany; Climate Risk Analysis, 06114 Halle, Germany
e-mail: mudelsee@climate-risk-analysis.com

J.P. Kropp, H.-J. Schellnhuber (eds.), *In Extremis*,
DOI 10.1007/978-3-642-14863-7_2,

in the climate system by their ability to infer with the carbon cycle (CO_2 and CH_4 emissions). This may have been the case since the Industrial Revolution (via CO_2, see [2.21]) at around, say, AD 1800 or even earlier, since what might be termed the agricultural revolution (via CH_4, see [2.32]) at around the beginning of the Holocene climate stage (~10 ka ago).

Risk is in statistical science defined as "adverse probability" [2.15]. Climate risk may therefore be defined from an anthropocentric viewpoint as the probability that a climate variable takes values that lead to loss of human lives or damages to economies. Normally, such values are in the tails of the probability density function (PDF) of a variable, which means climate risk comes from extreme values. Examples are storms (extremely high wind speed), droughts (extremely low precipitation and available water resources) or river floods (extremely large runoff values).

Because climate changes [2.21], so can the various types of climate risk change (Fig. 2.1). It is of immediate socioeconomic value to analyse changes in climate risk. As regards past changes, these can be analysed from documented records of

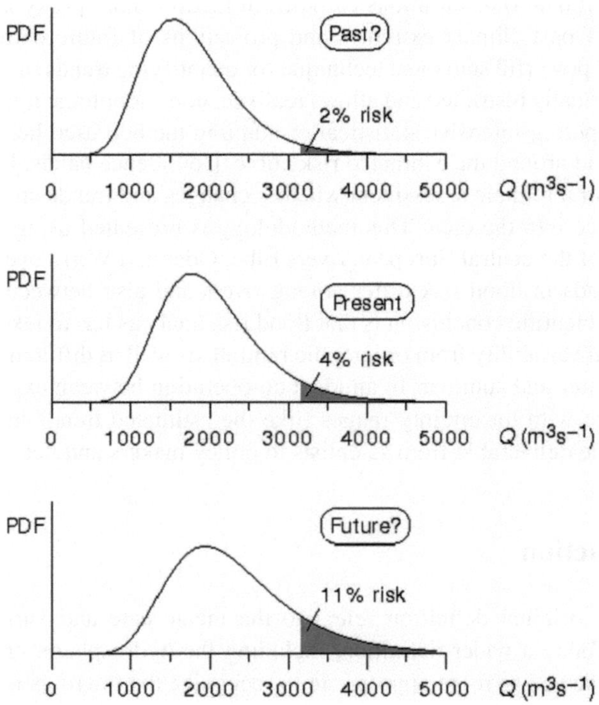

Fig. 2.1 Climate risk changes. This hypothetical example shows changes in the (right-skewed, non-Gaussian) PDF of maximum annual runoff, Q, at the station Dresden of the river Elbe. Climate risk is given by the integrated PDF. For example, in the hypothetical past the risk that within a year Q exceeds 3 110 m^3s^{-1} (the peak value of the March 1940 flood [2.26]) was 2%, it rose to 4% (*present*) and might rise to 11% (*future*). The question marks emphasize that knowing the past as well as the future situation is not certain but relies on imperfect data, coarse climate models and the validity of made assumptions (e.g. regarding the statistical method). In principle, one could assign a question mark also to the present situation

climate extremes. Such records come from more or less direct observations within the instrumental period (back to, say, AD 1850). Going further back in time, indirect measurements can be used, yielding so-called proxy records [2.1]. Examples are width of tree rings as indicator of summer temperature during the past centuries [2.4] and measured oxygen isotopic composition in a stalagmite from the Arabian peninsula as indicator of the strength of Indian Ocean monsoon rainfall during the Holocene [2.12]. Also written documents can be used to extend the time span from the beginning of the instrumental period back several centuries [2.2, 2.3, 2.26, 2.31]. Inevitably, records of past climate changes may contain measurement errors, proxy errors, textual interpretation errors or dating errors. As regards future changes, these can be inferred from mathematical models of the climate system. Although some of the most powerful computers are currently employed to do this task, owing to the limited computing power and the imperfect knowledge of the relevant processes acting in the climate system, also climate model projections are susceptible to uncertainties. One can try to improve the data situation by using better climate archives, measurement devices, model formulations and computers. In principle, however, our knowledge shall always be incomplete.

This means that estimated past or future climate risk changes have also errors. This is precisely the task of statistical analysis in climate risk research: to quantify the uncertainties, to give error bars or confidence intervals of our estimates and projections.

This chapter illustrates the bootstrap approach to quantify uncertainties in estimated climate risk changes. The bootstrap is a relatively new, computing-intensive, statistical resampling technique [2.10, 2.11]. Its advantage is that it is less restricted by parametric assumptions than more traditional approaches. For example, the assumption that proxy or measurement errors in climatology follow nicely a Gaussian distribution has been used for decades although it is generally appreciated that this assumption is often wrong (Fig. 2.1). The reason for making this simplistic assumption was to obtain analytically tractable statistical problems. With today's computing power, however, one need not rely on unrealistic assumptions and can instead use the bootstrap. The bootstrap approach is applied in this work to records of floods of European rivers from the past five centuries.

The major result is that changes in flood risk differ among the rivers Elbe, Oder and Werra. This reflects the spatial variability of rainfall [2.24], which in turn is partly owing to variations in orographic properties. The conclusion is that flood risk analysis has to take into account the high spatial variability and also the different hydrological regimes in winter and summer. It is useless when applied on a too large spatial scale. Rivers have to be analysed separately to obtain reliable results.

2.2 Method

Regression methods fail to detect changes in climate risk because they model the mean value and not the extremes. We [2.27] gave an example where the prescribed trend in climate risk is upward and a weak downward trend is superimposed as background, plus some noise. The regression line is downward, reflecting the tendency of the majority of data points and not the extremes. Taking the regression line as indicative for risk changes would thus give the wrong, opposite result.

The situation may be less severe when instead of the original data (background plus noise) some upper quantile data are available; this method is then called quantile regression [2.23]. Related is the case when, for example, the monthly runoff maxima instead of their means are taken. However, this still bears the possibility of two large events within a single month, of which one would not find entry into the flood risk analysis.

The alternative approach to regression, namely peak over threshold (POT), is preferred by us. For example, in runoff records one can find the POT data (flood dates) by applying a threshold (e.g. 50-year runoff level) and taking the data above that threshold. In documentary flood records as analysed here, the reported dates of floods are themselves already the POT data (time domain).

The simplest POT analysis technique compares two time intervals with respect to properties of the statistical distribution that describe the extreme values found inside. Typically chosen is the return period, which is the expected time for an extreme event to occur. That means the return period estimated from data of the first time interval is compared with the return period for the second interval. The problem with the interval comparison technique is that the time information is seriously degraded. For example, comparing Elbe floods between 1500 and 1750 with those between 1750 and 2000 would merely provide two estimates and miss the variability within each interval.

Frei and Schär [2.14] introduced the logistic model as a parametric description of the risk curve into climatology. (The term "risk curve" refers to the time-dependent occurrence rate defined in the following section. The logistic model is a parametric formulation of that time dependence.) This has the advantage of not degrading the time information. On the other hand, the logistic model is strictly monotonically increasing or decreasing. This means that it is not suited for analysing longer records (above a span of, say, 100 a) because on such timescales one cannot assume monotonic trends in climate risk but rather highs and lows, which might be related to the general climatic situation.

A powerful method to quantify time-dependent climate risk could be fitting a generalized extreme value (GEV) distribution to climate records, whereby the GEV parameters are allowed to vary with time [2.6, 2.22, 2.29]. Additional parameters are required for describing the time dependence. A challenge for this approach is adjusting the total number of parameters: allowing "enough" time variability while keeping the number of fit parameters low (Occam's razor). Because this nonstationary approach requires flood data measured with a reasonable accuracy and a sufficient size, it is hardly applicable to analysing flood risk in the pre-instrumental period.

The nonparametric kernel technique is therefore our method of choice. It analyses the extremes (POT advantage), does not degrade the time information and allows non-monotonic trends in climate risk. The kernel technique can be further combined with the bootstrap approach to yield confidence bands around the estimated risk curves. This helps the climate risk analyst to assess whether or not a high in flood risk is statistically significant or whether we can expect a significant increase in flood risk coming with global climate changes. We explain the kernel technique with bootstrap confidence band construction in the following section. A detailed description of the kernel technique with bootstrap confidence band construction is given elsewhere [2.27].

2.2.1 Kernel Risk Estimation

As indicated above and said previously [2.28], the simplest method to quantify flood risk over time is to form intervals (say, decades) and count the number of floods that occurred within each interval. The problem hereby is that only few estimation points would be produced. An improvement is to use quasi-continuously shifted intervals (as in running mean smoothing). The method is then called kernel smoothing, and the kernel function used is a uniform function [2.34], because all floods within an interval have same weight. Uniform kernel functions for flood risk estimation have been used in previous papers (e.g. [2.18]). The method can be further enhanced by adopting a smooth kernel function (that means, weighting) and using a mathematical method to solve the smoothing problem (choice of interval width). Finally, a confidence band around the estimated flood risk curve can be constructed using bootstrap simulations. See our previous paper [2.27] and the original work [2.8] for a detailed explanation of the method.

The kernel technique [2.9] estimates the occurrence rate as

$$\widehat{\lambda}(t) = h^{-1} \sum_{i=1}^{n} K\left(\left[t - T(i)\right] / h\right), \tag{2.1}$$

where $\lambda(t)$ is the time-dependent occurrence rate (probability of an extreme event per time unit), t is time, $T(i)$ are the flood dates, n is the total number of floods, K is the kernel function and h is the bandwidth. The "hat" denotes the estimate, reflecting that the true function $\lambda(t)$ is not known but has to be estimated from the data. A high time resolution is obtained by letting t run quasi-continuously within the observation interval of the data, $[t_1, t_2]$. Using a smooth kernel function yields a more realistic smooth estimate of the occurrence rate. A Gaussian kernel, $K(y) = \exp\left(-y^2/2\right)/(2\pi)^{1/2}$, is a convenient choice because it yields a smooth estimated occurrence rate and allows to calculate $\widehat{\lambda}(t)$ efficiently in Fourier space [2.33], leading to considerable computational acceleration.

Boundary effects (underestimation of $\widehat{\lambda}(t)$ near (i.e. within $\sim 3h$ distance) t_1 and t_2) can be reduced by generating pseudodata outside of $[t_1, t_2]$ before occurrence rate estimation [2.7]. Since pseudodata generation is equivalent to an extrapolation of the empirical distribution function, results at the boundaries should be judged cautiously. It is also advisable to slightly undersmooth, that is, to take a slightly smaller bandwidth than indicated by cross-validation (see next paragraph) to keep boundary effects small. Regarding boundary effects and confidence interval accuracy, see the original papers on the kernel occurrence rate estimation method [2.7, 2.8].

Bandwidth (h) selection determines bias and variance properties of the occurrence rate estimation and is therefore a crucial step. Small h leads to only few data points effectively contributing to the kernel estimation (2.1) and therefore a large variance of the estimate. But small h keeps bias low because data far away from the time point, t, have less influence on the estimate (2.1). On the other hand, large h leads to smaller estimation variance and higher estimation bias. The optimum bandwidth choice lies therefore somewhere in the middle, as the best compromise between statistical and systematic estimation uncertainties. One mathematical method for finding the solution to this smoothing problem is cross-validation [2.5]. Thereby, a cost function, determined by two terms describing variance and bias, is minimized. Cross-validated h depends, among other factors, also on the data size, n (see [2.27] for more details). In addition to cross-validation, bandwidth selection may be guided by the objective to reduce boundary effects. Another guide is to look at the confidence bands (next section) around the estimated occurrence rates and evaluate whether changes in risk are significant. A user-interactive analysis process is therefore most suited. Select a bandwidth; look at the risk curves and the significance of the highs and lows; if many, insignificant changes are found, then increase h; if one or no significant changes are found, then reduce h; etc.

2.2.2 Bootstrap Confidence Band Construction

A confidence band around $\widehat{\lambda}(t)$ is essential for interpreting results. For example, it might be asked if a low in $\widehat{\lambda}(t)$ is real or came instead by chance into the data. A confidence band can be obtained using bootstrap simulations [2.8] as follows:

1. From the set of data (augmented by pseudodata) draw one by one, with replacement, a simulated set of flood dates of same data size. This is the bootstrap resampling step.
2. Calculate $\widehat{\lambda}^*(t)$ after (2.1) using simulated data and same h.
3. Repeat the procedure simulation–estimation until 2 000 versions of $\widehat{\lambda}^*(t)$ are available.
4. A simple, percentile-based confidence interval (of level α) at time t is given by the central α values of ordered $\widehat{\lambda}^*(t)$. For example, for $\alpha = 90\%$, it is given by the interval between the 100th and 1 900th largest values.
5. The confidence band is given by the confidence intervals over time $t \in [t_1, t_2]$.
6. Cowling and co-workers [2.8] describe construction of a percentile-t type confidence band (denoted as "Type 1" and used by us), which has higher accuracy than the percentile-based band.

Note that the confidence band is "pointwise", it reflects the variability of the point estimate, $\widehat{\lambda}(t)$. The cited work [2.8] gives further bootstrap schemes and confidence band types, which have similar properties as the method shown here. This pioneering work also analyses the performance of kernel risk estimation by employing Monte Carlo simulations, that is, numerical experiments where the artificial data are generated from prescribed $\lambda(t)$ curves.

The method of kernel occurrence rate estimation with bootstrap confidence bands has been applied by us in the following studies: floods of the rivers Elbe and Oder [2.26, 2.27] and Werra [2.28] over the past 500–1 000 years, occurrence of wildfire events in the Canadian Boreal Shield since the end of the 18th century [2.16] and in climate model projections for the same region and the 21st century [2.17], North Sea storms since AD 1500 [2.30] and soil erosion events in Kenya over the past 300 years [2.13]. The methodology is currently being implemented into a user-friendly Windows version of the software.

2.3 Data

Table 2.1 shows the database of analysed river floods. The Elbe, Oder and Werra are rivers in central Europe. Their catchment areas (middle Elbe, 95 000 km^2; middle Oder, 54 000 km^2; middle and upper Werra, 5 505 km^2) are under low-mountainous climate. Floods in hydrological summer (May–October) are caused by heavy and/or prolonged rainfall and in the winter (November–April) also by thawing snow. Breaking river ice may function as water barrier and enhance a flood severely [2.19].

Documentary records of floods (Fig. 2.2) were consulted and analysed to construct the old parts of the flood database (Elbe, before 1850; Oder, before 1920; Werra, before 1900). Measured records of water stage and inferred runoff were used to complete the database to the present (Elbe, 1850–2002; Oder, 1854–2002; Werra, 1880–2003). Occasionally, the documentary entries contained information on the maximum flood water stage. Owing to overlapping documentary and instrumental periods, it was possible to quantify the size of a flood for number of events and to ensure data homogeneity across the boundary between documentary and instrumental periods [2.26]. The size of most of the flood events in the documentary periods could only roughly be quantified by means of information such as the duration of an event, the number of casualties, the economic damages caused. Following standard practice in documentary climatology [2.2, 2.26], the flood records were allowed to take only three values: 1 (minor flood event), 2 (strong flood event) and 3 (exceptionally strong flood event). In the present chapter, we focus on heavy floods (classes 2–3).

Table 2.1 Database (see also Fig. 2.3)

		Number of floods				
River	Interval	Total	Winter	Summer	Unknown	Reference
Elbe	1021–2002	328	208	117	3	[2.26]
Oder	1269–2002	218	108	106	4	[2.26]
Werra	1500–2003	143	111	32	0	[2.28]

Fig. 2.2 Pamphlet on the "Thuringian Flood". This catastrophe happened on 29 May 1613 (Julian calendar). This rendering (printed 1613 in Schmalkalden) is likely not realistic; Hellmann's work [2.20] contains a reproduction (p. 40) and further bibliographic details

The most severe problem when dealing with documentary data of climate extremes is to reduce the effects of data inhomogeneities in the form of document loss. In the earlier half of the last millennium, before the invention of printing in Europe, likely fewer documents (handwritings) survived, compared with the latter half, until they found entrance into permanent, secondary sources. Ignoring this type of data deficit could then lead to unjustified claims of increased flood risk in the second compared with the first half. In the case of Elbe and Oder floods (Table 2.1), the start of the observation intervals (1021 and 1269, respectively) likely is not the start of the interval in which homogeneous data can be assumed, which we set [2.26] to AD 1500. In the case of Werra floods (1500–2003), we tried by employing critical source interpretation and consultation of many documentary archives to achieve more or less complete information about floods within the relatively small area of the middle and upper Werra [2.28]. Despite this, the results (Sect. 2.4) suggest that

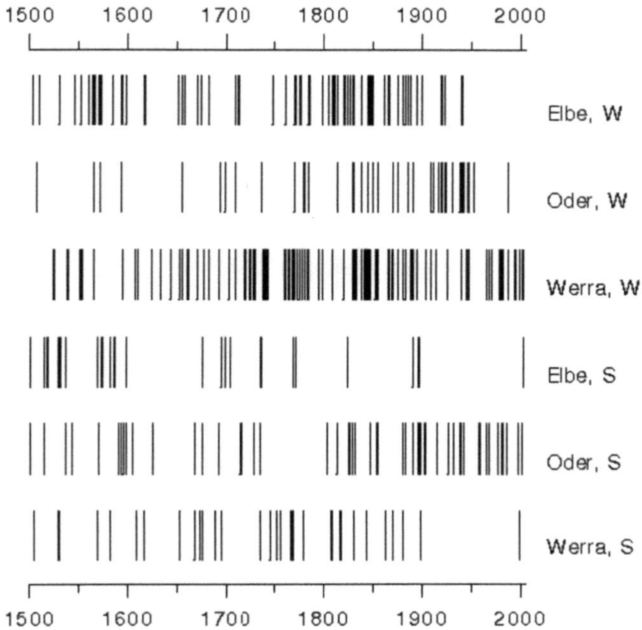

Fig. 2.3 Winter (W) and summer (S) floods of rivers Elbe, Oder and Werra since AD 1500. Floods with unknown season are few (Table 2.1); they are plotted here together with the winter floods. In the case of Elbe and Oder floods, only the heavy events (classes 2 and 3, see [2.26]) are shown

minor document loss could have occurred for data before the beginning of the 18th century.

See the data references [2.26–2.28] for more details on the rivers, river engineering work, orography, runoff–stage relations, critical source interpretation and comparisons of different sources.

2.4 Results

The results (Fig. 2.4) show that flood risk over the past five centuries varied significantly. There is further variation between winter and summer trends and also among the various rivers. These trends have been discussed in detail in the original publications [2.26–2.28]. Here we make some general comments and investigate selected points.

Werra flood risk (winter and summer) shows small, long-term increases in the early part ($\sim$1500–1700). It may be asked whether this reflects what really occurred in nature or instead results from a trend towards reduced document loss. Arguing from a historical–critical perspective, we believe that document loss played only a minor role in case of the Werra because historical information from that region is quite abundant and well preserved.

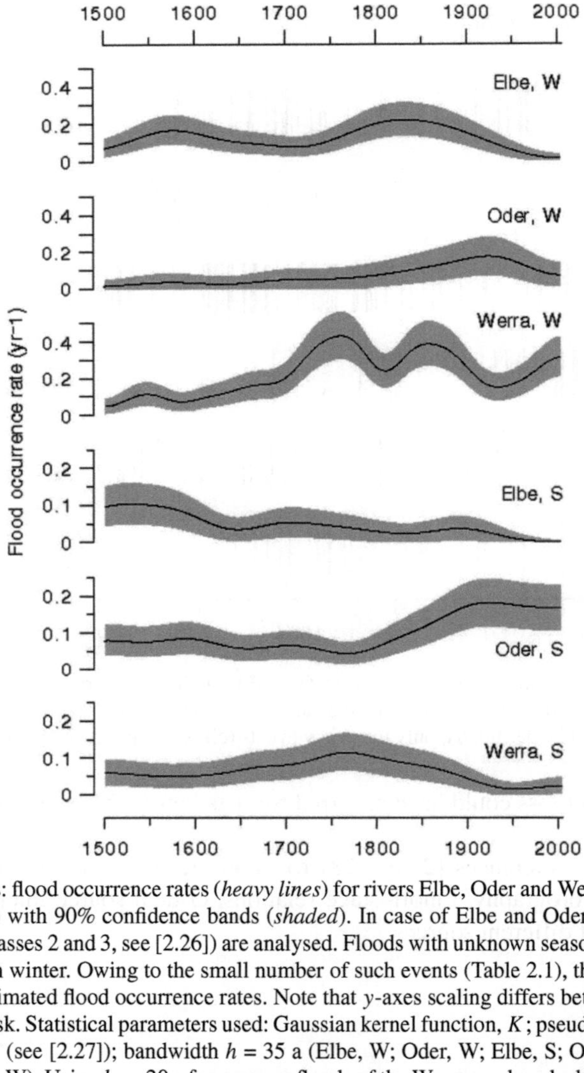

Fig. 2.4 Results: flood occurrence rates (*heavy lines*) for rivers Elbe, Oder and Werra in winter (W) and summer (S) with 90% confidence bands (*shaded*). In case of Elbe and Oder floods, only the heavy events (classes 2 and 3, see [2.26]) are analysed. Floods with unknown season are assumed to have occurred in winter. Owing to the small number of such events (Table 2.1), this has negligible effect on the estimated flood occurrence rates. Note that y-axes scaling differs between winter and summer flood risk. Statistical parameters used: Gaussian kernel function, K; pseudodata generation rule "reflection" (see [2.27]); bandwidth $h = 35$ a (Elbe, W; Oder, W; Elbe, S; Oder, S; Werra, S) and 20 a (Werra, W). Using $h = 20$ a for summer floods of the Werra produced additional, insignificant "wiggles" [2.28]

The high in Elbe winter flood risk in the latter half of the 16th century (Fig. 2.4) corresponds to increased risk levels in rivers in central and southwest Europe during that time, which were related to higher precipitation [2.2]. A low in Elbe flood risk at around 1700 could have been the effect of the cold and dry climate in Late Maunder Minimum [2.25], a period of nearly absent sunspots and reduced solar activity. However, the Oder does not exhibit such significant changes in the early period.

Length reductions of the middle Elbe (1740–1870) and middle Oder (1745–1850) did not leave a consistent imprint on winter and summer flood risk (Fig. 2.4) and were therefore only of minor influence.

The 18th and 19th centuries experienced strong, significant changes in flood occurrence rates (Fig. 2.4). The Elbe had a high in winter flood risk in 1800–1850, followed by a long-term decrease. Oder flood risk (winter and summer) increased, but this should be interpreted cautiously because the Oder flood record in the interval 1850–1920 is of reduced quality [2.26]. Werra winter flood risk peaked high at ~1760, then low at ~1810, then high again at ~1860 [2.28]. This pseudo-periodicity of 100 a is not the result of bandwidth selection because $h = 20$ a is clearly smaller. These Werra winter flood changes contrast markedly with the findings for the Elbe. Werra summer flood risk decreased gradually since ~1760 to the present.

In general, winter floods seem to have been more likely than summer floods over the past centuries, as is expressed most clearly by the Werra [2.28], the Elbe and, to a lesser degree, the Oder (Fig. 2.4).

The flood records for the past decades can be completely determined from instrumental observations. They have therefore an excellent degree of data homogeneity. Within the instrumental period, Elbe and Oder winter flood risk decreased significantly [2.26]. This is likely a climate signal from regional warming, which had reduced the probability of strong river freezing and, hence, the risk of "ice floods" [2.26, 2.27]. In this regard, the significant upward trend of Werra winter flood risk (Fig. 2.4) is interesting. We speculate that "ice flood" risk was reduced earlier (mid-19th century) for the Werra than for the other two rivers (mid-20th century [2.26, 2.27]). Contrarily, summer flood risk shows no trends (Elbe, Oder and Werra) over the past decades.

2.5 Conclusion

Producing large amounts of rainfall in the affected catchment areas requires a combination of several factors [2.27, 2.28]:

1. north-westerly to southerly, cyclonic airflow;
2. high atmospheric water vapour content;
3. low convective lability, preventing cell formation;
4. prolonged (at least half a day) flow against the orography (Fig. 2.5).

It would be naïve to assume that with climate changes only factor 2 would change (via the Clausius–Clapeyron equation). In particular, the role of factor 4, flow against orography, should be analysed in the case of flood risk changes in central European regions under low-mountainous climate. This is because of the differences among Werra, Elbe and Oder flood risk curves (Fig. 2.4), which indicate that the orographic differences among the catchment areas introduce a strong nonlinear component into the complex climate–hydrosphere system.

It is in our view required to carry out a large body of detailed scientific work to learn about flood risk and climate changes:

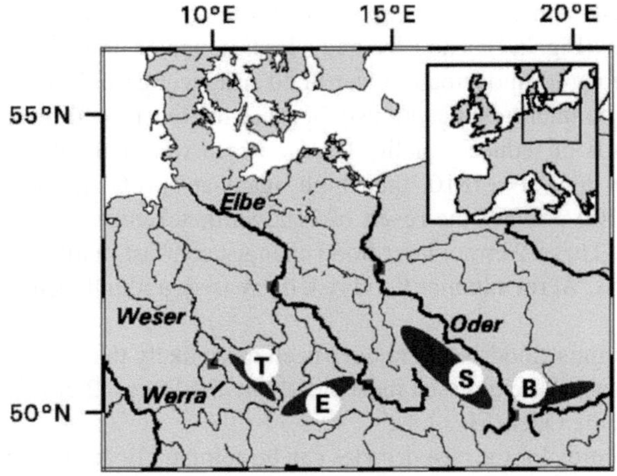

Fig. 2.5 Rivers Elbe, Oder and Werra in central Europe. Grey squares denote the places used [2.26, 2.28] to define the analysed river sections (middle Elbe, middle Oder and middle to upper Werra). Also shown are the mountain ranges relevant for orographically induced rainfall in the catchment areas of the river sections. T, Thüringer Wald; E, Erzgebirge; S, Sudeten; B, Beskids

1. Produce records of past floods in a hydrologically more or less homogeneous area, at monthly or at least seasonal resolution over the past centuries.
2. Combine documentary with instrumental evidence to achieve data homogeneity.
3. Use quantitative flood risk estimation methods with error band.
4. Use the results obtained from the analysis of past floods to train coupled climate models (global–regional–hydrological) that are used to make projections of future flood risk; trends from observed, past floods serve as targets for successful models.

As said in the introduction, estimated past or future climate risk changes have errors. It is the task of climate scientists to give error bars or confidence intervals of the estimates and projections. It is then the duty of policy makers to make, in this uncertain situation, decisions of sometimes strong impact. Luckily, politicians are trained to doing exactly that: making decisions in uncertain situations. This is a plea for a rational approach in this challenging socioeconomic situation: let the scientists do the science and the politicians make the decisions.

Acknowledgments We thank two anonymous reviewers for constructive remarks. We thank the following colleagues and institutions for data, discussions and information regarding extreme river floods and their statistical analysis: M. Börngen (Saxonian Academy of Sciences), N. Conrads (University of Stuttgart), M. Deutsch (University of Göttingen), H. Engel (Federal Institute of Hydrology Germany), W. Fröhlich (Federal Institute of Hydrology Germany), U. Grünewald (Technical University Cottbus), J. Jacobeit (University of Augsburg), B. Kowalski (Thuringian Environmental Agency), T. Lüllwitz (Federal Institute of Hydrology Germany), J. Luterbacher (University of Bern), T. Maurer (Federal Institute of Hydrology Germany), J. Munzar (Czech Academy of Sciences), C. Pfister (University of Bern), R. Oppermann (Federal Institute of Hydrology Germany), M. Schulz (University of Bremen), G. Tetzlaff (University of Leipzig), H. Wanner

(University of Bern), Q. Yao (London School of Economics), Global Runoff Data Centre (Federal Institute of Hydrology Germany), Thuringian State Archive Gotha, Thuringian State Archive Meiningen, Town Archive Meiningen and Town Archive Bad Salzungen. We thank M. Alkio (University of Massachusetts at Boston) for comments and proofreading. We especially thank H. Kroß (Samtgemeinde Bodenwerder, Sachgebiet Tourismus) for information on the painting shown in Fig. 1.5 and Bibliographisches Institut & F. A. Brockhaus AG for permission to reproduce it. Financial support by the German Science Foundation through research projects (MU 1595/1, TE 51/23) is acknowledged.

References

2.1. R.S. Bradley, *Paleoclimatology: Reconstructing Climates of the Quaternary*, 2nd edn. (Academic Press, San Diego, 1999)
2.2. R. Brázdil, R. Glaser, C. Pfister, P. Dobrovolný, J.-M. Antoine, M. Barriendos, D. Camuffo, M. Deutsch, S. Enzi, E. Guidoboni, O. Kotyza, F.S. Rodrigo, Clim. Change **43**(1), 239 (1999)
2.3. R. Brázdil, C. Pfister, H. Wanner, H. von Storch, J. Luterbacher, Clim. Change **70**(3), 363 (2005)
2.4. K.R. Briffa, T.J. Osborn, F.H. Schweingruber, I.C. Harris, P.D. Jones, S.G. Shiyatov, E.A. Vaganov, J. Geophys. Res. **106**(D3), 2929 (2001)
2.5. M.M. Brooks, J.S. Marron, Stoch. Process. Appl. **38**(1), 157 (1991)
2.6. S. Coles, *An Introduction to Statistical Modeling of Extreme Values* (Springer, London, 2001)
2.7. A. Cowling, P. Hall, J. R. Statist. Soc. B **58**(3), 551 (1996)
2.8. A. Cowling, P. Hall, M.J. Phillips, J. Am. Statist. Assoc. **91**(436), 1516 (1996)
2.9. P. Diggle, Appl. Stat. **34**(2), 138 (1985)
2.10. B. Efron, Ann. Statist. **7**(1), 1 (1979)
2.11. B. Efron, R.J. Tibshirani, *An Introduction to the Bootstrap* (Chapman and Hall, London, 1993)
2.12. D. Fleitmann, S.J. Burns, M. Mudelsee, U. Neff, J. Kramers, A. Mangini, A. Matter, Science **300**(5626), 1737 (2003)
2.13. D. Fleitmann, R.B. Dunbar, M. McCulloch, M. Mudelsee, M. Vuille, T.R. McClanahan, J.E. Cole, S. Eggins, Geophys. Res. Lett. **34**(4), L04401 (2007). [doi:10.1029/2006GL028525, electronic version freely available from `www.climate-risk-analysis.com`]
2.14. C. Frei, C. Schär, J. Climate **14**(7), 1568 (2001)
2.15. J.S. Gardenier, T.K. Gardenier, in: *Encyclopedia of statistical sciences*, Vol. 8, ed. by S. Kotz, N.L. Johnson, C.B. Read (Wiley, New York, 1988) p. 141.
2.16. M.P. Girardin, Y. Bergeron, J.C. Tardif, S. Gauthier, M.D. Flannigan, M. Mudelsee, Int. J. Wildland Fire **15**(3), 375 (2006)
2.17. M.P. Girardin, M. Mudelsee, Ecological Appl. **18**(2), 391 (2008)
2.18. R. Glaser, H. Stangl, Surv. Geophys. **25**(5–6), 485 (2004)
2.19. U. Grünewald et al., *Ursachen, Verlauf und Folgen des Sommer-Hochwassers 1997 an der Oder sowie Aussagen zu bestehenden Risikopotentialen. Eine interdisziplinäre Studie — Langfassung* (Deutsches IDNDR-Komitee für Katastrophenvorbeugung e. V., Bonn, 1998)
2.20. G. Hellmann, Veröffentlichungen des Königlich Preußischen Meteorologischen Instituts **256**, 21 (1913) [title *Die "Thüringische Sündflut" vom Jahre 1613*]
2.21. J.T. Houghton, Y. Ding, D.J. Griggs, M. Noguer, P.J. van der Linden, X. Dai, K. Maskell, C.A. Johnson (eds.), *Climate Change 2001: The Scientific Basis. Contribution of Working Group I to the Third Assessment Report of the Intergovernmental Panel on Climate Change* (Cambridge University Press, Cambridge, 2001)
2.22. R.W. Katz, M.B. Parlange, P. Naveau, Adv. Water Resour. **25**(8–12), 1287 (2002)

2.23. R. Koenker, K.F. Hallock, J. Econ. Perspect. **15**(4), 143 (2001)
2.24. G.H. Liljequist, K. Cehak, *Allgemeine Meteorologie*, 3rd edn. (Vieweg, Braunschweig, 1984)
2.25. J. Luterbacher, R. Rickli, E. Xoplaki, C. Tinguely, C. Beck, C. Pfister, H. Wanner, Clim. Change **49**(4), 441 (2001)
2.26. M. Mudelsee, M. Börngen, G. Tetzlaff, U. Grünewald, Nature **425**(6954), 166 (2003); [electronic version freely available from `www.climate-risk-analysis.com`]
2.27. M. Mudelsee, M. Börngen, G. Tetzlaff, U. Grünewald, J. Geophys. Res. **109**(D23), D23101 (2004); [doi:10.1029/2004JD005034, electronic version freely available from `www.climate-risk-analysis.com`]
2.28. M. Mudelsee, M. Deutsch, M. Börngen, G. Tetzlaff, Hydrol. Sci. J. **51**(5), 818 (2006) [electronic version freely available from `www.climate-risk-analysis.com`]
2.29. P. Naveau, M. Nogaj, C. Ammann, P. Yiou, D. Cooley, V. Jomelli, C.R. Geosci. **337**(10–11), 1013 (2005)
2.30. J. Neubauer, F. Rohrbeck, M. Mudelsee, *Occurrence of major windstorms in the North Sea region over the past decades to centuries* (Risk Prediction Initiative, Hamilton, Bermuda, 2004); [electronic version available from `www.bios.edu/rpi`]
2.31. C. Pfister, R. Brázdil, R. Glaser, M. Barriendos, D. Camuffo, M. Deutsch, P. Dobrovolný, S. Enzi, E. Guidoboni, O. Kotyza, S. Militzer, L. Rácz, F.S. Rodrigo, Clim. Change **43**(1), 55 (1999)
2.32. W.F. Ruddiman, Clim. Change **61**(3), 261 (2003)
2.33. B.W. Silverman, Appl. Stat. **31**(1), 93 (1982)
2.34. B.W. Silverman, *Density Estimation for Statistics and Data Analysis* (Chapman and Hall, London, 1986)

The Elbe flood 2002 which was classified as a 100-year event. Only four years later, in 2006, the Elbe discharge exceeded this event at several locations. The concept of a 100-year event was put into question by the media (Photo courtesy of Enrico Bär, creator)

Chapter 3
Confidence Intervals for Flood Return Level Estimates Assuming Long-Range Dependence

Henning W. Rust, Malaak Kallache, Hans Joachim Schellnhuber, and Jürgen P. Kropp

Standard flood return level estimation is based on extreme value analysis assuming independent extremes, i.e. fitting a model to excesses over a threshold or to annual maximum discharge. The assumption of independence might not be justifiable in many practical applications. The dependence of the daily run-off observations might in some cases be carried forward to the annual maximum discharge. Unfortunately, using the autocorrelation function, this effect is hard to detect in a short maxima series. One consequence of dependent annual maxima is an increasing uncertainty of the return level estimates. This is illustrated using a simulation study. The confidence intervals obtained from the asymptotic distribution of the maximum likelihood estimator (MLE) for the generalized extreme value distribution (GEV) turned out to be too small to capture the resulting variability. In order to obtain more reliable confidence intervals, we compare four bootstrap strategies, out of which one yields promising results. The performance of this semi-parametric bootstrap strategy is studied in more detail. We exemplify this approach with a case study: a confidence limit for a 100-year return level estimate from a run-off series in southern Germany was calculated and compared to the result obtained using the asymptotic distribution of the MLE.

3.1 Introduction

Many achievements regarding extreme value statistics and the assessment of potential climate change impacts on frequency and intensity of extreme events have been made in the past years, summarized for instance in [3.34], [3.49] or [3.45]. The IPCC stated that it is very likely for the frequency of intense precipitation to increase [3.32] with increasing global mean temperature. This implies changes in precipitation patterns, a major factor – among some others – for the intensity and frequency of floods, which ultimately can cause tremendous consequences for

H.W. Rust (✉)
Laboratoire des Sciences du climat et de l'Environnement, 91191 Gif-sur-Yvette, France
e-mail: henning.rust@lsce.ipsl.fr

J.P. Kropp, H.-J. Schellnhuber (eds.), *In Extremis*,
DOI 10.1007/978-3-642-14863-7_3,

nature and societies in a catchment area. This has been already observed in many regions of the world [3.37].

A pressing question is thus whether heavy rain or severe floods become more frequent or intense. Some concepts make use of non-stationary models, e.g. Chap. 9, [3.16, 3.33], or try to identify flood-producing circulation patterns [3.2]. A variety of approaches assess changes by comparing windows covering different time spans [3.3]. This procedure is especially useful for getting an impression of possible further developments by comparing GCM control and scenario runs [3.38, 3.58]. A useful indicator for changes in flood frequency and magnitude is the comparison of return level estimates. For this purpose a reliable quantification of the uncertainty of return level estimates is crucial.

Alerted by a seemingly increasing flood risk, decision makers demand for quantitative and explicit findings for readjusting risk assessment and management strategies. Regional vulnerability assessments can be one strategy to deal with the threat of extremes, such as floods or heat waves (e.g. [3.36]). Other approaches try to anticipate extreme scenarios using GCM runs. The development of risk assessment concepts, however, has still a long way to go, since forecasting of extreme precipitation or floods is highly uncertain (cf. for instance [3.42]). Another potential problem in the risk assessment framework is the quantification of uncertainty in extreme value statistics. In situations where common statistical approaches might not be applicable as usual, e.g. dependent records, specification of uncertainty bounds for a return level estimate cannot be made on the basis of the mathematically founded asymptotic theory. The simplified assumption of independent observations usually implies an underestimation of this uncertainty [3.4, 3.13, 3.35, 3.51]. The estimation of return levels and their uncertainty plays an important role in hydrological engineering and decision making. It forms the basis of setting design values for flood protection buildings like dikes. Since those constructions protect facilities of substantial value or are by themselves costly objects, it is certainly of considerable importance to have appropriate concepts of estimation and uncertainty assessment at hand. Otherwise severe damages, misallocation of public funds, or large claims against insurance companies might be possible. Thus, the approach presented in this contribution focuses on an improvement of common statistical methods used for the estimation of return levels with non-asymptotic bootstrap method.

In the present chapter, we focus on the block maxima approach and investigate the maximum likelihood estimator for return levels of autocorrelated run-off records and its uncertainty. In a simulation study, the increase in uncertainty of a return level estimate due to dependence is illustrated. As a result of comparing four strategies based on the bootstrap, we present a concept which explicitly takes autocorrelation into account. It improves the estimation of confidence intervals considerably relative to those provided by the asymptotic theory. This strategy is based on a semi-parametric bootstrap approach involving a model for the autocorrelation function (ACF) and a resampling strategy from the maxima series of the observations. The approach is validated using a simulation study with an autocorrelated process. Its applicability is exemplified in a case study: we estimate a 100-year return level and a related 95% upper confidence limit under the different assumptions of independent and dependent observations. The empirical run-off series was measured at the gauge Vilsbiburg at the river Große Vils in the Danube catchment.

The chapter is organized as follows: Sect. 3.2 describes the basic theory of the block maxima approach of extreme value statistics and the associated parameter estimation. Section 3.3 illustrates the effect of dependence on the variability of the return level estimator. In Sect. 3.4 the bootstrap strategies are presented including the methodological concepts they require. The performance of the most promising approach is evaluated in Sect. 3.5, followed by a case study in Sect. 3.6. A discussion and conclusions in Sects. 3.7 and 3.8, respectively, complete the chapter. Details regarding specific methods used are deferred to the Appendix 3.

3.2 Basic Theory

3.2.1 The Generalized Extreme Value Distribution

The pivotal element in extreme value statistics is the three types theorem, discovered by Fisher and Tippett [3.22] and later formulated in full generality by Gnedenko [3.23]. It motivates a family of probability distributions, namely the general extreme value distributions (GEV), as models for block maxima from an observed record, e.g. annual maximum discharge. We denote the maxima out of blocks of size n as M_n. According to the three types theorem, for n large enough the maxima distribution can be approximated by

$$\Pr\{M_n \leq z\} \approx G(z), \tag{3.1}$$

where $G(z)$ is a member of the GEV family (cf. Appendix).

The quality of the approximation in Eq. (3.1) depends in the first place on the block size n, which in hydrologic applications is usually chosen as 1 year. Further influencing factors are the marginal distribution of the observed series and – a frequently disregarded characteristic – its autocorrelation. Fortunately, the three types theorem holds also for correlated records under certain assumptions (cf. Appendix). The quality of approximation, however, is affected by the correlation as demonstrated in the following.

We compare records of white noise and a simple correlated process (AR[1], cf. Sect. 3.4.3) with the same (Gaussian) marginal distribution. For different block sizes n, we extract 2 000 block maxima from a sufficiently long record. Subsequently, the maxima are modelled with a Gumbel distribution being the appropriate limiting distribution in the Gaussian case [3.21]. We measure the quality of approximation for different n using the negative log-likelihood l (cf. Sect. 3.2.2). Figure 3.1 shows a decreasing negative log-likelihood with increasing block sizes n for the uncorrelated and the correlated record. This implies that the approximation in general ameliorates with block size n. However, for all n the approximation is better for the uncorrelated series than for the AR[1] series. This finding is consistent with

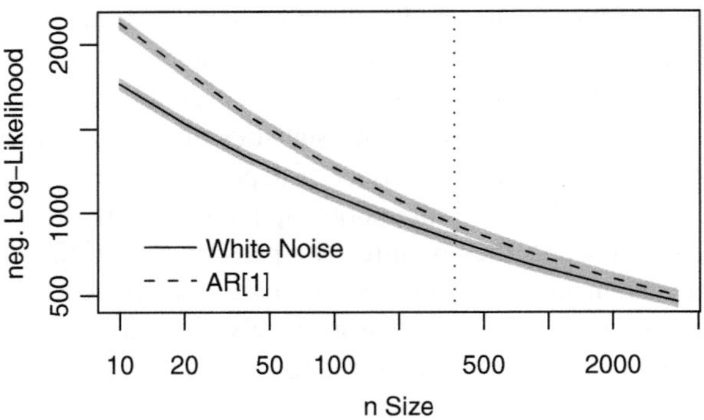

Fig. 3.1 Quality of approximation of a Gumbel fit to 2 000 maxima of realizations of a white noise and an AR[1] process for different block sizes n. The *lines* connect the means of 1 000 realizations, the *shadows* mark the mean ±1 SD. The *vertical line* marks a block size of $n = 365$

dependency reducing the effective number of data points [3.63], which in this case translates into a reduction of effective block size. The difference in approximation between the correlated and the uncorrelated case vanishes with increasing n [3.51].

3.2.2 GEV Parameter Estimation

To fully specify the model for the extremes, we estimate the GEV parameters from the data. Estimates can be obtained in several ways: probability weighted moments [3.30, 3.31], maximum likelihood (ML) [3.12, 3.54], or Bayesian methods [3.12, 3.14, 3.55]. These different approaches have advantages and drawbacks which are discussed in, e.g. [3.13, 3.31] and [3.56]. In the following we focus on ML estimation as the most general method. Within this framework models can be easily extended, for example, to non-stationary distributions [3.34].

Let $\hat{\theta} = (\hat{\mu}, \hat{\sigma}, \hat{\xi})$ be the maximum likelihood estimate (cf. Appendix) for the location (μ), scale (σ), and form (ξ) parameter of the GEV. For large block sizes n approximate $(1-\alpha)100\%$ confidence intervals for these estimates can be obtained from the Fisher information matrix I_E as $\hat{\theta}_j \pm z_{\frac{\alpha}{2}}\sqrt{\beta_{j,j}}$, with $\beta_{j,k}$ denoting the elements of the inverse of I_E and $z_{\frac{\alpha}{2}}$ the $(1-\frac{\alpha}{2})$-quantile of the standard normal distribution (cf. Appendix).

The m-year return level can be calculated straightforwardly once the location, scale, and shape parameter are estimated. In case of the Gumbel distribution the equation reads

$$\hat{r}_m = \hat{\mu} - \hat{\sigma}\log(y), \tag{3.2}$$

with $y = -\log(1-\frac{1}{m})$. An approximated confidence interval for $\hat{r}_m$ can be obtained using the delta method described in the Appendix [3.12].

For maximum likelihood estimation of the GEV parameters, we use the package `evd` [3.57] for the open source statistical language environment R [3.50].[1]

3.3 Effects of Dependence on Confidence Intervals

Annual maxima from river run-off frequently appear uncorrelated from an investigation of the empirical ACF. The left panel in Fig. 3.2 shows the ACF of the annual maxima series from the gauge Vilsbiburg (solid) and of a simulated record (dotted). Both records contain 62 values and their ACF estimates basically do not exceed the 95% significance level for white noise. If a longer series was available, as is the case for the simulated record, significant autocorrelations of the annual maxima are revealed by the ACF, Fig. 3.2 (right). This implies that considering annual maxima from run-off records a priori as uncorrelated can be misleading.

The ML estimator relies on the assumption of independent observations and is thus, strictly speaking, not correct for dependent observations. The main effect is that standard errors are underestimated if obtained from the Fisher information matrix [3.15]. In the following we illustrate this effect by a Monte Carlo (MC) simulation study using realizations of a long-range[2] dependent process (FAR$[1, d]$, cf. Sect. 3.4.3) with Hurst exponent $H = 0.75$ (or, equivalently, fractional differencing parameter $d = H - 0.5 = 0.25$). To ameliorate resemblance to a daily run-off series, we transform the Gaussian series X_t with an exponential function. The resulting record $Z_t = \exp(X_t)$ is then log-normally distributed.

Considering Z_t as 100 years of daily run-off ($N = 36\,500$), we perform an extreme value analysis, i.e. we model the annual maxima series by means of a GEV. Since the marginal distribution is by construction log-normal, we restrict

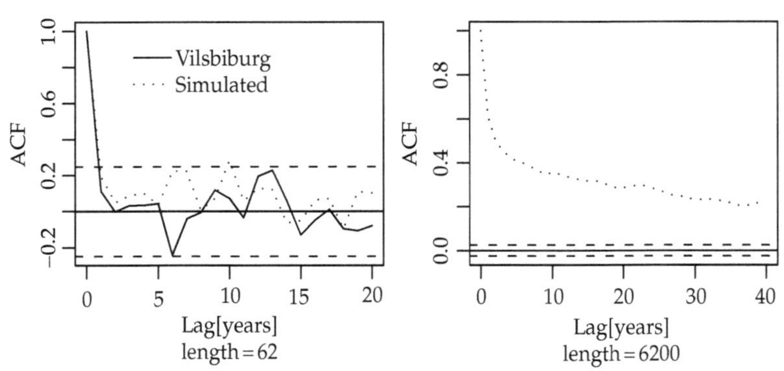

Fig. 3.2 Autocorrelation of the empirical maxima series and a section of same length cut out of the simulated series (*left*). The *right panel* shows the ACF of the full simulated maxima series (length = 6 200). The 95% significance levels are marked as *dashed lines*

[1] Both are freely available from `http://cran.r-project.org`

[2] A process is long-range dependent if its autocorrelation function is not summable, cf. Sect. 3.4.3.

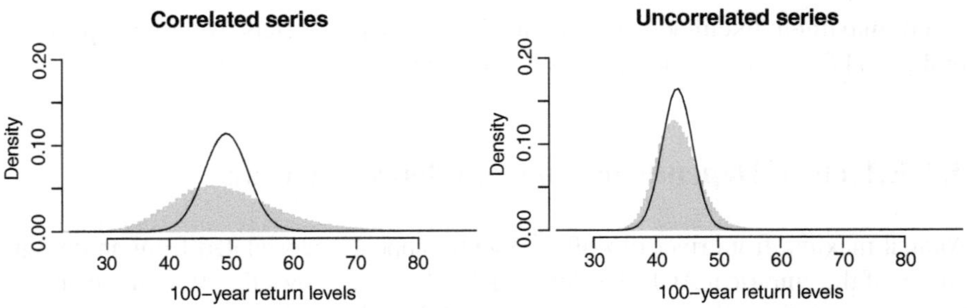

Fig. 3.3 Histogram (*grey*) of the estimated 100-year return levels from the MC ensemble of 10 000 realization of the FAR[1, d] process with fractional difference parameter $d = 0.25$ (or equivalently $H = 0.75$) and AR parameter $\phi_1 = 0.9$ (*left*). The *right panel* shows the result for a white noise process. The realizations contain $N = 36\,500$ data points. The *solid line* shows a Gaussian density function representing the limiting distribution of the 100-year return level estimator derived from the Fisher information matrix for one ensemble member

the extreme value analysis to a Gumbel distribution which is the proper limiting distribution in this case. Exemplarily, a 100-year return level is estimated using the MLE (cf. Sect. 3.2.2). Repeating this for 10 000 realizations of Z_t yields a frequency distribution representing the variability of the return level estimator for the FAR[1, d] process, shown as histogram in Fig. 3.3 (left panel). An analogous simulation experiment has been carried out for an uncorrelated series with a log-normal distribution (Fig. 3.3, right panel). Both histograms (grey) are compared with the limiting distribution (solid line) of the MLE (Eq. (3.19)) evaluated for an ensemble member with return level estimate close to the ensemble mean. For the uncorrelated series the limiting distribution provides a reasonable approximation in the sense that it roughly recovers the variability of the estimator. In the presence of correlation, the estimator variability is underestimated. This indicates that confidence intervals derived from the MLE's limiting distribution are not appropriate here.

Alternatively, confidence intervals can be obtained using the profile likelihood which is frequently more accurate [3.12]. Table 3.1 compares the upper limits of two-sided confidence intervals for three α-levels obtained using profile likelihood to the limiting distribution and the Monte Carlo simulation. The limits from the profile likelihood are indeed for the correlated and the uncorrelated process closer to the limits of the Monte Carlo ensemble. For the correlated process this improvement is not satisfying, since the difference to the Monte Carlo limit is still about 20% of the estimated return level.

Table 3.1 Upper limits of two-sided confidence intervals for various confidence levels obtained from the asymptotic distribution (Asympt.), the profile likelihood approach (Profile), and the Monte Carlo ensemble (MC)

Uncorrelated series				Correlated series			
Level	Asympt.	Profile	MC	Level	Asympt.	Profile	MC
0.68	45.97	46.10	46.69	0.68	52.72	53.00	56.91
0.95	48.33	48.90	50.42	0.95	56.21	57.19	67.26
0.99	49.84	50.87	53.10	0.99	58.43	60.12	75.34

To facilitate the presentation in the following, we compare the results from the bootstrap approaches to the confidence intervals obtained using the asymptotic distribution.

3.4 Bootstrapping the Estimators Variance

We discuss non-asymptotic strategies to more reliably assess the variability of the return level estimator. These strategies are based on the bootstrap, i.e. the generation of an ensemble of artificial maxima series. These series are simulated using a model which has been motivated by the data [3.17]. In the given setting we estimate the return levels for each ensemble member and study the resulting frequency distribution.

There are various strategies to generate ensembles. Four of them will be briefly introduced and, as far as necessary, described in the following.

***bootstrap*$_{\mathrm{cl}}$.** The first approach is a classical bootstrap resampling of the maxima [3.17, 3.19], denoted in the following as *bootstrap*$_{\mathrm{cl}}$: one ensemble member is generated by sampling with replacement from the annual maxima series. Autocorrelation is not taken into account here.

***iaaft*$_{\mathrm{d}}$.** We denote the second strategy as *iaaft*$_{\mathrm{d}}$, which makes use of daily observations. Ensemble members are obtained using the iterative amplitude adjusted Fourier transform (IAAFT) – a surrogate method described in Sect. 3.4.4. The IAAFT generates artificial series (so-called surrogates) preserving the distribution and the correlation structure of the observed daily record. Subsequently, we extract the maxima series to obtain an ensemble member. Linear correlation is thus accounted for in this case.

***bootstrap*$_{\mathrm{fp}}$.** The third strategy is a full-parametric bootstrap approach denoted as *bootstrap*$_{\mathrm{fp}}$. It is based on parametric models for the distribution and the autocorrelation function of the yearly maxima. This approach operates on the annual maxima in order to exploit the Fisher–Tippett theorem motivating a parametric model for the maxima distribution.

***bootstrap*$_{\mathrm{sp}}$.** The fourth strategy is a semi-parametric approach, which we call *bootstrap*$_{\mathrm{sp}}$. It similarly uses a parametric model for the ACF of the maxima series, but instead of the GEV we choose a nonparametric model for the distribution.

While the first two strategies are common tools in time series analysis and are well described elsewhere [3.17, 3.53], we focus on describing only the full-parametric and semi-parametric bootstrap strategy.

3.4.1 Motivation of the Central Idea

A return level estimate is derived from an empirical maxima series which can be regarded as a realization of a stochastic process. The uncertainty of a return level estimate depends on the variability among different realizations of this process. As a measure of this variability, we consider the deviation of a realization's empirical distribution function from the true distribution function. If this variability is low, it is more likely to have obtained a good representative for the true maxima distribution from one sample. For a high variability instead, it is harder to get a representative picture of the underlying distribution from one sample. We illustrate this effect using long realizations ($N = 10\,000$) from the correlated and the uncorrelated process introduced in Sect. 3.3. We compare the difference between the distributions $\hat{F}_s(x)$ of a short section ($N = 100$) of a realization and the entire realization's distribution $\hat{F}_0(x)$ by means of the Kolmogorov–Smirnov distance $D = \max_x |\hat{F}_s(x) - \hat{F}_0(x)|$ [3.18]. Smaller distances D indicate a larger similarity between $\hat{F}_s$ and $\hat{F}_0$. Figure 3.4 shows the cumulative distribution function $\hat{F}(D)$ of these distances D for an uncorrelated (circles) and a correlated process (triangles). For the correlated process, we find a distribution of distances D located at larger values. This implies that the sections are more diverse in their distribution. Thus for correlated processes the variability in short realization's maxima distribution is larger than for the uncorrelated process. Realizations of correlated processes are therefore not as likely to yield as representative results for the underlying distribution as a comparable sample of an uncorrelated process.

Since the variability of the return level estimator is a result of the variability of realization's maxima distribution, we employ this illustrative example and study the estimator's variability among sections of a long record. Ideally, the properties of this long record should be close to the underlying properties of the process under consideration. This requires a satisfying model for the maxima series' distribution and the autocorrelation function. In the approach pursued here, we initially provide two separate models for the two characteristics. Realizations of these two models are then combined to obtain one realization satisfying both the desired distribution and the ACF.

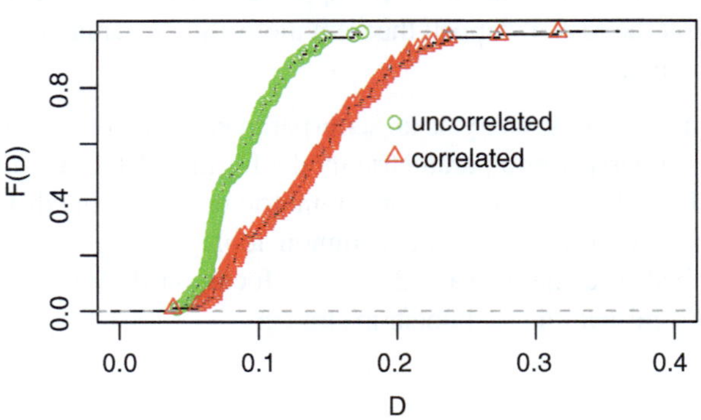

Fig. 3.4 Empirical cumulative distribution function $\hat{F}(D)$ of the Kolmogorov–Smirnov distances D between sections of a 100-year annual maxima series and the entire 10 000-year annual maxima series. For the uncorrelated record (○) distances D are located at smaller values than for the correlated record (△)

3.4.2 Modelling the Distribution

The aim of modelling the distribution is to provide means for generating realizations used in a later step of the bootstrap procedure.

For the semi-parametric approach, the distribution of the maxima is modelled by the empirical cumulative distribution function from the observed series. This means realizations from this model can be obtained simply by sampling with replacement from the observed maxima series [3.17].

The full-parametric approach exploits the Fisher–Tippett theorem for extreme values (Sect. 3.2.1). It uses the parametric GEV family as a model for the maxima distribution. Realizations are then obtained directly by sampling from the parametric model fitted to the empirical maxima series.

3.4.3 Modelling the ACF

At this point, we are mainly interested in an adequate representation of the empirical ACF. Such a representation can be achieved by modelling the series under investigation with a flexible class of linear time series models. We do not claim that these models are universally suitable for river run-off series. However, together with a static nonlinear transformation function and a deterministic description of the seasonal cycle, they capture the most dominant features regarding the stochastic variability. Especially, an adequate representation of the ACF can be achieved, which is the objective of this undertaking. These models are obviously not adequate for studying, e.g. the effect of changing precipitation patterns on run-off or other external influences. This is, however, not the focus of this chapter.

3.4.3.1 ARMA Processes

A simple and prevailing correlated stochastic process is the autoregressive first-order (AR[1]) process (or red noise) frequently used in various geophysical contexts, e.g. [3.26, 3.41, 3.64]. For a random variable X_t, it is a simple and intuitive way to describe a correlation with a predecessor in time by

$$X_t = \phi_1 X_{t-1} + \sigma_\eta \eta_t, \tag{3.3}$$

with η_t being a Gaussian white noise process ($\eta_t \sim W\mathcal{N}(0, 1)$) and ϕ_1 the lag-one autocorrelation coefficient. This approach can be extended straightforwardly to include regressors X_{t-k} with lags $1 \leq k \leq p$ leading to AR[p] processes allowing for more complex correlation structures, including oscillations. Likewise lagged instances of the white noise process $\psi_l \eta_{t-l}$ (moving average component) with lags $1 \leq l \leq q$ can be added leading to ARMA[p,q] processes, a flexible family of models for the ACF [3.9, 3.10]. There are numerous applications of ARMA models, also in the context of river run-off, e.g. [3.7, 3.27, 3.60].

3.4.3.2 FARIMA Processes

Since ARMA processes are short-range dependent, i.e. have a summable ACF [3.4], long-range dependence, which is frequently postulated for river run-off [3.40, 3.43, 3.47], cannot be accounted for. It is thus desirable to use a class of processes able to model this phenomenon. Granger [3.24] and Hosking [3.29] introduced fractional differencing to the concept of linear stochastic models and therewith extended the ARMA family to fractional autoregressive integrated moving average (FARIMA) processes. A convenient formulation of such a long-range-dependent FARIMA process X_t is given by

$$\Phi(B)(1 - B)^d X_t = \Psi(B)\eta_t, \tag{3.4}$$

with B denoting the back-shift operator ($BX_t = X_{t-1}$), $\eta_t \sim W\mathcal{N}(0, \sigma_\eta)$ a white noise process, and $d \in \mathbb{R}$ the fractional difference parameter. The latter is related to the Hurst exponent, frequently used in hydrology, $H = d + 0.5$ [3.4]. The autoregressive and moving average components are described by polynomials of order p and q

$$\Phi(z) = 1 - \sum_{i=1}^{p} \phi_i z^i, \quad \Psi(z) = 1 + \sum_{j=1}^{q} \psi_j z^j, \tag{3.5}$$

respectively. In practice, the fractional difference operator $(1 - B)^d$ has to be expanded in a power series, cf. [3.24]. A FARIMA$[p, d, q]$ process is stationary if $d < 0.5$ and all solutions of $\Phi(z) = 0$ in the complex plane lie outside the unit circle. It exhibits long range dependence or long-memory for $0 < d < 0.5$. Processes with $d < 0$ are said to possess intermediate memory; in practice this case is rarely encountered, it is rather a result of "over-differencing" [3.4]. Recently, the FARIMA model class has been used as a stochastic model for river run-off, e.g. [3.4, 3.33, 3.40, 3.43, 3.44]. Also self-similar processes are sometimes used in this context. The latter provide a simple model class, but are not as flexible as FARIMA processes and are thus appropriate only in some specific cases. FARIMA models, however, can be used to model a larger class of natural processes including self-similar processes. (For a comprehensive overview of long-range dependence and FARIMA processes refer to [3.4, 3.46] and references therein.)

3.4.3.3 Parameter Estimation

The model parameters $(\phi_1, \ldots, \phi_p, d, \psi_1, \ldots, \psi_q)$ are estimated using Whittle's approximation to the ML estimator [3.4]. It operates in Fourier space – with the spectrum being an equivalent representation of the autocorrelation function [3.48] – and is computationally very efficient due to the use of the fast Fourier transform. Therefore, the Whittle estimator is especially useful for long records where exact MLE is not feasible due to computational limits.

The model orders p and q are a priori unknown and can be determined using the Hannan–Quinn information criterion (HIC) which is advocated for FARIMA processes:

$$\mathrm{HIC} = N \log \hat{\sigma}_\eta^2 + 2c \log\log N(p+q+1), \tag{3.6}$$

with $c > 1$, $\hat{\sigma}_\eta^2$ the ML estimate of the variance of the driving noise η_t, and $p+q+1$ being the number of parameters [3.5, 3.6, 3.25]. We choose the model order p and q such that the HIC takes a minimum.

3.4.3.4 Indirect Modelling of the Maxima Series' ACF

Modelling the ACF of a run-off maxima series is usually hampered by the shortness of the records. For a short time series it is often difficult to reject the hypothesis of independence, cf. Sect. 3.3. To circumvent this problem, we model the daily series and assume that the resulting process adequately represents the daily series' ACF. It is used to generate long records whose extracted maxima series are again modelled with a FARIMA$[p, d, q]$ process. These models are then considered as adequate representatives of the empirical maxima series (ACF). This indirect approach of modelling the maxima series' ACF relies on the strong assumption that the model for the daily series and also its extrapolation to larger time scales is adequate.

3.4.3.5 Modelling a Seasonal Cycle

The seasonal cycle found in a daily river run-off series has a fixed periodicity of 1 year. It can be modelled as a deterministic cycle $C(t)$ which is periodic with period T (1 year): $C(t+T) = C(t)$. Combined with the stochastic model $X(t)$ this yields the following description:

$$Y(t) = C(t) + X(t). \tag{3.7}$$

In the investigated case studies $C(t)$ is estimated by the average yearly cycle obtained by averaging the run-off $Q(t)$ of a specific day over all M years, i.e. $\hat{C}(t) = 1/M \sum_j Q(t + jT), t \in [1, T]$, cf. [3.28].

3.4.3.6 Including a Static Nonlinear Transformation Function

River run-off is a strictly positive quantity and the marginal distribution is in general positively skewed. This suggests to include an appropriate static nonlinear transformation function in the model. Let $Z = T(Y)$ denote the Box–Cox transformation (cf. Sect. 3.10.3, [3.8]) of the random variable Y (Eq. (3.7)). Then Z is a positively skewed and strictly positive variable, suitable to model river run-off. One can think of this static transformation function as a change of the scale of measurement. This transformation has been suggested for the modelling of river run-off by Hipel and McLeod [3.28].

The full model for the ACF can be written as

$$\Phi(B)(1-B)^d X_t = \Psi(B)\eta_t, \tag{3.8}$$

$$Y(t) = C(t) + X(t), \tag{3.9}$$

$$Z(t) = T(Y). \tag{3.10}$$

3.4.3.7 Simulation of FARIMA Processes

Several algorithms are known to simulate data from a FARIMA process (for an overview refer to [3.1]). Here, we use a method based on the inverse Fourier transform described in [3.61]. It was originally proposed for simulating self-similar processes but can be straightforwardly extended to FARIMA processes.[3]

3.4.4 Combining Distribution and Autocorrelation

Having a model for the distribution and for the ACF we can generate realizations, i.e. a sample $\{W_i\}_{i=1,\ldots,N}$, from the distribution model and a series $\{Z_i\}_{i=1,\ldots,N}$ from the FARIMA model including the BOX–COX transformation and, if appropriate, the seasonal cycle. To obtain a time series $\{Q_i\}_{i=1,\ldots,N}$ with distribution equal to the one of $\{W_i\}_{i=1,\ldots,N}$ and ACF comparable to that of $\{Z_i\}_{i=1,\ldots,N}$, we employ the iterative amplitude adjusted Fourier transform (IAAFT).

The IAAFT was developed by Schreiber and Schmitz [3.52] to generate surrogate time series used in tests for nonlinearity [3.59]. The surrogates are generated such that they retain the linear part of the dynamics of the original time series including a possible nonlinear static transfer function. This implies that the power spectrum (or ACF, equivalently) and the frequency distribution of values are conserved. The algorithm basically changes the order of the elements of a record in a way that the periodogram stays close to a desired one [3.53].

Besides using the IAAFT on the daily series to generate an ensemble of surrogates (denoted as $iaaft_{\mathrm{d}}$), we employ this algorithm also to create records $\{Q_i\}_{i=1,\ldots,N}$ with a periodogram prescribed by a series $\{Z_i\}_{i=1,\ldots,N}$ and a frequency distribution coming from $\{W_i\}_{i=1,\ldots,N}$.

3.4.5 Generating Bootstrap Ensembles

With the described methods we are now able to build a model for the distribution and ACF of empirical maxima series and combine them to obtain long records. This

[3] An R package with the algorithms for the FARIMA parameter estimation (based on the code from Beran [3.4]), the model selection (HIC), and the simulation algorithm can be obtained from the author.

provides the basis of the full-parametric bootstrap ensemble $bootstrap_{\mathrm{fp}}$ and the semi-parametric ensemble $bootstrap_{\mathrm{sp}}$.

In detail, the strategies to obtain the ensembles $bootstrap_{\mathrm{fp}}$ and $bootstrap_{\mathrm{sp}}$ can be outlined as follows:

1. Model the correlation structure of the maxima
   (a) If necessary, transform the daily run-off data to follow approximately a Gaussian distribution using a log or BOX–COX transform [3.8, 3.28], cf. Sect. 10.3.
   (b) Remove periodic cycles (e.g. annual, weekly).
   (c) Model the correlation structure using a FARIMA$[p, d, q]$ process, select the model orders p and q with HIC (Sect. 3.4.3).
   (d) Generate a long series from this model ($N_{\mathrm{long}} \gtrsim 100 N_{\mathrm{data}}$).
   (e) Add the periodic cycles from 1(a). The result is a long series sharing the spectral characteristics, especially the seasonality, with the empirical record.
   (f) Extract the annual maxima series.
   (g) Model the correlation structure of the simulated maxima series using a FARIMA$[p_{\mathrm{max}}, d, q_{\mathrm{max}}]$ process, with orders p_{max} and q_{max} selected with HIC.
2. Model the distribution of the maxima according to the approach used:
   $bootstrap_{\mathrm{fp}}$: Estimate the parameters of a GEV model from the empirical maxima series using MLE (Sect. 3.2.2).
   $bootstrap_{\mathrm{sp}}$: Use the empirical maxima distribution as model.
3. Generate an ensemble of size N_{ensemble} of maxima series with length N_{max} with correlation structure and value distribution from the models built in 1 and 2:
   (a) Generate a series $\{Z_i\}$ with the FARIMA$[p_{\mathrm{max}}, d, q_{\mathrm{max}}]$ model from step 1(f) of length $N_{\mathrm{ensemble}} N_{\mathrm{max}}$ back-transform according to 1(a).[4]
   (b) Generate a sample $\{W_i\}$ with length $N_{\mathrm{ensemble}} N_{\mathrm{max}}$
   $bootstrap_{\mathrm{fp}}$ from the GEV model specified in step 2(a).
   $bootstrap_{\mathrm{sp}}$ from sampling with replacement from the empirical maxima series.
   (c) By means of IAAFT $\{W_i\}$ is reordered such that its correlation structure is similar to that of $\{Z_i\}$. This yields the run-off surrogates $\{Q_i\}_{i=1,\ldots,N_{\mathrm{ensemble}} N_{\mathrm{max}}}$.
   (d) Splitting $\{Q_i\}_{i=1,\ldots,N_{\mathrm{ensemble}} N_{\mathrm{max}}}$ into blocks of size N_{max} yields the desired ensemble.

Estimating the desired return level from each ensemble member as described in Sect. 3.2.2 yields a frequency distribution of return level estimates which can be used to assess the variability of this estimator.

[4] Instead of back-transforming here, one can BOX–COX transform the outcome of step 3(b), combine the results with IAAFT as in step 3(c), and back-transform afterwards. This procedure turned out to be numerically more stable.

3.5 Comparison of the Bootstrap Approaches

A comparison of the four different bootstrap approaches *bootstrap*$_{\rm cl}$, *iaaft*$_{\rm d}$, *bootstrap*$_{\rm fp}$, and *bootstrap*$_{\rm sp}$ is carried out on the basis of a simulation study. We start with a realization of a known process, chosen such that its correlation structure and its value distribution are plausible in the context of river run-off. Here, this is a FARIMA process similar to those used in [3.43] with a subsequent exponential transformation to obtain a log-normal distribution. This process is used to generate a Monte Carlo ensemble of simulated daily series. For each realization we extract the maxima series and estimate a 100-year return level. We obtain a distribution of 100-year return level estimates which represent the estimators' variability for this process. In the following, this distribution is used as a reference to measure the performance of the bootstrap approaches. A useful strategy should reproduce the distribution of the return level estimator reasonably well.

We now take a representative realization out of this ensemble and consider it as a record, we possibly could have observed. From this record, we generate the four bootstrap ensembles according to the approaches presented in Sect. 3.4.5. The resulting four frequency distributions of the 100-year return level estimates are then compared to the distribution of the reference ensemble and to the asymptotic distribution of the ML estimator.

3.5.1 Monte Carlo Reference Ensemble

We simulate the series for the reference ensemble with a FARIMA$[1, d, 0]$ process with parameters $d = 0.25$ (or $H = 0.75$), $\phi_1 = 0.9$, variance $\sigma^2 \approx 1.35$, Gaussian driving noise η, and length $N = 36\,500$ (100 years of daily observations). The skewness typically found for river run-off is achieved by subsequently transforming the records to a log-normal distribution. To resemble the procedure of estimating a 100-year return level we extract the annual maxima and estimate a 0.99 quantile using a Gumbel distribution as parametric model. From an ensemble of 100 000 runs, we obtain a distribution of 100-year return levels (0.99 quantiles) serving as a reference for the bootstrap procedures.

3.5.2 The Bootstrap Ensembles

Taking the "observed" series, we generate the four bootstrap ensembles according to Sect. 3.4.5. Since *iaaft*$_{\rm d}$ and *bootstrap*$_{\rm cl}$ are well described in the literature, we focus on the semi-parametric and full-parametric approach.

Following the outline in Sect. 3.4.5, we start modelling the correlation structure of the "observed" series using a FARIMA process as described in Sect. 3.4.3. As a transformation, we choose a log-transform as a special case of the BOX–COX. We treat the process underlying the sample as unknown and fit FARIMA$[p, d, q]$

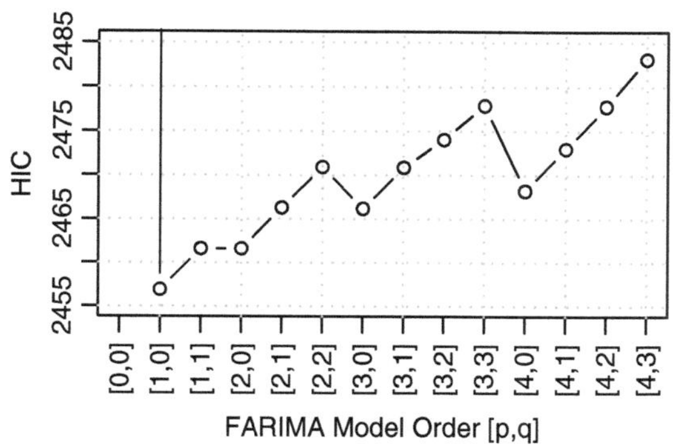

Fig. 3.5 Comparison of the HIC of different FARIMA$[p, d, q]$ models. Smaller values indicate a better model. On the abscissa different model orders $[p, q]$ are plotted ($0 \leq q \leq 3$, $0 \leq p \leq 4$, and $q \leq p$). The model orders are discrete, the *lines* connecting the discrete orders $[p, q]$ are drawn to enhance clarity

models with $0 \leq q < p \leq 4$. Figure 3.5 shows the result of the model selection criterion (HIC) (Eq. (3.6)) for these fits. The FARIMA$[1, d, 0]$ with parameters and asymptotic standard deviation $d = 0.250 \pm 0.008$, $\phi = 0.900 \pm 0.005$, $\sigma_\eta^2 = 0.0462$ yields the smallest HIC and is chosen to model the record. Thus, the proper model structure is recovered and step 1(b) is completed.

With this model we generate an artificial series longer than the original one ($N_{\mathrm{long}} = 100 N_{\mathrm{data}}$) according to step 1(c). The extracted annual maxima series contains $N_{\mathrm{max}} = 10\,000$ data points. Since we do not expect the ACF of this maxima series to require a more complex model (in terms of the number of parameters) than the daily data, we use orders $p \leq 1$ and $q = 0$, leaving only two models. The FARIMA$[0, d, 0]$ has a slightly smaller HIC value ($\mathrm{HIC}_{[0,d,0]}$=21 792.06) than FARIMA$[1, d, 0]$ ($\mathrm{HIC}_{[1,d,0]}$=21 792.40) and will thus be chosen in the following, step 1(f). The resulting parameters are $d = 0.205 \pm 0.008$ and $\sigma_\eta^2 = 0.535$.

Having modelled the correlation structure we now need a representation for the distribution. For the full-parametric bootstrap ensemble *bootstrap*$_{\mathrm{fp}}$, we get back to the "observed" series and model the annual maxima with a parametric Gumbel distribution, step 2(a). This results in ML estimates for the location and scale parameters: $\mu = 10.86$, $\sigma = 8.35$. Since the semi-parametric approach *bootstrap*$_{\mathrm{sp}}$ does not need a model but uses a classical bootstrap resampling from the empirical annual maxima series, we now can generate the desired bootstrap ensembles *bootstrap*$_{\mathrm{fp}}$ and *bootstrap*$_{\mathrm{sp}}$ both with 1 000 members according to step 3.

Figure 3.6 compares the frequency distributions of estimated return levels from the four bootstrap ensembles to the reference distribution (grey filled) and to the asymptotic distribution of the ML estimator (dotted). The left plot shows the result of the *bootstrap*$_{\mathrm{cl}}$ (solid) and the *iaaft*$_{\mathrm{d}}$ (dashed) ensembles. While *bootstrap*$_{\mathrm{cl}}$ accounts for more variability than the asymptotic distribution, *iaaft*$_{\mathrm{d}}$ exhibits less variability, although it takes autocorrelation of the daily data into account. This might be due to the fact that the records in the *iaaft*$_{\mathrm{d}}$ ensemble consist of exactly

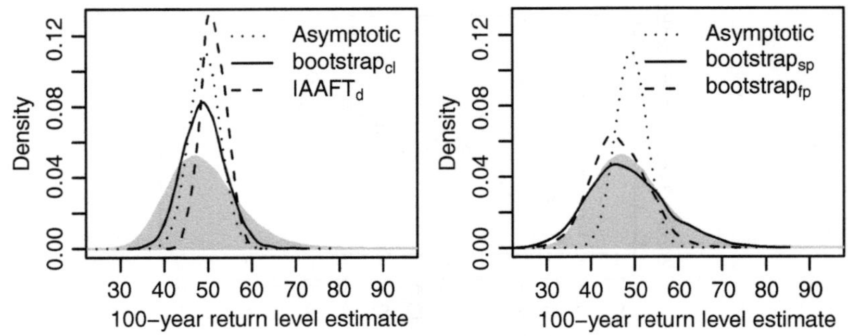

Fig. 3.6 Comparison of the result for different bootstrap ensembles to the MC reference ensemble (*grey area*) and the asymptotic distribution of the ML estimator (*dotted*). The bootstrap ensembles consist each of 1 000 members. The *left plot* shows the results of the nonparametric bootstrap resampling and the daily IAAFT surrogates. The full-parametric and semi-parametric bootstrap strategies are shown in the *right plot*

the same daily run-off values arranged in a different order. While this allows for some variability on the daily scale, an annual maxima series extracted from such a record is limited to a much smaller set of possible values. Since the temporal order of the maxima series does not influence the return level estimation, the variability of the estimates is reduced.

The right panel in Fig. 3.6 shows the result from the *bootstrap*$_{sp}$ (solid) and the *bootstrap*$_{fp}$ (dashed) ensembles. The latter strategy is better than the *bootstrap*$_{cl}$ but still yields a too narrow distribution. In contrast, the result from the *bootstrap*$_{sp}$ ensemble gets very close to the reference ensemble. Thus, this approach is a promising strategy to improve the uncertainty analysis of return level estimates and is studied in more detail in the following section.

3.5.3 Ensemble Variability and Dependence on Ensemble Size

We investigate the potential of the semi-parametric bootstrap approach by studying its inter-ensemble variability and the dependence on ensemble size. This can be achieved by performing an extensive simulation study, i.e. generating different sets of *bootstrap*$_{sp}$ ensembles, each set containing 100 ensembles of a fixed size. We are interested in the variability of the ensemble runs within one set of fixed size as well as in the effect of the ensemble size. The ensemble size varies between $N_{\text{ensemble}} = 50$ and $N_{\text{ensemble}} = 6\,000$. To facilitate the representation of the result, we do not consider the entire distribution of the return level estimates for each ensemble, but rather five selected quantiles with relative frequencies of 5, 25, 50, 75, and 95%. Figure 3.7 shows these quantiles estimated from the ensembles of different ensemble sizes as grey dots. Due to the large number of grey dots, they cannot be perceived as individual dots but rather as grey clouds.

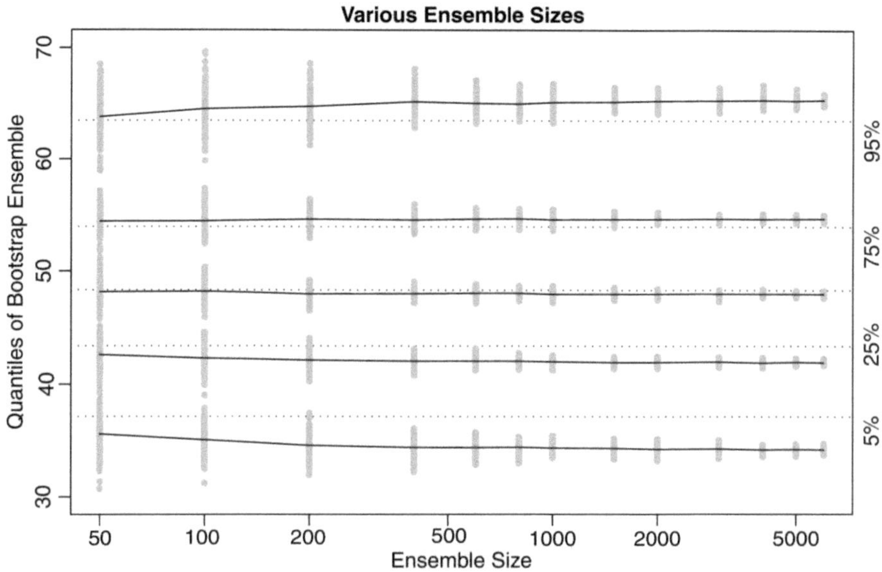

Fig. 3.7 The quantiles of the semi-parametric bootstrap ensembles of different size. One hundred ensembles of the same size are grouped in a set. The 5, 25, 50, 75, and 95% quantiles of each ensemble in a set is marked with a *grey dot*. This results in *grey areas* representing the variability within a set. The *solid lines* connect the sets' mean values for each quantile. The quantiles from the reference ensemble are represented as a *dotted line*

The variability of the quantile estimates decreases with increasing ensemble size indicated by the convergence of the grey clouds for each of the five quantiles. Consequently, the ensemble size should be chosen according to the accuracy needed. For small sizes, the means of the selected quantiles are close to the values from the reference ensemble, especially for the three upper quantiles. With an increasing size, difference to the reference ensemble increases for the extreme quantiles until they stagnate for ensembles with more than about 2 000 members. For the 5 and 95% quantiles, the difference between the bootstrap and the Monte Carlo is less than 6% of the return level estimates.

3.6 Case Study

To demonstrate the applicability of the suggested semi-parametric bootstrap approach ($bootstrap_{\mathrm{sp}}$), we exemplify the strategy with a case study. We consider the run-off record from the gauge Vilsbiburg at the river Große Vils in the Danube River catchment. Vilsbiburg is located in the south-east of Germany about 80 km north-east of Munich. The total catchment area of this gauge extends to 320 km^2. The mean daily run-off has been recorded from 01 Nov 1939 to 07 Jan 2002 and thus comprises $N_{\mathrm{years}} = 62$ full years or $N = 22\,714$ days. The run-off averaged over the whole observation period is about 2.67 m^3/s.

3.6.1 Extreme Value Analysis

First, we perform an extreme value analysis as described in Sect. 3.2.2, i.e. extracting the annual maxima and determining the parameters of a GEV distribution by means of ML estimation. In order to test whether the estimated shape parameter $\hat{\xi} = 0.04$ is significantly different from zero, we compare the result to a Gumbel fit using the likelihood-ratio test [3.12]. With a p-value of $p = 0.74$ we cannot reject Gumbel distribution as a suitable model on any reasonable level. The resulting location and scale parameters with asymptotic standard deviation are $\mu = (28.5 \pm 2.0)\text{m}^3/\text{s}$ and $\sigma = (15.0 \pm 1.5)\text{m}^3/\text{s}$. The associated quantile and return level plots are shown in Fig. 3.8 together with their 95% asymptotic confidence limits.

According to Eq. (3.2) we calculate a 100-year return level ($m = 100$) and use the delta method (Eq. (3.21)) to approximate a standard deviation under the hypothesis of independent observations: $r_{100} = 97.7 \pm 7.9$.

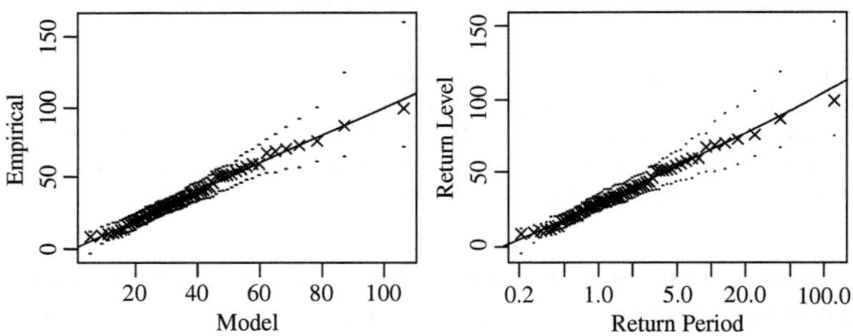

Fig. 3.8 Result of the ML estimation of the Gumbel parameters for the Vilsbiburg yearly maxima series compared to the empirical maxima series in a quantile plot (*left panel*) and return level plot (*right panel*)

3.6.2 Modelling the ACF of the Daily Series

In the second step, we model the correlation structure which requires preprocessing of the run-off series: To get closer to a Gaussian distribution, a BOX–COX transformation (Sect. 3.8) is applied to the daily run-off. The parameter $\lambda = -0.588$ is chosen such that the unconditional Gaussian likelihood is maximized. Subsequent to this static transformation, we calculate the average annual cycle in the mean and the variance, cf. Sect. 3.4.3. The cycle in the mean is subtracted from the respective days and the result is accordingly divided by the square root of the variance cycle. As we find also indication for a weekly component in the periodogram, we subtract this component analogously. The BOX–COX transformation and the seasonal filters have been suggested by Hipel and McLeod [3.28] for hydrological time series. Although the proposed estimates of the periodic components in mean and variance

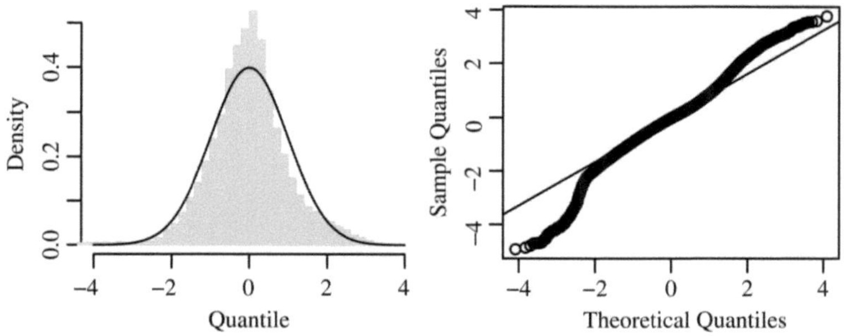

Fig. 3.9 Comparison of the BOX–COX transformed, deseasonalized, and mean-adjusted run-off values to a Gaussian distribution. Histogram (*left, grey*) with a density estimate (*solid line*) and quantile plot (*right*)

are consistent, especially the estimates of the annual cycle exhibit a large variance. Several techniques such as STL [3.11] are advocated to obtain smoother estimates. Studying those periodic components is, however, not the focus of this chapter.

Figure 3.9 shows a comparison of the transformed and mean-adjusted daily run-off record to a Gaussian distribution in the form of a density plot (left panel) and a plot of empirical versus theoretical quantiles (right panel). The transformed distribution is much less skewed than the original one. Differences to a Gaussian are mainly found in the tails.

We now fit FARIMA$[p, d, q]$ models of various orders with $0 \le q \le 5$, $0 \le p \le 6$, and $q \le p$ and compare the HIC of the different models in Fig. 3.10 (left). The smallest value for the HIC is obtained for the FARIMA$[3, d, 0]$ process, which is thus chosen to model the autocorrelation of the daily run-off. The parameters estimated for this process with their asymptotic standard deviation are $d = 0.439 \pm 0.016$, $\phi_1 = 0.415 \pm 0.017$, $\phi_2 = -0.043 \pm 0.007$, $\phi_3 = 0.028 \pm 0.008$, and $\sigma_\eta^2 = 0.2205$. Using the goodness-of-fit test proposed by Beran [3.4], we obtain a p-value of $p = 0.015$. The model thus cannot be rejected on a 1% level of significance. The result of this fit is shown in the spectral domain in Fig. 3.10 (right).

3.6.3 *Modelling the ACF of the Maxima Series*

Using this model, a long series is generated with $N_{\text{long}} = 100 N_{\text{data}}$. This simulated series is partially back-transformed: the overall mean and the seasonal cycles in mean and variance are added. Note that the BOX–COX transform is not inverted in this step. From the resulting record, we extract the annual maxima series. Figure 3.2 (left panel) shows the ACF of the original maxima series (solid) and compares it to the ACF of a section of the same length cut out of the maxima series gained from the simulated run (dotted). The original series has been BOX–COX transformed as well to achieve a comparable situation. The autocorrelation basically fluctuates within the 95% significance level (dashed) for a white noise. However, the full maxima

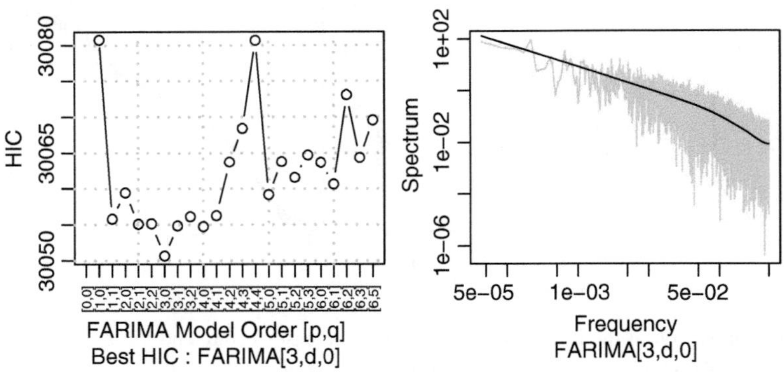

Fig. 3.10 Model selection for the BOX–COX transformed and deseasonalized daily run-off with mean removed. The spectral density of the model with smallest HIC is shown in double logarithmic representation together with the periodogram (*grey*) of the empirical series (*right panel*)

series from the simulated run of 6 200 data points exhibits prominent autocorrelation (Fig. 3.2, right panel). This indicates that, although an existing autocorrelation structure is not necessarily visible in a short sample of a process, it might still be present. Accounting for this dependence improves the estimation of confidence limits.

In the next step, we model the correlation structure of the maxima series with a FARIMA process. Again, we do not expect this series to be more adequately modelled by a more complex process than the daily data. We use HIC to choose a model among orders $p_{\max} \leq p = 3, q_{\max} \leq q = 1$. The smallest values for the HIC is attained for a FARIMA$[0, d, 0]$ model with $d = 0.398 \pm 0.010$. The goodness of fit yields a p-value of $p = 0.319$ indicating a suitable model.

3.6.4 Combining Distribution and ACF

Having the model for the ACF of the maxima series, we are now able to generate a bootstrap ensemble of artificial data sets according to the semi-parametric strategy *bootstrap*$_{\text{sp}}$ as described in Sect. 3.4.5. We use IAAFT to combine the results of the FARIMA simulation with the BOX–COX transformed resampled empirical maxima series. In the last step the ensemble is restored to the original scale of measurement by inverting the BOX–COX transform we started with.

3.6.5 Calculating the Confidence Limit

Subsequently, the 100-year return level $\hat{r}^*$ (or 0.99 quantile) is estimated for each ensemble member yielding the desired distribution for the 100-year return level estimator shown in Fig. 3.11. From this distribution we obtain an estimate for a

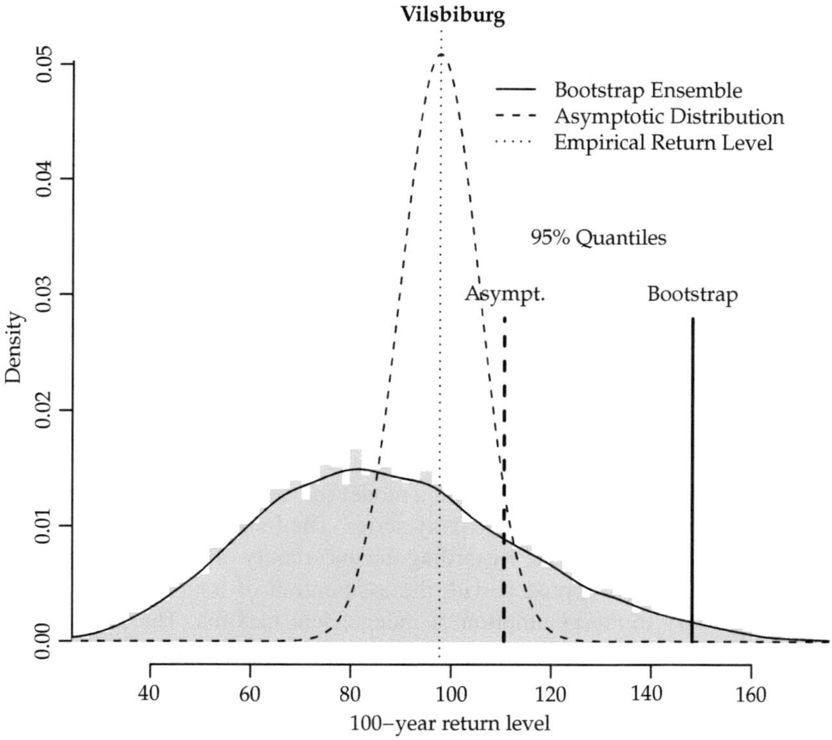

Fig. 3.11 Frequency distribution of the 100-year return level estimates from the *bootstrap*$_{sp}$ ensemble with 9 999 members for Vilsbiburg as histogram (*grey*) and density estimate (*solid line*) compared to the asymptotic distribution of the ML estimator derived from the Fisher information matrix (*dashed*). The 100-year return level estimate from the empirical maxima series is marked as *dotted vertical line*. The 95% quantiles of the asymptotic and bootstrap distributions are indicated as *dashed* and *solid vertical lines*, respectively

$(1-\alpha)\%$ one-sided upper confidence limit r^{α} using order statistics. r^{α} is calculated from order statistics as $r^{\alpha} = \hat{r}_{100} - (\hat{r}^*_{(N+1)\alpha} - \hat{r}_{100})$ [3.17], where the $\hat{r}^*_i$ are sorted in ascending order. With an ensemble size $N_{\text{ensemble}} = 9\,999$ we ensure $(N+1)\alpha$ being an integer for common choices of α.

To facilitate the comparison of the 95% confidence limits obtained from the bootstrap ensemble ($r^{0.95}_{\text{boot}} \approx 148\,\text{m}^3/\text{s}$) and the asymptotic distribution ($r^{0.95}_{\text{asymp}} \approx 110\,\text{m}^3/\text{s}$) they are marked as vertical lines in Fig. 3.11. The bootstrap 95% confidence level $r^{0.95}_{\text{boot}}$ clearly exceeds the quantile expected from the asymptotic distribution confirming the substantial increase in uncertainty due to dependence. Furthermore, the tails of the bootstrap ensemble decay slower than the tails of the asymptotic distribution. The interpretation of such a confidence level is the following: In 95% of 100-year return level estimates the expected ("true") 100-year return level will not exceed the 95% confidence limit.

3.7 Discussion

The approach is presented in the framework of GEV modelling of annual maxima using maximum likelihood. The concept can also be applied in the context of other models for the maxima distribution (e.g. log-normal) or also different parameter estimation strategies (e.g. probability weighted moments). Furthermore, it is conceivable to extend the class of models describing the dependence to FARIMA models with dependent driving noise (FARIMA-GARCH [3.20]) or seasonal models [3.40, 3.44].

The modelling approach using FARIMA[p,d,q] models and a subsequent adjustment of the values has been investigated in more detail by [3.62]. Using simulation studies, it was demonstrated that the combination of FARIMA models and the IAAFT is able to reproduce also other characteristics of time series than the distribution and power spectrum. Also the increment distribution and structure functions for river run-off are reasonably well recovered.

In the approach described, we obtain a model for the ACF of the maxima series only with the help of a model of the daily series. The longer this daily series is the more reliable the model will be. Regarding the uncertainty of return level estimates, we might consider this approach with the assumption of long-range dependence as complementary to the assumption of independent maxima. The latter assumption yields the smallest uncertainty while the assumption of long-range dependence yields larger confidence intervals which can be considered as an upper limit of uncertainty. The actual detection of long-range dependence for a given time series is by no means a trivial problem [3.41] but it is not the focus of this chapter.

It is also possible to include available annual maxima in the procedure for periods where daily series have not been recorded. This enhances the knowledge of the maxima distribution but cannot be used for the modelling of the ACF.

3.8 Conclusion

We consider the estimation of return levels from annual maxima series using the GEV as a parametric model and maximum likelihood (ML) parameter estimation. Within this framework, we explicitly account for autocorrelation in the records which reveals a substantial increase in uncertainty of the flood return level estimates. In the standard uncertainty assessment, i.e. the asymptotic confidence intervals based on the Fisher information matrix or the profile likelihood, autocorrelations are not accounted for. For long-range-dependent processes, this results in uncertainty limits being too small to reflect the actual variability of the estimator. On the way to fill this gap, we study and compare four bootstrap strategies for the estimation of confidence intervals in the case of correlated data. This semi-parametric bootstrap strategy outperforms the three other approaches. It showed promising results in the validation study using an exponentially transformed FARIMA[1,d,0] process. The main idea involves a resampling approach for the annual maxima and a parametric model (FARIMA) for their autocorrelation function. The combination of the resampling and the FARIMA model is realized with the iterative amplitude adjusted Fourier transform, a resampling method used in nonlinearity testing. The results of the semi-parametric bootstrap approach are substantially better than those based on the standard asymptotic approximation for MLE using the Fisher information matrix. Thus this approach might be of considerable value for flood risk assessment of water management authorities to avoid floods or misallocation of public funds. Furthermore, we expect the approach to be applicable also in other sectors where an extreme value analysis with dependent extremes has to be carried out.

The practicability is illustrated for the gauge Vilsbiburg at the river Vils in the Danube catchment in southern Germany. We derived a 95% confidence limit for the 100-year flood return level. This limit is about 38% larger than the one derived from the asymptotic distribution, a dimension worth being considered for planning options. To investigate to what extent this increase in uncertainty depends on catchment characteristics, we plan to systematically study other gauges. Furthermore, a refined model selection strategy and the accounting for instationarities due to climate change is the subject of further work.

Acknowledgments This work was supported by the German Federal Ministry of Education and Research (grant number 03330271) and the European Union Baltic Sea INTERREG III B neighbourhood program (ASTRA, grant number 0085). We are grateful to the Bavarian Authority for the Environment for the provision of data. We further wish to thank Dr. V. Venema, Dr. Boris Orlowsky, Dr. Douglas Maraun, Dr. P. Braun, Dr. J. Neumann, and H. Weber for several fruitful discussions.

Appendix

General Extreme Value Distribution

Consider the maximum

$$M_n = \max\{X_1, \dots, X_n\} \tag{3.11}$$

of a sequence of n independent and identically distributed (iid) variables $X_1, \dots, X_n$ with common distribution function F. This can be, for example, daily measured run-off at a gauge; M_n then represents the maximum over n daily measurements, e.g. the annual maximum for $n = 365$. The three types theorem states that

$$\Pr\{(M_n - b_n)/a_n \leq z\} \rightarrow G(z), \text{ as } n \rightarrow \infty, \tag{3.12}$$

with $\{a_n\}$ and $\{b_n\}$ being series of normalization constants and $G(z)$ a non-degenerate distribution function known as the general extreme value distribution (GEV)

$$G(z) = \exp\left\{-\left[1+\xi\left(\frac{z-\mu}{\sigma}\right)\right]^{-1/\xi}\right\}. \tag{3.13}$$

z is defined on $\{z|1+\xi(z-\mu)/\sigma > 0\}$. The model has a location parameter μ, a scale parameter σ, and a form parameter ξ. The latter decides whether the distribution is of type II (Fréchet, $\xi > 0$) or of type III (Weibull, $\xi < 0$). The type I or Gumbel family

$$G(z) = \exp\left[-\exp\left\{-\left(\frac{z-\mu}{\sigma}\right)\right\}\right], \qquad \{z| -\infty < z < \infty\} \tag{3.14}$$

is obtained in the limit $\xi \to 0$ [3.12].

It is convenient to transform Eq. (3.12) into

$$\Pr\{M_n \le z\} \approx G((z-b_n)/a_n) = G^*(z). \tag{3.15}$$

The resulting distribution $G^*(z)$ is also a member of the GEV family and allows the normalization constants and the location, scale, and shape parameter to be estimated simultaneously.

We consider an autocorrelated stationary series $\{X_1, X_2, \ldots\}$ and define a condition of near-independence: For all $i_1 < \cdots < i_p < j_1 < \cdots < j_q$ with $j_1 - i_p > l$,

$$|\Pr\{X_{i_1} \le u_n, \ldots, X_{i_p} \le u_n, X_{j_1} \le u_n, \ldots, X_{j_q} \le u_n\} - \tag{3.16}$$

$$\Pr\{X_{i_1} \le u_n, \ldots, X_{i_p} \le u_n\}\Pr\{X_{j_1} \le u_n, \ldots, X_{j_q} \le u_n\}| \le \alpha(n,l), \tag{3.17}$$

where $\alpha(n, l_n) \to 0$ for some sequence l_n, with $l_n/n \to 0$ as $n \to \infty$. It can be shown that the three types theorem holds also for correlated processes satisfying this condition of near-independence [3.12, 3.39]. This remarkable result implies that the limiting distribution of the maxima of uncorrelated and (a wide class) of correlated series belongs to the GEV family.

Maximum Likelihood Parameter Estimation of the GEV

Let $\{M_{n,1}, M_{n,2}, \ldots, M_{n,m}\}$ be a series of independent block maxima observations, where n denotes the block size and m the number of blocks available for estimation. We denote $M_{n,i}$ as z_i. The likelihood function now reads

$$L(\mu,\sigma,\xi) = \prod_{i=1}^{m} g(z_i;\mu,\sigma,\xi), \tag{3.18}$$

where $g(z) = \mathrm{d}G(z)/\mathrm{d}z$ is the probability density function of the GEV. In the following, we consider the negative log-likelihood function $l(\mu,\sigma,\xi|z_i) = -\log L(\mu,\sigma,\xi|z_i)$.

Minimizing the negative log-likelihood with respect to $\theta = (\mu, \sigma, \xi)$ leads to the ML estimate $\hat{\theta} = (\hat{\mu}, \hat{\sigma}, \hat{\xi})$ for the GEV. Under suitable regularity conditions – among them independent observations z_i – and in the limit of large block sizes $(n \to \infty)$ $\hat{\theta}$ is multivariate normally distributed:

$$\hat{\theta} \sim \mathrm{MVN}_d(\theta_0, I_E(\theta_0)^{-1}), \tag{3.19}$$

with $I_E(\theta)$ being the expected information matrix (or Fisher information matrix) measuring the curvature of the log-likelihood. Denoting the elements of the inverse of I_E evaluated at $\hat{\theta}$ as $\beta_{j,k}$ we can approximate an $(1-\alpha)100\%$ confidence interval for each component j of $\hat{\theta}$ by

$$\hat{\theta}_j \pm z_{\frac{\alpha}{2}} \sqrt{\beta_{j,j}}, \tag{3.20}$$

with $z_{\frac{\alpha}{2}}$ being the $(1-\alpha/2)$ quantile of the standard normal distribution [3.12].

The m-year return level can be easily calculated as specified in Eq. (3.2). An approximated confidence interval for $\hat{r}_m$ can be obtained under the hypothesis of a normally distributed estimator $\hat{r}_m$ and making use of the standard deviation $\sigma_{\hat{r}_m}$. The latter can be calculated from the information matrix using the delta method [3.12]. For the Gumbel distribution we obtain

$$\sigma^2_{\hat{r}_m} = \beta_{11} - (\beta_{22} + \beta_{21}) \log\left(-\log\left(1 - \frac{1}{m}\right)\right) + \beta_{22}\left(\log\left(-\log\left(1 - \frac{1}{m}\right)\right)\right)^2. \tag{3.21}$$

BOX–COX Transform

The BOX–COX transformation can be used to transform a record $\{x_i\}$ such that its distribution is closer to a Gaussian. For records $\{x_i\}$ with $x_i > 0$ for all i, it is defined as[3.8]

$$y = \begin{cases} \frac{(x^\lambda - 1)}{\lambda}, & \lambda \neq 0 \\ \log(x), & \lambda = 0 \end{cases}.$$

We choose the parameter λ such that the unconditional Gaussian likelihood is maximized. Hipel also advocates the use of this transformation for river run-off [3.28].

References

3.1. J.-M. Bardet, G. Lang, G. Oppenheim, A. Philippe, M.S. Taqqu, Generators of long-range dependent processes: A survey. In ed. by P. Doukhan, G. Oppenheim, M.S. Taqqu, *Theory and Applications of Long-Range Dependence*, (Birkhäuser, Boston, 2003) pp. 579–623.

3.2. A. Bárdossy, F. Filiz, Identification of flood producing atmospheric circulation patterns. J. Hydrol. **313**(1–2), 48–57 (2005)

3.3. A. Bárdossy, S. Pakosch, Wahrscheinlichkeiten extremer Hochwasser unter sich ändernden Klimaverhältnissen. Wasserwirtschaft **7–8**, 58–62 (2005)
3.4. J. Beran, *Statistics for Long-Memory Processes*. Monographs on Statistics and Applied Probability (Chapman & Hall, New York, NY 1994)
3.5. J. Beran, R.J. Bhansali, D. Ocker, On unified model selection for stationary and nonstationary short- and long-memory autoregressive processes. Biometrika **85**(4), 921–934 (1998)
3.6. L. Bisaglia, Model selection for long-memory models. Quaderni Statistica **4**, 33–49 (2002)
3.7. P. Bloomfield, Trends in global temperature. Clim Change **21**(1), 1–16 (1992)
3.8. G.E.P. Box, D.R. Cox, An analysis of transformations. J. R. Statist. Soc. **62**(2), 211–252 (1964)
3.9. G.E.P. Box, G.M. Jenkins, *Time Series Analysis: Forecasting and Control* (Prentice Hall, New Jersey, 1976)
3.10. P.J. Brockwell, R.A. Davis, *Time Series: Theory and Methods*. Springer Series in Statistics (Springer, Berlin, 1991)
3.11. R.B. Cleveland, W.S. Cleveland, J.E. McRae, I. Terpenning, Stl: A seasonal-trend decomposition procedure based on loess. J. Official Statist., **6**(1), 3–73 (1990)
3.12. S.G. Coles, *An Introduction to Statistical Modelling of Extreme Values* (Springer, London, 2001)
3.13. S.G. Coles, M.J. Dixon, Likelihood-based inference for extreme value models. Extremes **2**(1), 5–23 (1999)
3.14. S.G. Coles, J.A. Tawn, A Bayesian analysis of extreme rainfall data. Appl. Stat. **45**, 463–478 (1996)
3.15. S.G. Coles, J.A. Tawn, Statistical methods for extreme values. Internet (1999)
3.16. J.M. Cunderlik, T.B. Ouarda, Regional flood-duration-frequency modeling in the changing environment. J. Hydrol., **318**(1–4), 276–291 (2006)
3.17. A.C. Davison, D.B. Hinkley, *Bootstrap Methods and Their Application*. Cambridge series in Statistical and Probabilistic Mathematics. (Cambridge University Press, Cambridge, 1997)
3.18. M.H. DeGroot, *Probability and Statistics*. Series in Statistics. Reading, UK, 2nd ed. (Addison-Wesley, 1975)
3.19. B. Efron, Bootstrap methods: Another look at the jackknife. Ann. Stat. **7**(1), 1–26 (1979)
3.20. P. Elek, L. Márkus. A long range dependent model with nonlinear innovations for simulating daily river flows. Nat. Hazards Earth Syst. Sci. **4**, 227–283 (2004)
3.21. P. Embrechts, C. Klüppelberger, T. Mikosch, *Modelling Extremal Events for Insurance and Fincance* (Springer, Berlin, 1997)
3.22. R.A. Fisher, L.H.C. Tippett, Limiting forms of the frequency distribution of the largest or smallest member of a sample. Proc. Cambridge Phil. Soc. **24**, 180–190 (1928)
3.23. B.V. Gnedenko, Sur la distribution limite du terme maximum d'une série aléatoire. Ann. Math. (2) **44**, 423–453 (1943)
3.24. C.W. Granger, R. Joyeux, An introduction to long-memory time series models and fractional differences. J. Time Ser. Anal. **1**(1), 15–29 (1980)
3.25. E.J. Hannan, B.G. Quinn, The determination of the order of an autoregression. J. R. Statist. Soc. **41**, 190–195 (1979)
3.26. H. Held, T. Kleinen, Detection of climate system bifurcations by degenerate fingerprinting. Geophys. Res. Lett. 31(L23207) (2004)
3.27. K.W. Hipel, A.I. McLeod, Preservation of the rescales adjusted range 2. Simulation studies using Box-Jenkins models. Water Resour. Res. **14**(3), 509–516 (1978)
3.28. K.W. Hipel, A.I. McLeod, *Time Series Modelling of Water Resources and Environmental Systems*. (Development in Water Science, Elsevier, Amsterdam, 1994)
3.29. J.R.M. Hosking, Fractional differencing. Biometrika **68**, 165–176 (1981)
3.30. J.R.M. Hosking, L-moments: Analysis and estimation of distributions using linear combinations of order statistics. J. R. Statist. Soc. **B** 105–124 (1990)
3.31. J.R.M. Hosking, J.R. Wallis, E.F. Wood, Estimation of the generalized extreme-value distribution by the method of probability-weighted moments. Technometrics **27**(3), 251–261 (1985)

3.32. Climate change 2001, The scientific basis, contribution of working group I to the 3 assessment report (2001). Intergovernmental Panel on Climate Change (IPCC). (Cambridgde University Press, Cambridge, UK)
3.33. M. Kallache, H.W. Rust, J. Kropp, Trend assessment: Applications for hydrology and climate. Nonlin. Proc. Geoph. **2**, 201–210 (2005)
3.34. R.W. Katz, M.B. Parlange, P. Naveau, Statistics of extremes in hydrology. Adv. Water Resour. **25**, 1287–1304 (2002)
3.35. D. Koutsoyiannis, Climate change, the Hurst phenomenon, and hydrological statistics. Hydrol. Sci. J. **48**(1), 3–24 (2003)
3.36. J.P. Kropp, A. Block, F. Reusswig, K. Zickfeld, H.J. Schellnhuber, Semiquantitative assessment of regional climate vulnerability: The North Rhine – Westphalia study. Clim Change **76**(3–4), 265–290 (2006)
3.37. Z.W. Kundzewicz, Is the frequency and intensity of flooding changing in Europe. In ed. by W. Kirch, B. Menne, R. Bertollini *Extreme Weather Events and Public Health Responses*, (Springer, Berlin, 2005)
3.38. Z.W. Kundzewicz, H.-J. Schellnhuber, Floods in the IPCC TAR perspective. *Nat. Hazards* **31**, 111–128 (2004)
3.39. M.R. Leadbetter, G. Lindgren, H. Rootzén, *Extremes and Related Properties of Random Sequences and Processes*. Springer Series in Statistics (Springer, New York, NY, 1983)
3.40. M. Lohre, P. Sibbertsen, T. Könnig, Modeling water flow of the Rhine River using seasonal long memory. Water Resour. Res. **39**(5), 1132 (2003)
3.41. D. Maraun, H.W. Rust, J. Timmer, Tempting long-memory – on the interpretation of DFA results. Nonlin. Proc. Geophys. **11**, 495–503 (2004)
3.42. B. Merz, A.H. Thieken, Separating natural and epistemic uncertainty in flood frequency analysis. J. Hydrol. **309**, 114–132 (2005)
3.43. A. Montanari, R. Rosso, M.S. Taqqu, Fractionally differenced ARIMA models applied to hydrologic time series: Identification, estimation and simulation. Water Resour. Res. **33**(5), 1035 (1997)
3.44. A. Montanari, R. Rosso, M.S. Taqqu, A seasonal fractional ARIMA model applied to the Nile river monthly flows. Water Resour. Res. **36**(5), 1249–1259 (2000)
3.45. P. Naveau, M. Nogaj, C. Ammann, P. Yiou, D. Cooley, V. Jomelli, Statistical methods for the analysis of climate extremes. C.R. Geosci. **377**, 1013–1022 (2005)
3.46. M. Ooms, J. Doornik, Inference and forecasting for fractional autoregressive integrated moving average models, with an application to US and UK inflation. Technical report, Econometric Institute, Erasmus University, Rotterdam (1999). Econometric Institute Report 9947/A.
3.47. M. Ooms, P.H. Franses, A seasonal periodic long memory model for monthly river flows. Environ. Model. Softw. **16**, 559–569 (2001)
3.48. M.B. Priestley, *Spectral Analysis and Time Series* (Academic Press, London, 1992)
3.49. C. Prudhomme, D. Jakob, C. Svensson, Uncertainty and climate change impact on the flood regime of small UK catchments. J. Hydrol. **277**, 1–23 (2003)
3.50. R Development Core Team, *R: A Language and Environment for Statistical Computing*. R Foundation for Statistical Computing (Vienna, Austria, 2004) ISBN 3-900051-07-0
3.51. H. Rust, The effect of long-range dependence on modeling extremes with the generalized extreme value distribution. Eur. Phys. J. Special Topics **174**, 91–97 (2009)
3.52. T. Schreiber, A. Schmitz, Improved surrogate data for nonlinearity tests. Phys. Rev. Lett. **77**, 635 (1996)
3.53. T. Schreiber, A. Schmitz, Surrogate time series. Physica D **142**, 346–382 (2000)
3.54. R.L. Smith, Maximum likelihood estimation in a class of nonregular cases. Biometrika **72**(1), 67–90 (1985)
3.55. R.L. Smith, Comparison of maximum likelihood and Bayesian estimators for the three-parameter Weibull distribution. Appl. Stat. **36**(3), 358–369 (1987)
3.56. R. L. Smith, Environmental statistics. Internet (2001). Version 5.0.

3.57. A. Stephenson, C. Ferro, *Evd: Functions for Extreme Value Distributions* (2004). R package version 2.1–6
3.58. P.A. Stott, D.A. Stone, M.R. Allen, Human contribution to the European heatwave of 2003. Nature **432**, 600–614 (2004)
3.59. J. Theiler, S. Eubank, A. Longtin, B. Galdrikian, J.D. Farmer, Testing for nonlinearity in time series: The method of surrogate data. Physica D **58**(77) (1992)
3.60. J. Timmer, Modeling noisy time series: Physiological tremor. *Int. J. Bifurcation Chaos* 8(7), 1998.
3.61. J. Timmer, M. König, On generating power law noise. A&A **300**(3), 707–710 (1995)
3.62. V. Venema, S. Bachner, H. W. Rust, C. Simmer, Statistical characteristics of surrogate data based on geophysical measurements. *Nonlin. Proc. Geophys.* **13**, 449–466 (2006)
3.63. H. von Storch, F.W. Zwiers, *Statistical Analysis in Climate Research* (Cambridge University Press, Cambridge, New York, NY 1999)
3.64. T.M.L. Wigley, P.D. Jones, Detecting CO_2-induced climatic change. Nature **292**, 205–208 (1981)

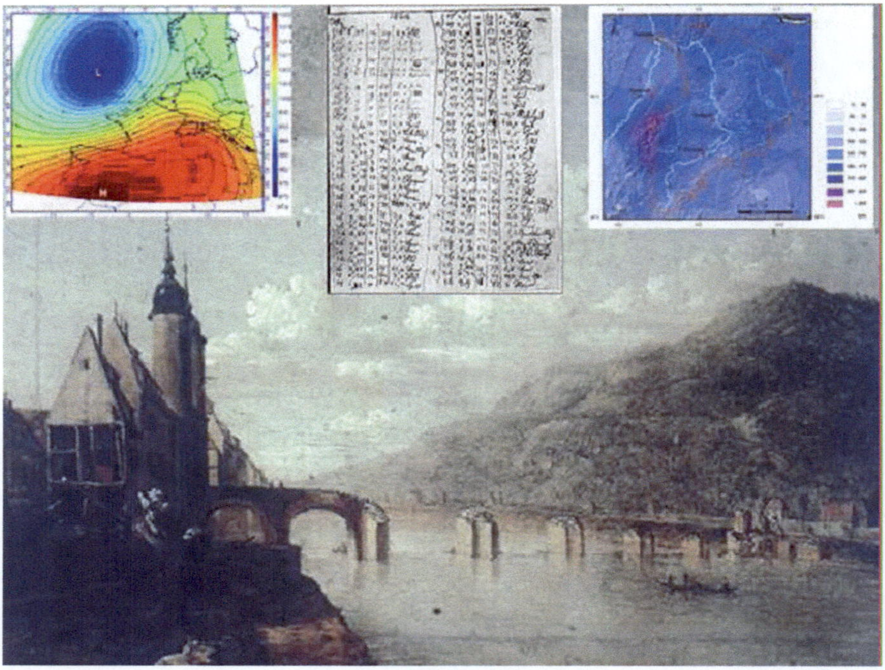

Big picture: Historical painting from the destroyed Neckar bridge in Heidelberg after the flood of October 1824. *Small picture left*: Reconstructed atmospheric circulation pattern for the 30th of October 1824. *Small picture middle*: Meteorological measurements from the scientist Bohnenberger in Tübingen for October 1824. *Small picture right*: Reconstructed area precipitation for the end of October 1824 that induced the extreme flood in the Neckar catchment

Chapter 4
Regional Determination of Historical Heavy Rain for Reconstruction of Extreme Flood Events

Paul Dostal, Florian Imbery, Katrin Bürger, and Jochen Seidel

The reconstruction of historical extreme hydrometeorological events contributes to a validation of extreme value statistics. This can mitigate several uncertainties in the flood risk analysis, e.g. in calculating possible discharges with extreme value statistics which are based on short reference data series [4.5]. The presented case study of the extreme flood of 1824 in the Neckar catchment and their triggering precipitation patterns can take place in a recent flood risk management and can be used to validate the results in trend and extreme value analysis of hydrometeorological time series.

4.1 Introduction

Climate variability, floods and their impacts in central Europe have received increasing attention in Germany in recent years [4.7]. In particular, the Elbe flood in 2002 showed that changes in the frequency and characteristics of floods may reflect climatic transitions. This led to an initiative by the German Federal Ministry of Education and Research (BMBF) for a "Risk Management of Extreme Flood Events" to avoid or limit the adverse impact of floods in Germany. The research group Xfloods from the Department of Physical Geography and the Meteorological Institute of the University of Freiburg is part of this initiative with a special focus on past extreme floods in the Federal State of Baden-Württemberg, southwest Germany. The research project integrates the information of historical data to identify and quantify extreme flood events of the past 500 years as a basis for flood risk management as well as for the calculation of record regional precipitation. The data will be extracted from historical records such as local annuals and chronicles from 1500 to 1900 and supplemented by instrumental observations available since the middle of the 18th century. The project can contribute to a safer handling of extreme floods in the future. This applies in particular to the analysis and modelling of millenarian flood events, which have been little considered previously. In such a way, the knowledge of the past can be integrated in the flood protection for tomorrow.

P. Dostal (✉)
Environmental Modelling Group, University Mainz, 55099, Mainz, Germany
e-mail: dostal@uni-mainz.de

J.P. Kropp, H.-J. Schellnhuber (eds.), *In Extremis*,
DOI 10.1007/978-3-642-14863-7_4, © Springer-Verlag Berlin Heidelberg 2011

4.2 Methodological Concept and Data Overview

Extreme flood events in the context of the climate change discussion have received increasing attention in Germany in recent years. The research project integrates the information of historical data to identify and quantify extreme flood events of the past 500 years as a basis for flood risk management as well as for the calculation of maximum regional precipitation. Different data were extracted to gain area precipitation of the Neckar catchment for the 1824 flood from historical records as well as local annuals and chronicles from 1500 to 1900 and supplemented by instrumental observations available since the middle of the 18th century. This applies in particular to the analysis and modelling of millenarian flood events, which have been little considered previously. For the reconstruction of the historical strong precipitation event of 1824 a model with various regression tools and geostatistical methods were developed. An overview of the Neckar catchment is shown in Fig. 4.1.

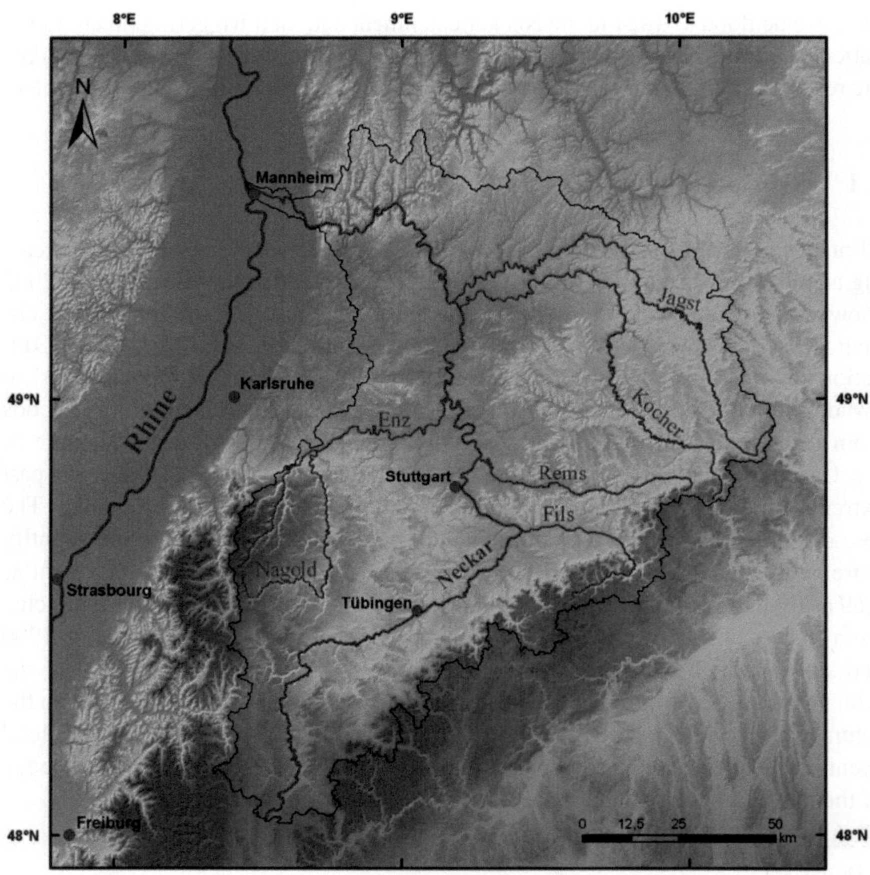

Fig. 4.1 Neckar catchment with its main tributaries Nagold, Enz, Kocher and Jagst

4.2.1 Meteorological Data from October 1824

For the calculation of the historical area precipitation (AP1824), sets of historical meteorological data, like precipitation (N), air temperature (T) sea level pressure (SLP), wind speed (u) and wind direction (dd), were used. With a European wide data set of historical meteorological data [4.2] synoptic weather charts were developed for October 1824. With the synoptic weather charts it was possible to find similar weather situations in the reference data set with the modern analogue method [4.1]. Weather charts help to understand the superior weather situation for this hydrometeorological extreme event and improves the results of the kriging calculation for the spatial rainfall [4.8, 4.9]. Table 4.1 shows the data for the event of 1824 which were used for the synoptic weather reconstruction and the area precipitation calculation. To calculate the spatial precipitation, reference data from the German Weather Service (DWD) were used. The reference data are meteorological data from 1900 to 2006 provided from the DWD. As a reference basis the whole meteorological data amount of south Germany (Rheinland-Pfalz, Hessen, Saarland, Bavaria and Baden-Württemberg) were used with a total of more than 1 000 meteorological stations. The meteorological data were available in several resolutions, e.g. 10 min mean and from the climate stations with the daily standard periods of 7, 14 and 21 h.

Table 4.1 Meteorological stations since the 18th century as data basis for the 1824 flood analysis

Location	Measured time series	Country
Freiburg	1780–2006	Germany
Karlsruhe	1799–2006	Germany
Stuttgart	1800–2006	Germany
Freudenstadt	1824–2006	Germany
Rottweil	1818–1827	Germany
Ellwangen	1818–1825	Germany
Giengen	1820–2006	Germany
Genkingen	1820–1826	Germany
Augsburg	1812–2006	Germany
Basel	1755–2006	Switzerland
Bern	1824–2006	Switzerland
Straßburg	1800–2006	France
Paris	1824–2006	France
Armagh	1795–1880	Ireland
Edinburgh	1785–2006	Scotland
Barcelona	1780–2006	Spain
Càdiz	1786–2006	Spain
Madrid	1786–2006	Spain
Milano	1763–2006	Italy
Padova	1766–2006	Italy
Palermo	1790–2006	Italy
Prague	1781–2006	Czech Rep.
Reykjavik	1816–2006	Iceland
Stockholm	1756–2006	Sweden
Uppsala	1722–2006	Sweden

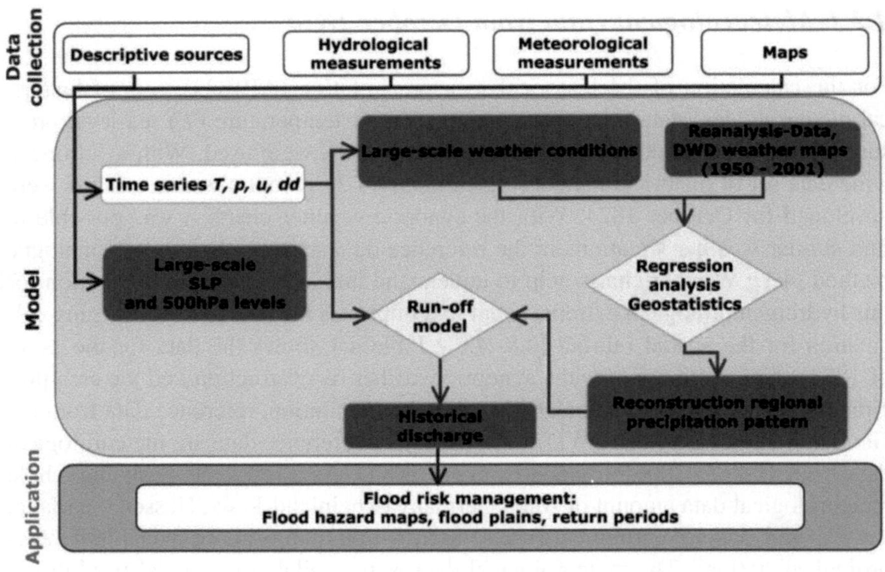

Fig. 4.2 Flow chart for the reconstruction of historical floods and their integration into a contemporary flood risk management

With a European-wide data set of historical meteorological data synoptic weather charts were developed for October 1824. With the synoptic weather charts it is possible to find similar weather situations in the reference data set. The weather charts help to understand the superior weather situation for this hydrometeorological extreme event and improve the results of the kriging calculation for the spatial rainfall [4.8, 4.9] (Fig. 4.2).

Table 4.1 shows the data for the event of 1824 which were used for the synoptic weather charts and the area precipitation calculation.

4.2.2 Methods

4.2.2.1 Synoptic Situation

On the basis of historical measurements of air pressure (see Table 4.1) the synoptic situation for middle Europe and parts of the northern Atlantic was reconstructed for October 26th until November 1st. These weather maps were compared with recent weather data from the DWD, based on the monthly weather bulletins of the DWD and the heavy precipitation statistics for southwest Germany [4.3] with the aim to find a contemporary similar weather condition, for which a sufficient database is available. The best agreement could be determined for the period 26–28 October 1998. The weather situation from October 1998 also caused strong precipitation and floods in the Neckar catchment.

4.2.2.2 Regression 1998–1824

Assuming that the regional distribution of precipitation in October 1824 and 1998 has a high similarity, the regional precipitation pattern in the Neckar catchment for the flood event of 1824 was developed as a function of the 1998 precipitation and the orographic circumstances of the investigation area [4.10]. A simple linear regression between historical precipitation measurements and precipitation data from 1998 could not give a sufficient result. Hence, a linear-logarithmic regression model in the form

$$N_{1824} = \left(1.5 + \frac{4}{\ln\left(\mathrm{DEM}\right)^4}\right) \cdot N_{1998}, \tag{4.1}$$

with N for the rainfall amount for each event and DEM as the altitude (m a.s.l.) of the measuring point, derived from a digital elevation model, was used. With this function, a good correlation could be gained between measured and modelled precipitation values ($r^2 = 0.88$). Assuming a high distribution of the historical measuring points in the Neckar catchment, the total rainfall amount of 220 points in the Neckar catchment was calculated on the basis of this equation.

4.2.2.3 Kriging

The spatial variation of short-term precipitation is highly variable in space. For the spatial interpolation of the precipitation event of October 1824 in the Neckar catchment a kriging ruler was used. Ordinary kriging calculates a weighted sum of known observed precipitation amounts to estimate an unknown value using the equation

$$\hat{z}_0 = \sum_{i=1}^{n} \omega_i z_i \,. \tag{4.2}$$

The observed values are selected from the sample data set within a user-specified neighbourhood. Covariances are derived for observations within the neighbourhood using the model variogram. The kriging process calculates a weight for each sample datum in the neighbourhood based on its statistical distance, as opposed to Euclidean distance, to each and every datum in the neighbourhood as well as to the estimate location. Then it employs a weighted linear combination of the data in the neighbourhood to estimate the new value.

The ordinary kriging system combines a set of matrices as a system of $n+1$ linear equations that must be solved so that for each neighbourhood of n values, the condition of unbiasedness holds. It is enforced by requiring the weights derived for each sample in the neighbourhood to sum to 1:

$$\sum_{i=1}^{n} \omega_i = 1. \tag{4.3}$$

As summarized by Isaaks and Srivastava [4.9], for each estimated location, weights are derived for every sample in the neighbourhood in the following manner:

$$\omega = C^{-1} \cdot D, \tag{4.4}$$

where C is the covariance between sample data and D is the covariance between sample data and estimated location.

The covariances are stored in matrices that, when multiplied together, produce a system of equations. In actuality, another equation is added to the set using a technique known as the Lagrange parameter, because the weights are constrained to sum to 1. The estimated value and its error variance become a function of $n + 1$ variables, n weights and the Lagrange multiplier – a dummy variable that forces the sum of weights to equal 1. The solution of $n + 1$ equations produces a set of weights that minimizes the modelled error variance under the constraint that the weights must sum to 1. C and D are determined for every estimation location.

The C matrix takes into account the effects of clustering and redundancy in data samples. If two or more points are close together in the inverse distance method, the number of points in the cluster adversely affects the weighting. If the covariances between these clustered samples are large, the effect of the clustering in D is lessened by the inverse of the covariances in C. Kriging, therefore, takes into account distance, clustering or redundancy, and spatial dependency of samples when estimating each new location. Ordinary kriging tries to create a mean residual or error that is equal to zero and as well, minimizes the variance of the errors. In fact, the mean error and the true variance of the errors for the estimates are always unknown, though it is possible to work with an average error and minimize the variance of the modelled error. The error variance is derived using the variances calculated during the weighted linear combination. The minimized error variance for an estimated location is related by

$$\mathrm{var}^2 = C^2 - \omega \cdot D. \tag{4.5}$$

The mean-squared error is a prediction error. It is probabilistic in nature based on the relationship of the variogram model and its parameters modelled from samples of the stationary random process [4.1]. Prediction variance, or kriging variance, is generated simultaneously for each kriged location. Both the kriged and variance surfaces can be visualized side by side. The kriging variance is likely to be lower when the sampling density is higher, although highly variable covariances in clustered areas do occur. Such a surface provides invaluable information about the fit of the model to the sample data. It also provides information on the effects of the distribution of the sampling scheme relative to the applied model (see Fig. 4.3).

4.2.2.4 Runoff simulation

LARSIM (large area runoff simulation model, [4.4]) is a river basin model which was developed for the systematic modelling of runoff generation and flood routing.

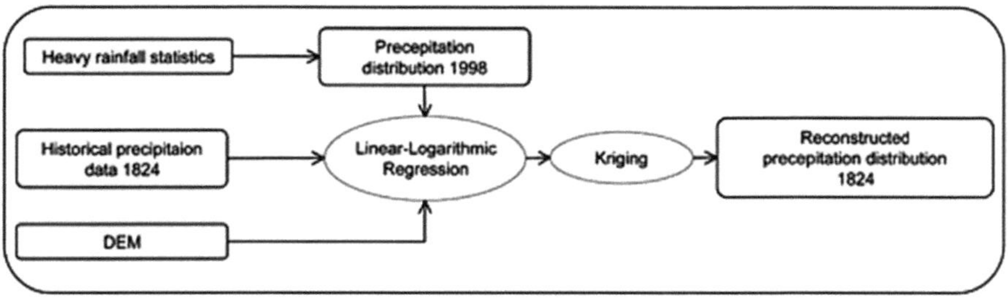

Fig. 4.3 Flow chart of spatial modelling of heavy rainfall at the Neckar catchment in October 1824

LARSIM can be applied both as a water balance model for continuous simulation and as an event-based flood forecast model. It uses proven and comparatively simple model components which are satisfying for practical purposes. For example, LARSIM is used for the flood prediction at the Neckar catchment.

For validation, the modelled spatial precipitation pattern for the flood event of October 1824 was used for the input in LARSIM for comparing modelled to historical runoff data.

4.3 Results

To analyse the flood triggering weather situation of October 1824 and therefore the AP1824, a similar weather situation had to be found in the reference data. Examinating the reference data, the best correspondence was the weather situation of 27th and 28th October 1998.

This was a weather situation which brought nearly in the total Neckar catchment a great amount of rain with devastating consequences at the Neckar and the Upper Rhine tributaries due to heavy runoffs. In comparison to the historical flood event of 1824 the flood of 1998 was not so severe because of different weather and environmental conditions (e.g. minor precipitation totals) in the run up of this extreme event.

4.3.1 The Extreme Weather Situation of October 1824

The weather situation of October 1824 can be described as follows:

At the beginning of October 1824, the weather was predominantly cloudy. Around 20th October warmer air temperatures brought fair weather. Starting on 26th October, a dramatic change in the weather was signalled when thunderstorms arose. The weather process causing the flood is described in the historical sources as follows: The evening of 26th October thunderstorms, accompanied by rain, broke out in several regions of southwest Germany, but no unusual rise of the water levels in the rivers could be noticed. On the 27th October, there was some rainfall in the

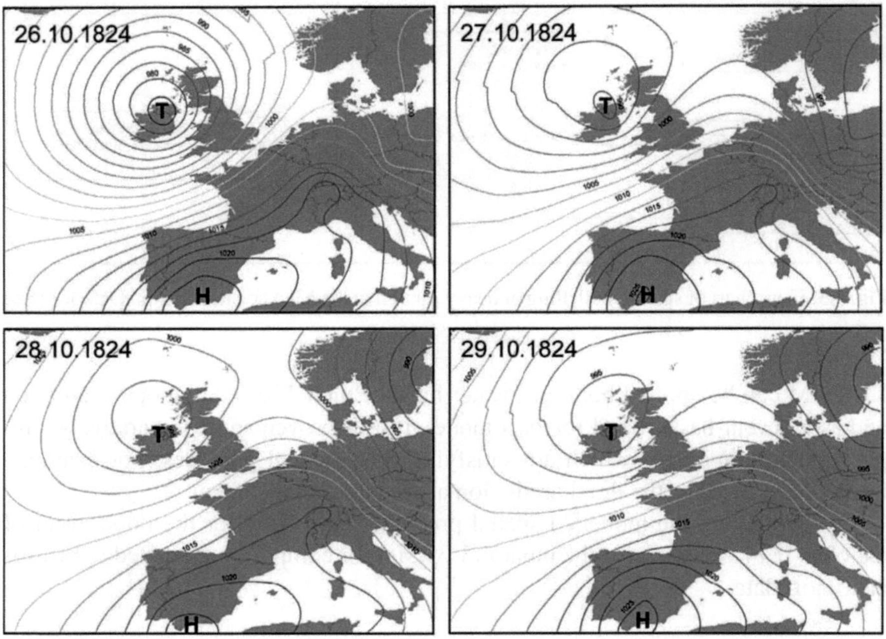

Fig. 4.4 Atmospheric circulation pattern on 29th of October 1824: CW (cyclonic westerly)

upper Neckar valley. In the evening of 28th October unusually strong rain weather began, which lasted 36 h with few interruptions up to the early morning hours of 30th October and caused this exceptional inundation. The rain quantity was very high in the Black Forest. The regions of the river Enz and the lower Neckar were especially affected by this inundation. In many places, the rainy weather with exception of smaller interruptions continued till 2nd November [4.11].

The reconstructed weather charts start at the 24th of October 1824 and shows the development of the weather situation until the end of this hydrometeorological extreme event (see Fig. 4.4).

4.3.2 Calculation of the 1824 Precipitation

In the Neckar catchment and its vicinity are 220 rainfall stations located. The precipitation of October 1824 was modelled referring to all these rainfall stations. The modelling was done with a linear regression. The $r^2 = 0.88$ shows a very good comparability to the historical measured data. In combination with the reference data of the flood event of 1998 and the data of the maximum precipitation [4.3] for the Neckar catchment, the precipitation sums are shown in Table 4.2.

Table 4.2 Precipitation sums for the year 1824 and modelled precipitation data

Location	1824 measured	1824 modelled
Freudenstadt	197	194
Wangen/Neckar	121	149
Hohenheim	128	127
Stuttgart	111	124
Genkingen	92	92
Tübingen	103	89
Giengen	92	89

4.3.3 Modelling the Area Precipitation of 1824

Aim of the presented project was to calculate the area precipitation for a 1 km^2 grid, to fit the modelled data in the water balance model LARSIM which is also based on a 1 km^2 grid. As seen in Fig. 4.5, the highest rainfall amounts with values up to 230 mm in 36 h occurred in the northern parts of the Black Forest. Secondary maxima were located in the northern and eastern parts of the Neckar catchment. This spatial rainfall distribution can be explained by the atmospheric circulation pattern and local orographic features. The location of the high- and low-pressure cells in the large-scale synoptic reconstruction implicates southwesterly warm and humid air mass flow to central Europe as a causal mechanism. Southwesterly large-scale

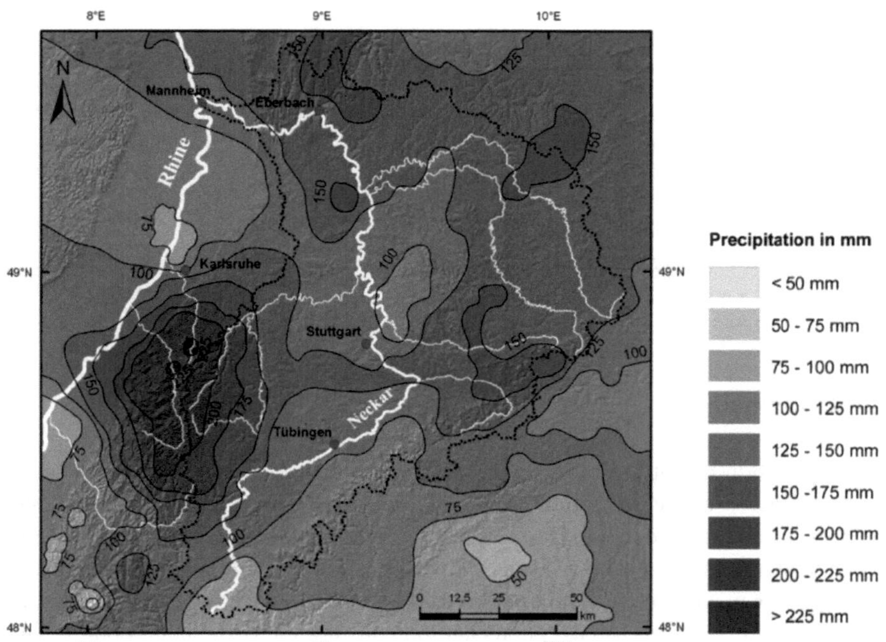

Fig. 4.5 Modelled precipitation pattern for the flood event at the end of October 1824 (with *isohyetal lines* showing the rainfall amount in millimetres)

air flows are typically modified by the mountain ranges of the Vosges mountains and the Black Forest. Particularly the northern parts of the Black Forest frequently receive high rainfall amounts during such weather conditions due to the orographic features on the western site of the Upper Rhine Valley (see Fig. 4.5).

4.3.4 Modelled Neckar Discharges for the Flood Event of October 1824

Table 4.3 and Fig. 4.6 show the modelled discharges of the river Neckar and its tributaries derived from the calculated precipitation data that were integrated in the waterbalance model LARSIM. The discharges show values which are significantly higher than $Q_{extreme}$ discharges for the Neckar [4.12, 4.13]. In all probability the flood event of October 1824 was one of the most extreme flood events during the last 300–500 years.

Table 4.3 Discharges for several locations at the Neckar in October 1824 modelled with LARSIM

Location	River	Q_{1824} (m^3/s)	Q_{100} (m^3/s)	$Q_{extreme}$ (m^3/s)
Rottweil	Neckar	137	260	390
Oberndorf	Neckar	223	522	522
Horb	Neckar	416	800	800
Plochingen	Fils	1400	1145	1600
Pforzheim	Enz	580		670
Besigheim	Enz	612	586	821
Lauffen	Neckar	2398	1877	2550
Gaildorf	Kocher	411	350	490
Stein	Kocher	1002	709	993
Elpershofen	Jagst	461	387	582
Untergriesheim	Jagst	984	525	771
Rockenau	Neckar	4185	2665	3600
Heidelberg	Neckar	4320	2806	3700
Mannheim	Neckar	4264	2833	3750

4.4 Conclusion

The application of data from historical sources, like in the actual flood risk management, increases the protection of potential inundation areas. This chapter shows the possibilities and the methods to reconstruct hydrometeorological extreme events which lay far in the past. These results show the potential of historical flood analysis for flood risk management. Better understanding of extreme flood events and better protection of endangered areas, goods and humans are the tangible outcomes. The importance of analysing historical floods is the possibility to demonstrate the consequences of such extreme events in a selected river catchment, such as the Neckar. This knowledge can be incorporated into flood risk management. By combining

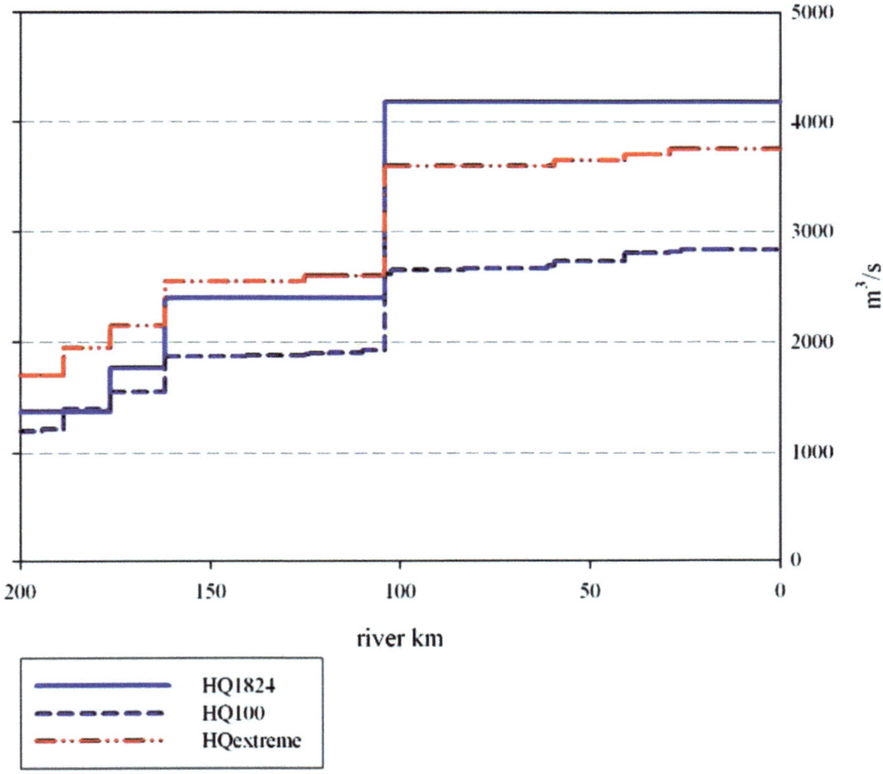

Fig. 4.6 Modelled discharge of 1824 flood compared with reference discharges, e.g. Q_{extreme}, of the Neckar

the historical precipitation and flood data with contemporary river channel morphology, current hazards and impacts can be predicted. This will lead to a better understanding of flood processes, as well as their characteristics. Improvements in the analysis of historical extreme events should be extended to other river catchments in order to mitigate the consequences of catastrophes such as the Elbe flood in 2002.

Acknowledgments The authors thank the RIMAX project, funded by the German Federal Ministry of Education and Research for financial support of their work (under grant no. 0330685).

References

4.1. A. Bárdossy, F. Filiz, *Identification of Flood Producing Atmospheric Circulation Patterns* in ed. by V.R. Thorndycraft, M. Benito, M. Barriendos, M.C. Llasat, Paleofloods, Historical Data & Climatic Variability (Proc. of the PHEFRA Int. Workshop, 16. – 19. October 2202, Barcelona, Spain), 307 – 312 (2003)

4.2. M. Barriendos, D. Coeur, M. Lang, M.C. Llasat, R. Naulet, F. Lemaitre and A. Barrera, Stationarity Analysis of Historical Flood Series in France and Spain (14th–20th centuries). *Natural Hazards and Earth System Sciences* **3**, 583–592 (2003).
4.3. H. Bartels, *Starkniederschlagshöhen für Deutschland: KOSTRA* (Selbstverlag des Deutschen Wetterdienstes, Offenbach, 1997) p. 23
4.4. M. Bremicker, A. Luce, I. Haag and A. Sieber, *Das Wasserhaushaltsmodell LARSIM - Modellgrundlagen* (Hochwasser Vorhersagezentrale Baden- Württemberg, 2005); [available from `www.hvz.baden-wuerttemberg.de/pdf/LARSIM_DE_2005-06-24`] (Accessed in the year 2006)
4.5. K. Bürger, J. Seidel, F. Imbery, P. Dostal, *RIMAX-Projekt Xfloods "Analyse historischer Hochwassereignisse für ein integratives Konzept zum vorbeugenden Hochwasserschutz"* (in: Umweltwissenschaften und Schadstoff-Forschung (UWSF), **18** (1), 27–29 (2006).
4.6. K. Bürger, J. Seidel, P. Dostal, F. Imbery, *Historic Flood as a Basis for Flood Risk Management in River Valleys* (in: GR international, **2**(2), (2006)).
4.7. R. Glaser, H. Stangl, *Climate and Floods in Central Europe since AD 1000: Data, Methods, Results and Consequences* (in: Surveys in Geophysics **25** (5–6), 495 – 510 (2004))
4.8. A. Hinterding: *InterNied - A Geostatistical Interpolation Procedure Four Hourly Measured Precipitation Data* (in: P. Krahe and D. Herpertz (eds.), International Commission for the Hydrology of the Rhine Basin (CHR)- Report I-20, 41–47 (2001))
4.9. E.H. Isaaks, R.M. Srivastava: *An introduction to applied geostatistics* (Oxford University Press, New York, 1989) p. 561
4.10. J. Jacobeit, R. Glaser, J. Luterbacher, M. Nonnebacher, H. Wanner: *Links Between Flood Events in central Europe since AD 1500 and the Large-scale Atmospheric Circulation* (in: V.R. Thorndycraft, M. Benito, M. Barriendos and M.C. Llasat (eds.), Paleofloods, Historical Data & Climatic Variability: Applications in Flood Risk Assessment, 269–274, 2003)
4.11. G. Schuebler *Über die ungewöhnliche Überschwemmung zu Ende Octobers des vorherigen Jahres und die dabei in verschiedenen Gegenden Württembergs Gefallene Regenmenge.* (Annalen der Physik und Chemie. Jg. 1825 Dritter Band, zweites Stück, 1825)
4.12. Landesanstalt für Umweltschutz (LfU): *Das Hochwasser vom Oktober/November 1998 in Baden-Württemberg* (in: Oberirdische Gewässer/Gewässerökologie, Karlsruhe, 65, 2000)
4.13. Landesanstalt für Umweltschutz (LfU): *Abflusskennwerte in Baden-Württemberg* (in: Oberirdische Gewässer/Gewässerökologie, Karlsruhe, CD-ROM, 94, 2005)

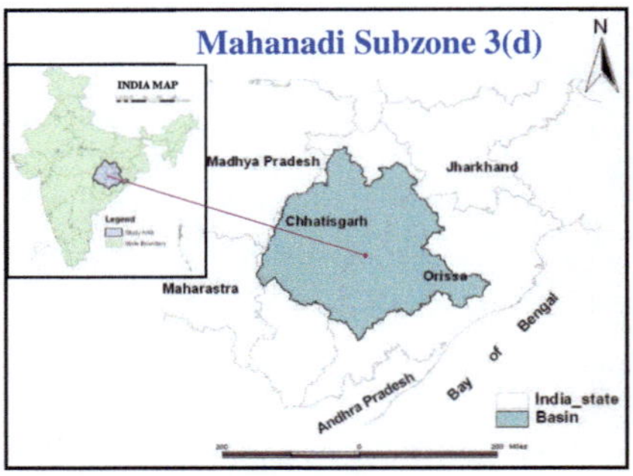

$$Q_T = C_T * A^{0.863}$$

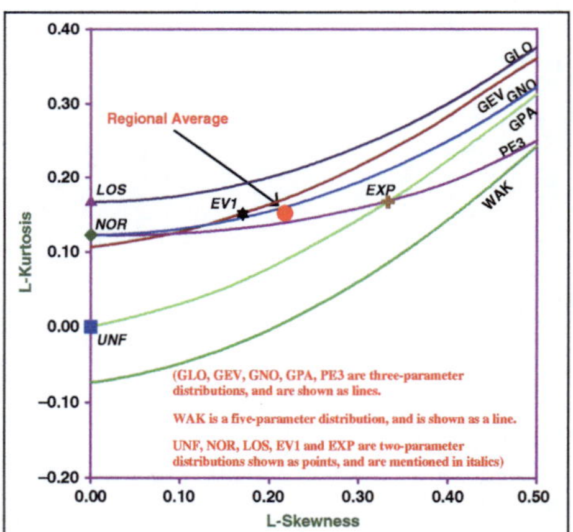

L-moment Ratio Diagram for Mahanadi Subzone 3(d)

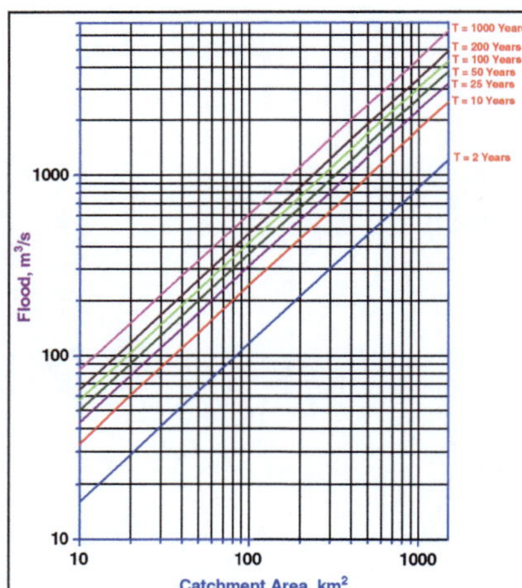

Variation of floods of various return periods with catchment area for Mahanadi Subzone 3(d)

Ungauged catchments need strong analytical concepts for a provision of suitable management advice

Chapter 5
Development of Regional Flood Frequency Relationships for Gauged and Ungauged Catchments Using L-Moments

Rakesh Kumar and Chandranath Chatterjee

For planning, development, and sustainable management of water resources, applications of new and advanced methodologies are essential for design flood estimation. The L-moments are a recent development within statistics and offer significant advantages over ordinary product moments. Regional flood frequency relationships are developed based on the L-moment approach. The annual maximum peak floods data are screened using the discordancy measure (D_i), and homogeneity of the region is tested employing the L-moment-based heterogeneity measure (H). For computing heterogeneity measure H, 500 simulations are performed using the κ-distribution. Twelve frequency distributions namely extreme value (EV1), generalized extreme value (GEV), logistic (LOS), generalized logistic (GLO), normal (NOR), generalized normal (GNO), uniform (UNF), Pearson type-III (PE3), exponential (EXP), generalized Pareto (GPA), κ- (KAP), and five-parameter Wakeby (WAK) are employed. Based on the L-moment ratio diagram and $|Z_i^{\text{dist}}|$-statistic criteria, GNO is identified as the robust frequency distribution for the study area. For estimation of floods of various return periods for gauged catchments of the study area, the regional flood frequency relationship is developed using the L-moment-based GNO distribution. Also, for estimation of floods of various return periods for ungauged catchments, the regional flood frequency relationships developed for gauged catchments is coupled with the regional relationship between mean annual maximum peak flood and catchment area.

5.1 Introduction

Information on flood magnitudes and their frequencies is needed for design of various types of water resources projects/hydraulic structures such as dams, spillways, road and railway bridges, culverts, urban drainage systems as well as for taking up various non-structural measures such as flood plain zoning, economic evaluation

R. Kumar (✉)
Surface Water Hydrology Division, National Institute of Hydrology, Roorkee-247667, Uttaranchal, India
e-mail: rakesh@nih.ernet.in

J.P. Kropp, H.-J. Schellnhuber (eds.), *In Extremis*,
DOI 10.1007/978-3-642-14863-7_5,

of flood protection projects. Since scientific hydrology began in the seventeenth century, one of the most difficult problems facing engineers and hydrologists is how to predict flow in basins with no records. Whenever rainfall or river flow records are not available at or near the site of interest, it is difficult for hydrologists or engineers to derive reliable design flood estimates directly. In such a situation, regional flood frequency relationships developed for the region are one of the alternative methods for prediction of design floods, especially for small-to medium-size catchments.

The approaches for design flood estimation may be broadly categorized as follows: (i) deterministic approach using design storm and (ii) probabilistic approach involving flood frequency analysis. The deterministic and probabilistic methods, which have been used for design flood estimation, are empirical methods, rational method, flood frequency analysis methods, unit hydrograph techniques, and watershed models. Pilgrim and Cordery (1992) [5.16] mention that estimation of peak flows on small- to medium-sized rural drainage basins is probably the most common application of flood estimation as well as being of greatest overall economic importance. In almost all cases, no observed data are available at the design site, and little time can be spent on the estimate, precluding use of other data in the region. The authors further state that hundreds of different methods have been used for estimating floods on small drainage basins, most involving arbitrary formulas. The three most widely used types of methods are the rational method, the US Soil Conservation Service method, and regional flood frequency methods. Regional flood frequency analysis resolves the problem of short data records or unavailability of data by "trading space for time"; as the data from several sites are used in estimating flood frequencies at any site. The choice of method primarily depends on design criteria applicable to the structure and availability of data.

Considering the importance of prediction in ungauged catchments, the International Association of Hydrological Sciences (IAHS) launched "Prediction of Ungauged Basins (PUBs)" as one of its initiatives and declared the current decade as "Decade of PUBs." As per the Bureau of Indian Standards (BIS) hydrologic design criteria, frequency-based floods find their applications in estimation of design floods for almost all the types of hydraulic structures, viz., small-size dams, barrages, weirs, road and railway bridges, cross-drainage structures, flood control structures, excluding large- and intermediate-size dams. For design of large- and intermediate-size dams probable maximum flood and standard project flood are adopted, respectively. Most of the small-size catchments are ungauged or sparsely gauged. To overcome the problems of prediction of floods of various return periods for ungauged and sparsely gauged catchments, a robust procedure of regional flood frequency estimation is required to be developed.

In this study, regional flood frequency relationships are developed based on the L-moment approach for the gauged and ungauged catchments of Mahanadi subzone 3(d) in India. For this purpose, various frequency distributions were employed. Regional flood frequency relationship is developed for gauged catchments based on the most robust identified frequency distribution. This relationship is coupled with the regional relationship between mean annual peak flood and catchment area, and a simple regional flood frequency relationship is also developed for ungauged catchments of the study area. The study presents regional flood frequency relationships for gauged catchments and in particular very simple and robust regional flood frequency relationships for ungauged catchments of the study area.

5.2 Regional Flood Frequency Analysis

Cunnane [5.5] mentions 12 different regional flood frequency analysis (RFFA) methods. Out of these methods, some of the commonly used methods are the following: (i) Dalrymple's index flood method, (ii) Natural Environmental Research Council (NERC) method, (iii) United States Water Resources Council (USWRC) method, (iv) Bayesian method, and (v) regional regression-based methods.

Based on data availability and record length of the data the following three types of approaches may be adopted for developing the flood frequency relationships: (a) at-site flood frequency analysis, (b) at-site and regional flood frequency analysis, and (c) regional flood frequency analysis. The steps involved in carrying out flood frequency analysis based on the above approaches are mentioned below.

5.2.1 At-Site Flood Frequency Analysis

(i) Fit various frequency distributions to the annual maximum peak flood data of a stream flow gauging site.
(ii) Select the best fit distribution based on the goodness-of-fit criteria.
(iii) Use the best fit distribution for estimation of T-year flood.

5.2.2 At-Site and Regional Flood Frequency Analysis

(i) Identify a hydrometeorologically homogeneous region.
(ii) Screen the observed annual maximum peak flood data of the stream flow gauging sites of the homogeneous region and test the regional homogeneity.
(iii) Develop flood frequency relationships for the region considering various frequency distributions.
(iv) Select the best fit distribution based on the goodness-of-fit criteria.
(v) Estimate the at-site mean annual peak flood.
(vi) Use the best fit regional flood frequency relationship for estimation of T-year flood.

5.2.3 Regional Flood Frequency Analysis

Steps (i)–(iv) mentioned under Sect. 5.2.2 remain unchanged here. Subsequently, the following steps are followed:

(i) Develop a regional relationship between mean annual peak flood and catchment and physiographic characteristics for the region.
(ii) Estimate the mean annual peak flood using the developed relationship.
(iii) Use the best fit regional flood frequency relationship for estimation of T-year flood.

Among some of the recent regional flood frequency analysis studies, Iacobellis and Fiorentino [5.11] presented a new rationale, which incorporates the climatic control for deriving the probability distribution of floods which is based on the assumption that the peak direct streamflow is a product of two random variates, namely the average runoff per unit area and the peak contributing area. The probability density function of peak direct streamflow can thus be found as the integral over total basin area, of that peak contributing area times the density function of average runoff per unit area. The model was applied to the annual flood series of eight gauged basins in Basilicata (southern Italy) with catchment area ranging from 40 to 1 600 km^2. The results showed that the parameter tended to assume values in good agreement with geomorphologic knowledge and suggest a new key to understand the climatic control of the probability distribution of floods.

Jingyl and Hall [5.12] applied the geographical approach (residual method), Ward's cluster method, the Fuzzy c-means method, and a Kohonen neural network [5.13] to 86 sites in the Gan-Ming river basin of China to delineate homogeneous regions based on site characteristics. The authors state that since the Kohonen neural network can be employed to identify the number of sub-regions as well as the allocation of the sites to sub-regions, this method is preferred over Ward's method and the Fuzzy c-means approach. The regional L-moment algorithm has been used both to take advantage of identifying an appropriate underlying frequency distribution and to construct sub-regional growth curves.

Chokmani and Quarda [5.3] proposed a physiographical space-based kriging method for regional flood frequency estimation. The methodology relies on the construction of a continuous physiographical space using physiographical and meteorological characteristics of gauging stations and the use of multivariate analysis techniques. Two multivariate analysis methods were tested: canonical correlation analysis (CCA) and principal component analysis (PCA). Ordinary kriging, a geostatistical technique, was then used to interpolate flow quantiles through the physiographical space. Data from 151 gauging stations across the southern part of the province of Quebec, Canada, were used to illustrate this approach. Results of the proposed method were compared to those produced by a traditional regional estimation method using the canonical correlation analysis. The proposed method estimated the 10-year return period specific flow with a coefficient of determination of 0.78. However, this performance decreases with the increase in quantile return period. It is also observed that the proposed method works better when the physiographical space is defined using canonical correlation analysis.

Merz and Blöschl [5.15] examined the predictive performance of various regionalization methods for the ungauged catchment case, based on a jack-knifing comparison of locally estimated and regionalized flood quantiles of 575 Austrian catchments. It is observed that spatial proximity is a significantly better predictor of regional flood frequencies than are catchment attributes. A method that combines spatial proximity and catchment attributes yields the best predictive performance. The authors have presented this novel method which is based on kriging and takes differences in the length of the flood records into account. It is shown that short flood records contain valuable information which can be exploited by the method proposed by the authors. A method that used only spatial proximity performs second best. The methods that only use catchment attributes perform significantly poorer than those based on spatial proximity. These are a variant of the region of influence (ROI) approach, applied in an automatic model and multiple regressions. The authors suggest that better predictive variables and similarity measures need to be found to make these methods more useful.

Kumar and Chatterjee [5.14] carried out regional flood frequency analysis for North Brahmaputra region of India. In the study, data of 13 streamflow gauging sites were screened using the discordancy measure (D_i) and homogeneity of the region is then tested employing the L-moment-based heterogeneity measure (H). Based on this test, it was observed that the data of 10 out of 13 gauging sites constituted a homogeneous region. Comparative regional flood frequency analysis studies were conducted employing the L-moment-based commonly used frequency distributions. Based on the L-moment ratio diagram and $|Z_i^{\mathrm{dist}}|$-statistic criteria, generalized extreme value (GEV) distribution was identified as the robust distribution for the study area. Regional flood frequency relationships were developed for estimation of floods of various return periods for gauged and ungauged catchments using the L-moment-based GEV distribution and a regional relationship between mean annual peak flood and catchment area. Flood frequency estimates of gauged and ungauged catchments were compared; when, without satisfying the criteria of regional homogeneity, data of all the 13 gauging sites were used instead of data of only 10 gauging sites constituting the homogeneous region.

Cunderlik and Burn [5.4] mention that because of sampling variability, catchment similarity in flood seasonality can significantly deviate from the true similarity. Therefore, sampling variability should be directly incorporated in the pooling algorithm to decrease the level of pooling uncertainty. The authors developed a new pooling approach that takes into consideration the sampling variability of flood seasonality measures used as pooling variables. A nonparametric resampling technique is used to estimate the sampling variability for the target site, as well as for every site that is a potential member of the pooling group for the target site. The variability is quantified by Mahalanobis distance ellipses. The similarity between the target site and the potential site is then assessed by finding the minimum confidence interval at which their Mahalanobis ellipses intersect. The confidence intervals can be related to regional homogeneity, which allows the target degree of regional homogeneity to be set in advance. The approach is applied to a large set of catchments from Great Britain, and its performance is compared with the performance of a previously used pooling technique based on Euclidean distance. The results demonstrated that the proposed approach outperforms the previously used approach in terms of the overall homogeneity of delineated pooling groups in the study area.

Zhang and Singh [5.21] mention that using the Gumbel–Hougaard copula, trivariate distributions of flood peak, volume, and duration were derived, and then conditional return periods were obtained. The derived distributions were tested using flood data from the Amite River Basin in Louisiana. The authors state that a major advantage of the copula method is that marginal distributions of individual variables can be of any form and the variables can be correlated. Griffis and Stedinger [5.7] presented evolution of flood frequency analysis with *Bulletin 17*. The current methodology recommended for flood frequency analyses by US Federal Agencies is presented in *Bulletin 17B*. *Bulletin 17* was first published in 1976, minor corrections were made in 1977 resulting in *Bulletin 17A*, which was later succeeded by *Bulletin 17B* published in 1982. The authors mention that the fields of hydrology and flood frequency analysis have substantially evolved since *Bulletin 17* was first published. New techniques are now available which should become part of these standard procedures. A comparison is provided which demonstrates how the standard and weighted *Bulletin 17B* quantile estimators perform relative to alternative LP3 quantile estimators that also make use of regional information. Chebana and Quarda [5.2] presented a multivariate L-moment homogeneity test with the aim to extend the statistical homogeneity test of Hosking and Wallis [5.9] to the multivariate case. The usefulness of the methodology is illustrated on flood events. Monte Carlo simulations are also performed for a bivariate Gumbel logistic model with Gumbel marginal distributions. Results illustrate the power of the proposed multivariate L-moment homogeneity test to detect heterogeneity on the whole structure of the model and on the marginal distributions. In a bivariate flood setting, a comparison is carried out with the classical homogeneity test of Hosking and Wallis based on several types of regions. In this study, regional flood frequency relationships have been developed based on the L-moment approach for estimation of floods of various return periods for gauged and ungauged catchments of the Mahanadi subzone 3(d) of India.

5.3 L-Moment Approach

Some of the commonly used parameter estimation methods for most of the frequency distributions include the following: (i) method of least squares, (ii) method of moments, (iii) method of maximum likelihoods, (iv) method of probability weighted moments, (v) method based on principle of maximum entropy, and (vi) method based on L-moments. L-moments are a recent development within statistics [5.8]. In a wide range of hydrologic applications, L-moments provide simple and reasonably efficient estimators of characteristics of hydrologic data and of a distribution's parameters [5.18]. Like the ordinary product moments, L-moments summarize the characteristics or shapes of theoretical probability distributions and observed samples. Both moment types offer measures of distributional location (mean), scale (variance), skewness (shape), and kurtosis (peakedness).

Recently a number of regional flood frequency analysis studies have been carried out based on the L-moment approach. The L-moment methods are demonstrably superior to those that have been used previously and are now being adopted by many organizations worldwide [5.10]. The L-moments offer significant advantages over ordinary product moments, especially for environmental data sets, because of the following [5.20].

1. L-moment ratio estimators of location, scale, and shape are nearly unbiased, regardless of the probability distribution from which the observations arise [5.8].
2. L-moment ratio estimators such as L-coefficient of variation, L-skewness, and L-kurtosis can exhibit lower bias than conventional product moment ratios, especially for highly skewed samples.
3. The L-moment ratio estimators of L-coefficient of variation and L-skewness do not have bounds which depend on sample size as do the ordinary product moment ratio estimators of coefficient of variation and skewness.
4. L-moment estimators are linear combinations of the observations and thus are less sensitive to the largest observations in a sample than product moment estimators, which square or cube the observations.
5. L-moment ratio diagrams are particularly good at identifying the distributional properties of highly skewed data, whereas ordinary product moment diagrams are almost useless for this task [5.19].

5.3.1 Probability Weighted Moments and L-Moments

The L-moments are an alternative system of describing the shapes of probability distributions [5.10]. They arose as modifications of probability weighted moments (PWMs) of Greenwood et al. [5.6]. Probability weighted moments is defined as

$$M_{p,r,s} = E\left(x^{p}\{F\}^{r}\{1-F\}^{s}\right) = \int_{0}^{1} \{x(F)\}^{p} F^{r}\{1-F\}^{s}\,\mathrm{d}F, \qquad (5.1)$$

where $F = F(x)$ is the cumulative distribution function (CDF) for x, $x(F)$ is the inverse CDF of x evaluated at the probability F, and p, r, and s are real numbers. If p is a nonnegative integer, $M_{p,0,0}$ represents the conventional moment of order p about the origin. If $p = 1$ and $s = 0$,

$$M_{1,r,0} = \beta_r = \int_{0}^{1} x(F)\,F^{r}\,\mathrm{d}F. \qquad (5.2)$$

For an ordered sample $x_1 \le x_2, \ldots, \le x_N$, $N > r$, the unbiased sample PWMs are given by

$$\hat{\beta}_r = \frac{1}{N}\frac{\sum_{i=1}^{N}\binom{i-1}{r}x_i}{\binom{N-1}{r}}. \qquad (5.3)$$

For any distribution the rth L-moment λ_r is related to the rth PWM [5.8], through

$$\lambda_{r+1} = \sum_{k=0}^{r} \beta_k (-1)^{r-k} \binom{r}{k} \binom{r+k}{k}. \tag{5.4}$$

These L-moments are linear functions of PWMs. For example, the first four L-moments are related to the PWMs using

$$\begin{aligned} \lambda_1 &= \beta_0, \\ \lambda_2 &= 2\beta_1 - \beta_0, \\ \lambda_3 &= 6\beta_2 - 6\beta_1 + \beta_0, \\ \lambda_4 &= 20\beta_3 - 30\beta_2 + 12\beta_1 - \beta_0. \end{aligned} \tag{5.5}$$

The L-moments are analogous to their conventional counterparts as they can be directly interpreted as measures of scale and shape of probability distributions and hence are more convenient than the PWMs. Hosking (1990) [5.8] defined L-moment ratios which are analogous to conventional moment ratios as

$$\begin{aligned} &L\text{-coefficient of variation: } L\text{-CV}(\tau_2) = \lambda_2/\lambda_1, \\ &L\text{-coefficient of skewness: } L\text{-skew}(\tau_3) = \lambda_3/\lambda_2, \\ &L\text{-coefficient of kurtosis: } L\text{-kurtosis}(\tau_4) = \lambda_4/\lambda_2. \end{aligned} \tag{5.6}$$

Analogous to the conventional moment ratios, λ_1 is a measure of location, τ_2 is a measure of scale and dispersion, τ_3 is a measure of skewness, and τ_4 is a measure of kurtosis. Hosking [5.8] showed that for $x \geq 0$, the value of τ_2 lies between 0 and 1, while the absolute values of τ_3 and τ_4 lie between 0 and 1. This restriction in the values of the L-coefficients works out to be an advantage in their interpretation as opposed to the conventional moments which do not have any bounds [5.17].

5.3.2 Screening of Data Using Discordancy Measure Test

The objective of screening of data is to check that the data are appropriate for performing the regional flood frequency analysis. In this study, screening of the data was performed using the L-moment-based discordancy measure (D_i). Discordancy is measured in terms of the L-moments of the sites' data, and the aim is to identify those sites that are grossly discordant with the group as a whole. The sample L-moment ratios (t_2, t_3, and t_4) of a site are considered as a point in a three-dimensional space. A group of sites form a cluster of such points in the three-dimensional space. A site is considered discordant if it is far from the center of the cluster. Hosking and Wallis [5.10] defined the discordancy measure (D_i) for a site i in a group of N sites. Let $u_i = [t_2^{(i)} t_3^{(i)} t_4^{(i)}]^T$ be a vector containing the sample L-moment ratios t_2, t_3, and t_4 values for site i, analogous to their regional values termed as τ_2, τ_3, and τ_4, expressed in Eq. (5.6). T denotes transposition of a vector or matrix. Let

$$\overline{u} = N^{-1} \sum_{i=1}^{N} u_i \tag{5.7}$$

be the (unweighted) group average. The sample covariance matrix is defined as

$$A_m = \sum_{i=1}^{N} (u_i - \overline{u})(u_i - \overline{u})^T. \tag{5.8}$$

The discordancy measure for site i is defined as

$$D_i = \frac{1}{3} N (u_i - \overline{u})^T A_m^{-1} (u_i - \overline{u}). \tag{5.9}$$

The site i is declared to be discordant, if D_i is greater than the critical value of the discordancy statistic D_i, given in a tabular form by Hosking and Wallis [5.10].

5.3.3 Test of Regional Homogeneity

For testing regional homogeneity, a test statistic H, termed as heterogeneity measure, was proposed by Hosking and Wallis [5.9]. It compares the "inter-site variations in sample L-moments for the group of sites" with "what would be expected of a homogeneous region." The inter-site variations in sample L-moments are evaluated based on any of the three measures of variability V_1 (based on L-CV), V_2 (based on L-CV and L-skew), and V_3 (based on L-skew and L-kurtosis). These measures of variability are computed as follows:

1. V_1 is the weighted standard deviation of at-site L-CV's ($t_2^{(i)}$):

$$V_1 = \left[\frac{\sum_{i=1}^{N} n_i \, (t_2^{(i)} - t_2^R)^2}{\sum_{i=1}^{N} n_i} \right]^{1/2}, \tag{5.10}$$

   where n_i is the record length at each site and t_2^R is the regional average L-CV weighted proportionally to the sites' record length as given below:

$$t_2^R = \frac{\sum_{i=1}^{N} n_i \, t_2^{(i)}}{\sum_{i=1}^{N} n_i}. \tag{5.11}$$

2. V_2 is the weighted average distance from the site to the group weighted mean on a graph of t_2 versus t_3:

$$V_2 = \frac{\sum_{i=1}^{N} n_i \left\{ (t_2^{(i)} - t_2^R)^2 + (t_3^{(i)} - t_3^R)^2 \right\}^{1/2}}{\sum_{i=1}^{N} n_i}, \tag{5.12}$$

where t_3^R is the regional average L-skew weighted proportionally to the sites' record length.

3. V_3 is the weighted average distance from the site to the group weighted mean on a graph of t_3 versus t_4:

$$V_3 = \frac{\sum_{i=1}^{N} n_i \left\{ (t_3^{(i)} - t_3^R)^2 + (t_4^{(i)} - t_4^R)^2 \right\}^{1/2}}{\sum_{i=1}^{N} n_i}, \tag{5.13}$$

where t_4^R is the regional average L-kurtosis weighted proportionally to the sites' record length. To establish *what would be expected of a homogeneous region*, first simulations are used to generate homogeneous regions with sites having same record lengths as those of the observed data. In order to generate the simulated data, a four-parameter κ-distribution is used. The four-parameter κ-distribution is chosen so as not to commit to a particular two or three-parameter distribution. Further, the four-parameter κ-distribution includes as special cases the generalized logistic, generalized extreme value, and generalized Pareto distributions and hence acts as a good representation of many of the probability distributions occurring in environmental sciences. The parameters of the κ-distribution are obtained using the regional average L-moment ratios t_2^R, t_3^R, t_4^R, and mean = 1. A large number of data regions are generated (say $N_{\text{sim}} = 500$) based on this κ-distribution. The simulated regions are homogeneous and have no cross-correlation or serial correlation. Further, the sites have the same record lengths as the observed data. For each generated region, V_j (i.e., any of V_1, V_2, or V_3) is computed using Eqs. (5.10), (5.11), (5.12), and (5.13). Subsequently, their mean (μ_v) and standard deviation (σ_v) are computed.

The heterogeneity measure $H(j)$ (i.e., $H(1)$, $H(2)$, or $H(3)$) is computed as

$$H(j) = \frac{V_j - \mu_v}{\sigma_v}. \tag{5.14}$$

If the heterogeneity measure is sufficiently large, the region is declared to be heterogeneous. Hosking and Wallis [5.10] suggested the following criteria for assessing heterogeneity of a region: if $H(j) < 1$, the region is acceptably homogeneous; if $1 \le H(j) < 2$, the region is possibly heterogeneous; and if $H(j) \ge 2$, the region is definitely heterogeneous. These boundary values of $H(j)$ being 1 and 2 were determined by Hosking and Wallis [5.10] by performing a series of Monte Carlo experiments in which the accuracy of quantile estimates corresponding to different values of $H(j)$ was computed. The authors further observed that for both real-world data and artificially simulated regions, $H(1)$ has much better power to discriminate between homogeneous and heterogeneous regions as compared to $H(2)$ and $H(3)$.

5.3.4 Identification of Robust Regional Frequency Distribution

The best fit frequency distribution for a homogeneous region is determined by how well the L-skewness and L-kurtosis of the fitted distribution match the regional average L-skewness and L-kurtosis of the observed data [5.10]. The procedure adopted is briefly stated below. Initially, several three-parameter distributions are fitted to the regional average L-moments t_2^R, t_3^R, and mean =1. Let τ_4^{Dist} be the L-kurtosis of the fitted distribution which may be GEV, GLO, GNO, PE3, etc. Using the N_{sim} number of simulated regions of the κ-distribution (as obtained for the heterogeneity measure described in Sect. 5.3.3), the regional average L-kurtosis, t_4^m, is computed for the mth simulated region. The bias of t_4^R is computed as

$$B_4 = N_{\text{sim}}^{-1} \sum_{m=1}^{N_{\text{sim}}} \left(t_4^m - t_4^R \right). \tag{5.15}$$

The standard deviation of t_4^R is computed as

$$\sigma_4 = \left[(N_{\text{sim}} - 1)^{-1} \left\{ \sum_{m=1}^{N_{\text{sim}}} \left(t_4^m - t_4^R \right)^2 - N_{\text{sim}} B_4^2 \right\} \right]^{1/2}. \tag{5.16}$$

The goodness-of-fit measure for each distribution is computed as [5.10]

$$Z^{\text{dist}} = \left(\tau_4^{\text{dist}} - t_4^R + B_4 \right) / \sigma_4. \tag{5.17}$$

The fit is considered to be adequate if $|Z^{\text{dist}}|$-statistic is sufficiently close to 0, a reasonable criterion being $|Z^{\text{dist}}|$-statistic less than 1.64. Hosking and Wallis [5.10] state that the $|Z^{\text{dist}}|$-statistic has the form of a normal distribution under suitable assumptions. Thus the criterion $|Z^{\text{dist}}|$-statistic less than 1.64 corresponds to acceptance of the hypothesized distribution at a confidence level of 90%.

5.4 Study Area and Data Availability

Mahanadi subzone 3(d) in India comprises Mahanadi, Brahmani, and Baitarani basins. It is located between longitudes of 80° 25′ to 87° east and latitudes 19° 15′ to 23° 35′ north (Fig. 5.1). Its total drainage area is about 195 256 km². About 50% of the area of this subzone is hilly varying from 300 to 1 350 m. Rest of the area lies in the elevation range of 0–300 m. The normal annual rainfall over the region varies from 1 200 to 1 600 mm. The subzone receives about 75–80% of the annual rainfall from Southwest monsoon during the monsoon season from June to September. The red and yellow soils cover major part of the subzone. The red sandy, submontane, and coastal alluvial soils cover the remaining part of the subzone. The

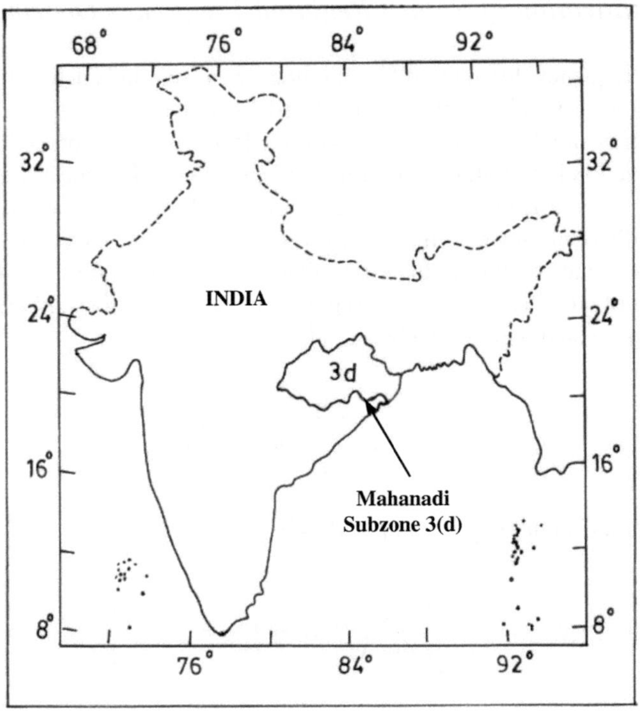

Fig. 5.1 Index map showing location of Mahanadi subzone 3(d) in India

subzone has an extensive area under forest. Paddy is the main crop grown on the cultivable land. Annual maximum peak flood data of 23 bridge sites lying in the hydrometeorologically homogeneous region [5.1] of Mahanadi subzone 3(d) as well as the catchment areas of the bridge sites were available for the study. These data were observed during the period 1957–1990 with record lengths varying from 11 to 30 years.

5.5 Analysis and Discussion of Results

Regional flood frequency analysis was performed using the various frequency distributions, viz., extreme value (EV1), generalized extreme value (GEV), logistic (LOS), generalized logistic (GLO), normal (NOR), generalized normal (GNO), uniform (UNF), Pearson type-III (PE3), exponential (EXP), generalized Pareto (GPA), κ-(KAP), and five-parameter Wakeby (WAK). Screening of the data, testing of regional homogeneity, identification of the regional distribution, and development of regional flood frequency relationships are described below.

5.5.1 Screening of Data Using Discordancy Measure Test

Values of discordancy statistic have been computed in terms of the L-moments for all the 23 gauging sites of the study area. It is observed that the D_i values for all the 23 sites vary from 0.04 to 3.06. As 3.06 is greater than the critical value of D_i, i.e., 3.00 for a region defined by 15 or more sites [5.10], the site having D_i value of 3.06 is discarded from the analysis. The D_i values for 22 sites vary from 0.04 to 2.32, and the data of 22 sites may be considered suitable for regional flood frequency analysis.

5.5.2 Test of Regional Homogeneity

The values of the heterogeneity measures $H(1)$, $H(2)$, and $H(3)$ were computed utilizing the data of 22 gauging sites by generating 500 regions using the fitted κ-distribution. Using the data of 22 sites, $H(1)$, $H(2)$, and $H(3)$ values are computed as 4.21, 1.79, and 0.16, respectively. As $H(1)$ is greater than 2.0 and $H(2)$ is greater than 1.0, the region defined by the 22 gauging sites is considered as heterogeneous. Thus, based on the statistical properties of the data of the gauging sites, one by one seven sites of the region are excluded till $H(1)$ value between 1.0 and 2.0; and both $H(2)$ and $H(3)$ values less than 1.0 are obtained. Further efforts to reduce the value of $H(1)$ led to significant loss of data, and hence, this was not attempted. The data sample comprising of 15 gauging sites and yielding $H(1)$, $H(2)$, and $H(3)$ values as 1.68, −0.71, and −1.98, respectively, is considered as reasonably homogeneous. The details of catchment data and statistical parameters including the discordancy measure for the 15 gauging sites are given in Table 5.1. The values of heterogeneity measures computed by carrying out 500 simulations using the κ-distribution based on the data of 15 sites are given in Table 5.2.

Table 5.1 Catchment area, sample statistics, sample size, and discordancy measure for 15 gauging sites of Mahanadi subzone 3(d)

Stream gauging site	Catchment area (km^2)	Mean annual peak flood (m^3/s)	Sample size (years)	L-CV (τ_2)	L-skew (τ_3)	L-kurtosis (τ_4)	Discordancy measure (D_i)
48	109	103.90	30	0.4020	0.2950	0.1658	0.46
93 K	74	153.07	28	0.2740	0.1235	0.1974	1.44
59 KGP	30	72.89	29	0.4079	0.2770	0.1780	0.74
308	19	41.22	27	0.3461	0.2339	0.0882	0.87
332 NGP	225	188.59	22	0.2899	0.2117	0.2020	1.23
59 BSP	136	196.23	22	0.4068	0.3471	0.2283	1.48
698	113	247.00	25	0.4240	0.3210	0.1356	1.09
121	1 150	1 003.86	21	0.2690	0.1622	0.0787	1.19
332 KGP	175	71.83	24	0.3102	0.1569	0.1647	0.51
40 K	115	260.67	21	0.3469	0.2328	0.1784	0.14
42	49	53.50	20	0.2260	0.0488	0.0530	1.92
69	173	238.89	19	0.3457	0.2392	0.1455	0.08
90	190	130.73	11	0.3570	0.1566	0.1335	2.11
195	615	963.77	13	0.2394	0.1305	0.1614	1.10
235	312	176.14	14	0.3128	0.2205	0.1130	0.63

Table 5.2 Heterogeneity measures for 15 gauging sites of Mahanadi subzone 3(d)

#	Heterogeneity measure	Value
	Heterogeneity measure H(1)	
	(a) Observed standard deviation of group L-CV	0.0615
1.	(b) Simulated mean of standard deviation of group L-CV	0.0472
	(c) Simulated standard deviation of group L-CV	0.0085
	(d) Standardized test value $H(1)$	1.68
	Heterogeneity measure H(2)	
	(a) Observed average of L-CV/L-skewness distance	0.0861
2.	(b) Simulated mean of average L-CV/L-skewness distance	0.0980
	(c) Simulated standard deviation of average L-CV/L-skewness distance	0.0167
	(d) Standardized test value $H(2)$	−0.71
	Heterogeneity measure H(3)	
	(a) Observed average of L-skewness/L-kurtosis distance	0.0821
3.	(b) Simulated mean of average L-skewness/L-kurtosis distance	0.1198
	(c) Simulated standard deviation of average L-skewness/L-kurtosis distance	0.0191
	(d) Standardized test value $H(3)$	−1.98

5.5.3 Identification of Robust Regional Frequency Distribution

The L-moment ratio diagram and $|Z_i^{\text{dist}}|$-statistic are used as the best fit criteria for identifying the robust distribution for the study area. The regional average values of L-skewness, i.e., τ_3 = 0.2180 and L-kurtosis, i.e., τ_4 = 0.1510 are obtained. Figure 5.2 shows the L-moment ratio diagram for the study area. The Z_i^{dist}-statistic

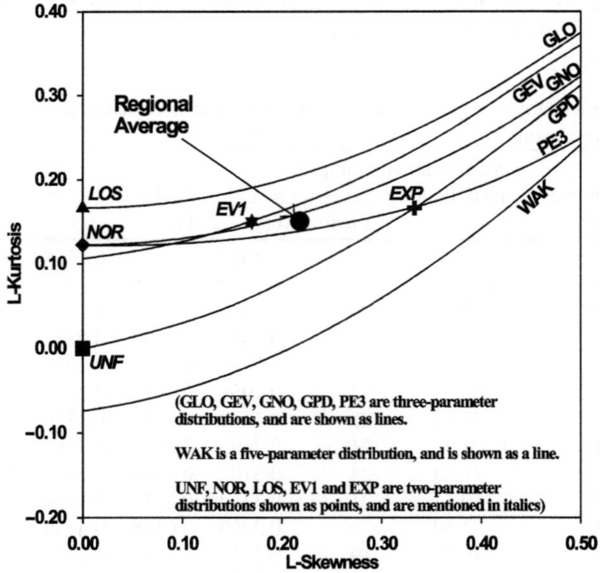

Fig. 5.2 L-moment ratio diagram for Mahanadi subzone 3(d) for various distributions

Table 5.3 Z_i^{dist}-statistic for various distributions for Mahanadi subzone 3(d)

S. No.	Distribution	Z_i^{dist}-statistic
1	Generalized logistic (GLO)	2.08
2	Generalized extreme value (GEV)	0.66
3	Generalized normal (GNO)	0.22
4	Pearson type-III (PE3)	−0.62
5	Generalized Pareto (GPD)	−2.68

Table 5.4 Regional parameters for various distributions for Mahanadi subzone 3(d)

Distribution	Parameters of the distribution				
GNO	$\xi = 0.870$	$\alpha = 0.548$	$k = -0.451$		
PE3	$\mu = 1.000$	$\sigma = 0.629$	$\gamma = 1.316$		
GEV	$\xi = 0.704$	$\alpha = 0.452$	$K = -0.073$		
WAK	$\xi = 0.100$	$\alpha = 1.985$	$\beta = 6.486$	$\gamma = 0.684$	$\delta = -0.078$

for various three-parameter distributions is given in Table 5.3. It is observed that the $|Z_i^{\text{dist}}|$-statistic values are lower than 1.64 for the three distributions, viz., GEV, GNO, and PE3. Further, the $|Z_i^{\text{dist}}|$-statistic is found to be the lowest for GNO distribution, i.e., 0.22. Thus, based on the L-moment ratio diagram and $|Z_i^{\text{dist}}|$-statistic criteria, the GNO distribution is identified as the robust distribution for the study area. The values of regional parameters for the various distributions which have Z^{dist}-statistic value less than 1.64 (i.e., distributions accepted at the 90% confidence level) as well as the five-parameter Wakeby distribution are given in Table 5.4. The regional parameters of the Wakeby distribution have been included in Table 5.4 because the Wakeby distribution has five parameters, more than most of the common distributions, and it can attain a wider range of distributional shapes than can the common distributions. This makes the Wakeby distribution particularly useful for simulating artificial data for use in studying the robustness, under changes in distributional form of methods of data analysis. It is preferred to use Wakeby distribution for heterogeneous regions.

5.5.4 Regional Flood Frequency Relationship for Gauged Catchments

For estimation of floods of various return periods for gauged catchments regional flood frequency relationship has been developed based on the robust identified GNO distribution. The cumulative density function of the three-parameter GNO distribution as parameterized by Hosking and Wallis [5.10] is given below:

$$F(x) = \phi\left[-k^{-1}\log\{1 - k(x-\xi)/\alpha\}\right] \qquad k \neq 0, \tag{5.18}$$

$$= \phi\left[(x-\xi)/\alpha\right] \qquad k = 0, \tag{5.19}$$

where ξ, α, and k are its location, scale, and shape parameters, respectively. When $k = 0$, it becomes normal distribution (NOR) with parameters ξ and α. This distribution has no explicit analytical inverse form. ϕ is the cumulative distribution function of the standard normal distribution given by

$$\phi(x) = \int_{-\infty}^{x} (2\pi)^{-1/2} \exp\left(-\frac{1}{2}x^2\right) \mathrm{d}x. \tag{5.20}$$

The quantile functions or the inverse form of the frequency distributions used in this study are given in Appendix.

Floods of various return periods may be computed by multiplying mean annual peak flood of a catchment by the corresponding values of growth factors of GNO distribution given in Table 5.5. The growth factor or site-specific scale factor $(Q_T/\overline{Q})$ is computed by dividing flood quantile (Q_T) by the annual mean peak flood of a gauging site $(\overline{Q})$. Monte Carlo simulation is used to estimate the accuracy of the quantiles of the GNO distribution. The region used in the simulation procedure contains 15 sites with record lengths same as that for Mahanadi subzone 3(d) data, the sites having GNO distributions with L-CV varying linearly from 0.2260 at site 1 to 0.4240 at site 15 and L-skewness 0.2180. The inter-site dependence is quantified in terms of an average correlation of about 0.32. Ten thousand regions were generated and a GNO distribution was fitted to each generated region. The regional average relative RMSE of the estimated growth curve was computed from the simulations. The 90% error bounds for the growth curves are also computed and are presented in Table 5.6.

Table 5.5 Values of growth factors $(Q_T/\overline{Q})$ for Mahanadi subzone 3(d)

	Return period (years)						
	2	10	25	50	100	200	1 000
Distribution	Growth factors						
GNO	0.870	1.821	2.331	2.723	3.125	3.538	4.552
PE3	0.866	1.842	2.329	2.683	3.028	3.366	4.134
GEV	0.872	1.809	2.332	2.745	3.175	3.627	4.767
WAK	0.865	1.848	2.353	2.712	3.052	3.374	4.058

Table 5.6 Accuracy measures for growth factors of GNO distribution for Mahanadi subzone 3(d)

Return period (years)	Growth factors	RMSE	90% Error bounds	
			Lower	Upper
2	0.870	0.035	0.829	0.889
10	1.821	0.076	1.789	1.905
25	2.331	0.101	2.248	2.517
50	2.723	0.116	2.587	3.008
100	3.125	0.128	2.924	3.526
200	3.538	0.140	3.267	4.075
1000	4.552	0.163	4.075	5.479

5.5.5 Regional Flood Frequency Relationship for Ungauged Catchments

For ungauged catchments the at-site mean cannot be computed in the absence of the observed flow data. Hence, a relationship between the mean annual peak flood of gauged catchments in the region and their pertinent physiographic and climatic characteristics is needed for estimation of the mean annual peak flood. Figure 5.3 shows a plot of the mean annual peak flood versus catchment area for the 15 gauging sites of Mahanadi subzone 3(d). The regional relationship in the form of a power law ($Y = aX^b$) is developed for the region using Levenberg–Marquardt iteration on the data of 15 gauging sites and is given below:

$$\overline{Q} = 2.519\,(A)^{0.863}, \tag{5.21}$$

where A is the catchment area in km^2, and $\overline{Q}$ is the mean annual peak flood in m^3/s. The best fit power equation is shown in Fig. 5.3 together with the error values of the parameter estimates. For Eq. 5.21, the coefficient of determination is $r^2 = 0.834$ and the standard error of estimates is 127.7.

For development of regional flood frequency relationship for ungauged catchments, the regional flood frequency relationship developed for gauged catchments is coupled with regional relationship between mean annual peak flood and catchment area, given in Eq. 5.21, and the following regional frequency relationship is developed:

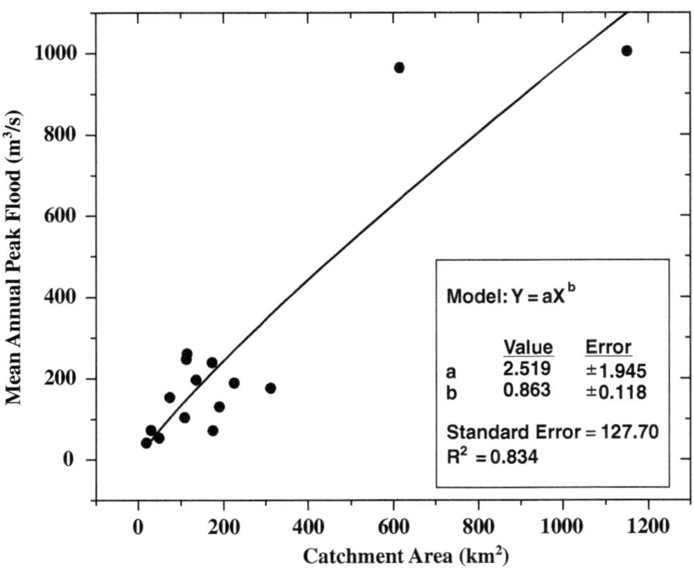

Fig. 5.3 Variation of mean annual peak flood with catchment area for various gauging sites of Mahanadi subzone 3(d)

$$Q_T = C_T \times A^{0.863}, \tag{5.22}$$

where Q_T is the flood estimate in m^3/s for T-year return period, A is the catchment area in km^2, and C_T is a regional coefficient. Values of C_T for some of the commonly used return periods are given in Table 5.7. The above regional flood formula (Eq. 5.22), its tabular form (Table 5.8), or graphical representation (Fig. 5.4) may be used for estimation of floods of desired return periods for ungauged catchments of the Mahanadi subzone 3(d). The conventional empirical flood formulae do not provide floods of various return periods for the ungauged catchments. However, the regional flood formula developed in this study (Eq. 5.22) is capable of providing flood estimates for desired return periods.

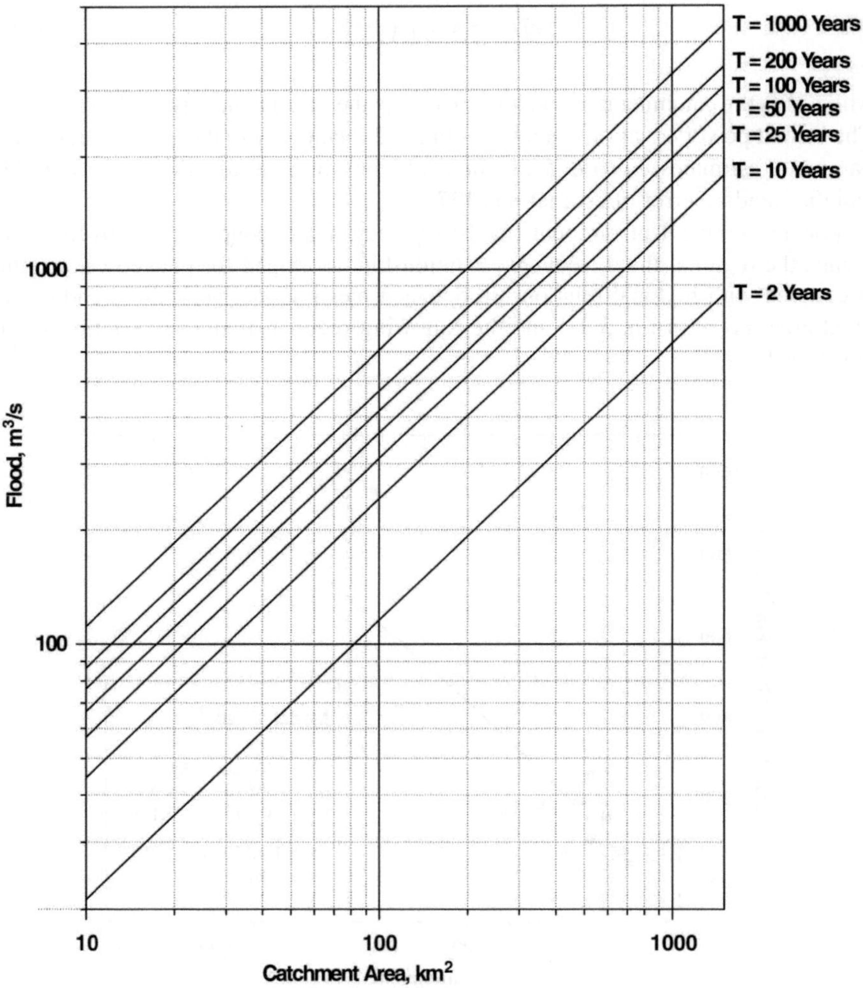

Fig. 5.4 Variation of floods of various return periods with catchment area for Mahanadi subzone 3(d)

Table 5.7 Values of regional coefficient C_T for Mahanadi subzone 3(d)

Distribution	Return period (years)						
	2	10	25	50	100	200	1 000
	Growth factors						
GNO	2.192	4.587	5.872	6.859	7.872	8.912	11.466

Table 5.8 Floods of various return periods for different catchment areas of Mahanadi subzone 3(d)

Catchment area (km^2)	Return period (years)						
	2	10	25	50	100	200	1000
	Floods of various return periods (m^3/s)						
10	16	33	43	50	57	65	84
20	29	61	78	91	104	118	152
30	41	86	111	129	148	168	216
40	53	111	142	166	190	215	277
50	64	134	172	201	230	261	335
60	75	157	201	235	270	305	393
70	86	179	230	268	308	349	448
80	96	201	258	301	346	391	503
90	107	223	285	333	382	433	557
100	117	244	312	365	419	474	610
150	166	346	443	518	594	673	866
200	212	444	568	664	762	863	1 110
250	257	538	689	805	924	1 046	1 345
300	301	630	806	942	1 081	1 224	1 575
350	344	720	921	1 076	1 235	1 398	1 799
400	386	807	1 034	1 207	1 386	1 569	2 018
450	427	894	1 144	1 337	1 534	1 737	2 234
500	468	979	1 253	1 464	1 680	1 902	2 447
600	548	1 146	1 467	1 713	1 966	2 226	2 864
700	625	1 309	1 675	1 957	2 246	2 543	3 271
800	702	1 469	1 880	2 196	2 520	2 853	3 671
900	777	1 626	2 081	2 431	2 790	3 159	4 064
1 000	851	1 780	2 279	2 662	3 056	3 459	4 451
1 100	924	1 933	2 475	2 891	3 317	3 756	4 832
1 200	996	2 084	2 668	3 116	3 576	4 049	5 209
1 300	1 067	2 233	2 858	3 339	3 832	4 338	5 581
1 400	1 137	2 380	3 047	3 559	4 085	4 625	5 950
1 500	1 207	2 526	3 234	3 778	4 336	4 908	6 315

5.6 Conclusion

Screening of the data conducted using the annual maximum peak flood data of the Mahanadi subzone 3(d) employing the discordancy measure (D_i) test reveals that 22 out of the 23 gauging sites are suitable for regional flood frequency analysis.

However, based on the heterogeneity measures, $H(j)$, the annual maximum peak flood data of 15 stream flow gauging sites are considered to constitute a homogeneous region and the same are used for regional flood frequency analysis. Various distributions, viz., EV1, GEV, LOS, GLO, UNF, PE3, NOR, GNO, EXP, GPA, KAP, and WAK have been employed. Based on the L-moments ratio diagram and $|Z_i^{\text{dist}}|$-statistic criteria, the GNO distribution has been identified as the robust distribution for the study area.

The developed regional flood frequency relationships may be used for the estimation of floods of desired return periods for gauged and ungauged catchments of the study area. As the regional flood frequency relationships have been developed using the data of catchments varying from 19 to 1150 km^2 in area; hence, these relationships are expected to provide estimates of floods of various return periods for catchments lying nearly in the same range of the areal extent. Further, the relationship between mean annual peak flood and catchment area is able to explain 83.4% of initial variance ($r^2 = 0.834$). Hence, in case of ungauged catchments the results of the study are subject to these limitations. However, the regional flood frequency relationships may be refined for obtaining more accurate flood frequency estimates; when the data for some more gauging sites become available and physiographic characteristics other than catchment area as well as some of the pertinent climatic characteristics are also used for development of the regional flood frequency relationships.

Appendix

The following commonly adopted frequency distributions have been used in this study. The details about these distributions and relationships among parameters of these distributions and L-moments are available in literature [5.10].

Generalized Extreme Value Distribution (GEV)

The generalized extreme value distribution (GEV) is a generalized three-parameter extreme value distribution. The quantile function or the inverse form of the distribution is expressed as

$$x(F) = u + \alpha\{1 - (-\ln F)^k\}/k \qquad k \neq 0, \tag{5.23}$$

$$= u - \alpha \ln(-\ln F) \qquad k = 0\,, \tag{5.24}$$

where u, α, and k are location, scale, and shape parameters of GEV distribution, respectively. EV1 distribution is the special case of the GEV distribution when $k = 0$.

Generalized Logistic Distribution (GLO)

The inverse form of the generalized logistic distribution (GLO) is expressed as

$$x(F) = u + [\alpha[1 - \{(1 - F)/F\}^k]/k \qquad k \neq 0, \tag{5.25}$$

$$= u - \alpha \ln \{(1 - F)/F\} \qquad k = 0, \tag{5.26}$$

where u, α, and k are location, scale, and shape parameters, respectively. The logistic distribution (LOS) is a special case of the generalized logistic distribution when $k = 0$.

Generalized Pareto Distribution (GPA)

The inverse form of the generalized Pareto distribution (GPA) is expressed as

$$x(F) = u + \alpha\{1 - (1 - F)^k\}/k \qquad k \neq 0, \tag{5.27}$$

$$= u - \alpha \ln(1 - F) \qquad k = 0, \tag{5.28}$$

where u, α, and k are location, scale, and shape parameters, respectively. The exponential distribution (EXP) is special case of generalized Pareto distribution when $k = 0$.

Pearson Type-III Distribution (PT-III)

The inverse form of the Pearson type-III distribution is not explicitly defined. Hosking and Wallis [5.10] mention that the Pearson type-III distribution combines Γ-distributions (which have positive skewness), reflected Γ-distributions (which have negative skewness), and the normal distribution (which has zero skewness). The authors parameterize the Pearson type-III distribution by its first three conventional moments, viz., mean μ, the standard deviation σ, and the skewness γ. The relationship between these parameters and those of the Γ-distribution is as follows.

Let X be a random variable with a Pearson type-III distribution with parameters μ, σ, and γ. If $\gamma > 0$, then $X - \mu + 2\sigma/\gamma$ has a Γ-distribution with parameters $\alpha = 4/\gamma^2$, $\beta = \sigma\gamma/2$. If $\gamma = 0$, then X has a normal distribution with mean μ and standard deviation σ. If $\gamma < 0$, then $-X + \mu - 2\sigma/\gamma$ has a Γ-distribution with parameters $\alpha = 4/\gamma^2$, $\beta = |\sigma\gamma/2|$.

If $\gamma \neq 0$, let $\alpha = 4/\gamma^2$, $\beta = (1/2)\sigma/\gamma$, and $\xi = \mu - 2\sigma/\gamma$ and $\Gamma(.)$ is Γ-function. If $\gamma > 0$, then the range of x is $\xi \leq x < \propto$ and the cumulative distribution function is

$$F(x) = G\left(\alpha, \frac{x - \xi}{\beta}\right) / \Gamma(\alpha). \tag{5.29}$$

If $\gamma < 0$, then the range of x is $-\infty < x \leq \xi$ and the cumulative distribution function is

$$F(x) = 1 - G\left(\alpha, \frac{\xi - x}{\beta}\right) / \Gamma(\alpha). \tag{5.30}$$

Here, $G(\alpha, x)$ is the incomplete gamma function expressed as

$$G(\alpha, x) = \int_0^x t^{\alpha-1} e^{-t}\, \mathrm{d}t. \tag{5.31}$$

κ-Distribution (KAP)

The κ-distribution is a four-parameter distribution that includes as special cases the generalized logistic (GLO), generalized extreme value (GEV), and generalized Pareto distribution (GPA):

$$x(F) = \xi + \alpha\left[1 - \left\{(1 - F)^h / h\right\}^k\right] / k, \tag{5.32}$$

where, ξ is the location parameter and α is the scale parameter.

When $h = -1$, it becomes generalized logistic (GLO) distribution; $h = 0$ is the generalized extreme value (GEV) distribution; and $h = 0$ is the generalized Pareto (GPA) distribution. It is useful as a general distribution with which to compare the fit of two- and three-parameter distributions and for use in simulating artificial data in order to assess the accuracy of statistical methods [5.10].

Wakeby Distribution (WAK)

Inverse form of the five-parameter Wakeby (WAK) distribution is expressed as.

$$x(F) = \xi + \frac{\alpha}{\beta}\{1 - (1 - F)^{\beta}\} - \frac{\gamma}{\delta}\{1 - (1 - F)^{-\delta}\}, \tag{5.33}$$

where ξ, α, β, γ, and δ are the parameters of the Wakeby distribution.

References

5.1. Central Water Commission, *Flood Estimation Report for Mahanadi Subzone 3 (d). Small Catchments* (Directorate of Hydrology, New Delhi, 1982)

5.2. F. Chebana, T.B.M.J. Ouarda, Multivariate L-moment homogeneity test. Water Resour. Res. **43**, W08406. doi:10.1029/2006WR005639 (2007)

5.3. K. Chokmani, T.B. M.J. Quarda. Physiographical space-based kriging for regional flood frequency estimation at ungauged sites. Water Resour. Res. **40**, W12514 (2004)
5.4. J. M. Cunderlik, D.H. Burn, Switching the pooling similarity distances: Mahalanobis for Euclidean. Water Resour. Res. **42**, W03409 (2006)
5.5. C. Cunnane, Methods and merits of regional flood frequency analysis. J. Hydrol. **100**, 269–290 (1988)
5.6. J.A. Greenwood, J.M. Landwehr, N.C. Matalas, J.R. Wallis, Probability weighted moments: Definition and relation to parameters of several distributions expressible in inverse form. Water Resour. Res. **15**(5), 1049–1054 (1979a)
5.7. V.W. Griffis, J.R. Stedinger, Evolution of flood frequency analysis with *Bulletin 17*. J. Hydrologic Eng. **12**(3), 283–297 (2007)
5.8. J.R.M. Hosking, L-moments: Analysis and estimation of distributions using linear combinations of order statistic. J. Royal Stat. Soc., Series B, **52**(2), 105–124 (1990)
5.9. J.R.M. Hosking, J.R. Wallis, Some statistics useful in regional frequency analysis. Water Resour. Res. **29**(2), 271–281 (1993)
5.10. J.R.M. Hosking, J.R. Wallis, *Regional frequency analysis-an approach based on L-moments* (Cambridge University Press, New York, 1997)
5.11. V. Iacobellis, M. Fiorentino, Derived distribution of floods based on the concept of partial area coverage with a climatic appeal. Water Resour. Res. **36**(2), 469–482 (2000)
5.12. Z. Jingyl, M.J. Hall, Regional flood frequency analysis for the gan-ming river basin in China. J. Hydrol. **296**, 98–117 (2004)
5.13. T. Kohonen. Self-organiszing maps. (Springer, New York, 2001)
5.14. R. Kumar, C. Chatterjee. Regional flood frequency analysis using L-moments for North Brahmaputra region of India. J. Hydrologic Eng. *ASCE*, **10**(1), 1–7 (2005)
5.15. R. Merz, G. Blöschl, Flood frequency regionalization – spatial proximity vs. catchment attributes. J. Hydrol. **302**, 283–306 (2005)
5.16. D. H. Pilgrim, I. Cordery, Flood runoff. In *Handbook of Hydrology*, ed. by D. R. Maidment (New York, Mc Graw-Hill, 1992)
5.17. A.R. Rao, K.H. Hamed, Flood frequency analysis (CRC Press, Washington, D.C., 2000)
5.18. J.R. Stedinger, R.M. Vogel, E. Foufoula-Georgiou, Frequency analysis of extreme events. In *Handbook of Hydrology*, ed. by D.R. Maidment, (New York, Mc Graw-Hill, 1992)
5.19. R.M. Vogel, N.M. Fennessey, L-moments should replace product moments diagrams. Water Resour. Res. **29**(6), 1745–1752 (1993)
5.20. A. Zafirakou-Koulouris, R.M. Vogel, S.M. Craig, J. Habermeier, L- moment diagrams for censored observations. Water Resour. Res. **34**(5), 1241–1249 (1998)
5.21. L. Zhang, V.P. Singh, Trivariate flood frequency analysis using the Gumbel–Hougaard Copula. J. Hydrol. Eng., **12**(4), 431–439 (2007)

Part II
Extremes, Trends, and Changes

Benign spring flooding of the Rver Biebrza (Poland) near to the Szostaki village. Courtesy of: http://mocar.obiezyswiat.org

Chapter 6
Intense Precipitation and High Floods – Observations and Projections

Zbigniew W. Kundzewicz

According to physical laws, the water-holding capacity of the atmosphere and hence the potential for intense precipitation increases with warming. Since a robust warming signal and a number of large rain-caused floods have been observed recently, it is of paramount importance to examine whether there has been an increasing trend in intense precipitation and flood flow. However, even if widespread increases in observed intense precipitation have been reported in many areas, the analysis of annual maximum river flow records does not detect an ubiquitous and coherent increasing trend. This is in disagreement with some projections for the future, where increasing intense precipitation and flood hazard are expected. One can conclude that flood process is complex, influenced by several non-climatic factors, and can be caused by several generating mechanisms, which are affected in different ways by climate change. Hence, issuing a flat-rate statement on change in flood hazard is not justified.

6.1 Introduction

Flooding has been a major concern in many regions of the world. There is a perception that flooding is increasing worldwide and part of this perception may result from increasing media coverage of flood events. However, in addition to the media attention, it is also clear that flood risk, understood as the probability of extreme event multiplied by a measure of its adverse consequences, has been on the rise. Among the factors contributing to the growth of flood risk are changes in socio-economic systems (population increase and accumulation of wealth in vulnerable areas), terrestrial systems, and climate [6.16].

In several recent flood events the material losses have exceeded US $10 billion (up to US $30 billion material damage during the summer 1998 floods in China).

Z.W. Kundzewicz (✉)
Research Centre for Agricultural and Forest Environment Polish Academy of Sciences, 60809 Poznan, Poland; Potsdam Institute for Climate Impact Research, 14412 Potsdam, Germany
e-mail: zkundze@man.poznan.pl; zbyszek@pik-potsdam.de

J.P. Kropp, H.-J. Schellnhuber (eds.), *In Extremis*,
DOI 10.1007/978-3-642-14863-7_6, © Springer-Verlag Berlin Heidelberg 2011

The death toll has remained high; in some events in less developed countries the number of flood fatalities has been greater than 1 000. In an average year, floods cause material damage of the order of tens of billions of dollars globally, in both developed and less developed countries, and more than 10 000 fatalities mostly in less developed countries. It is estimated that the costs of extreme weather events have exhibited a rapid upward trend. Yearly economic losses from weather extremes increased 10-fold (in inflation-adjusted dollars) between the 1950s and the 1990s, and flood material damages have changed in a similar way. The insured portion of these losses has grown even stronger.

Most destructive floods, in terms of human losses, occur outside of Europe (in particular in Asia: China, India, and Bangladesh), but in the last decades, floods have severely hit large parts of the European continent. There have been a number of large floods in Europe since the 1990s. For example, Poland suffered dramatic summer flood in 1997 (this event in the Odra basin also caused considerable damage in the Czech Republic and Germany), 1998, 2001, and 2010. The year 2002 was particularly abundant in large floods in Europe. It is estimated that the material flood damage recorded in the European continent in 2002 has been higher than in any single year before and exceeded 21 billion euros. The material flood damage in Central Europe (Czech Republic, Germany, Austria) in August 2002 alone on the rivers Danube, Labe/Elbe, and their tributaries, exceeded 15 billion euros. Not long after, in September 2002, during severe storms and floods in southern France (Rhone valley), people were killed and the total material losses exceeded 1 billion euros. Also in 2005, 2006, and 2007, a number of further destructive floods have been recorded in several countries of Europe.

Episodes of intense and/or long-lasting precipitation, which have occurred in Europe in recent years leading to high-impact floods, raise considerable concern. Based on climate change scenarios one may anticipate more severe problems of having destructive abundance of water in the future. Projections for the future indicate a potential for more intense precipitation in a warmer climate, further augmenting flood risk (cf. [6.15, 6.17]).

6.2 Observations of Intense Precipitation

Groisman et al. [6.9] observed widespread increase in heavy precipitation during the past 50–100 years in middle latitudes, even if, unfortunately, reliable estimates of very heavy and extreme precipitation are available only for regions with dense observation networks. The robust finding was disproportionate increase in precipitation extremes. Groisman et al. [6.9] detected changes in intense precipitation for more than half of the land area of the globe. They report increase in heavy precipitation in some areas (South Africa, Siberia, Central Mexico, Northeastern USA, Northern Japan), where the total precipitation and frequency of wet days did not change or decreased. As stated by Groisman et al. [6.9], "the empirical evidence from the period of instrumental observations [...] indicate an increasing probability of intense precipitation events for many extra-tropical regions, including the USA". This can be interpreted as effect of warming, accompanied by increase in water vapour, which leads to increased intensity but reduced duration of precipitation.

There are a number of studies, in which increase of intense precipitation in observed records from Europe is documented. In a comprehensive study, Klein Tank and Koennen [6.11] found that Europe average indices of wet extremes increase, although the spatial coherence of the trends is low. The spatial inhomogeneity of the trend patterns is largely influenced by the orography and subgrid-scale processes. Groisman et al. [6.9] observed that a clear majority of stations in Europe show increasing trends in the number of wet (75%) and very wet (95% quantile) days during the second half of the twentieth century. There have been a number of national studies reporting changes in extreme precipitation in Europe. Fowler and Kilsby [6.6] found that 10-day precipitation totals with 50-year return period in Northern England and Scotland increased by a factor of 2–5 by the 1990s. The low-frequency variability of heavy precipitation was investigated by Schmidli and Frei [6.29] for a dense data set (104 stations) of time series in Switzerland covering the period 1901–2000. The analysis of a wide range of daily and multi-day precipitation statistics encompassing basic characteristics, measures of heavy precipitation, and indices of typical and extreme wet spells led to the detection of a clear trend signal for winter and autumn. The centennial increase was found between 10 and 30% for the high quantiles and the seasonal 1- to 10-day extremes. These results corroborate the finding of a significant increase of frequency of intense precipitation event in winter and autumn for many stations for Alpine region of Switzerland in the time period 1901–1994 [6.7]. Increase in extreme rainfall in Italy and Spain (1951–1995) was reported by Alpert et al. [6.1]. Sometimes the reported change is more complex. For instance, for the 1961–2000 data in the UK, daily precipitation was found to be more intense in winter and less intense in summer [6.25]. However, several studies in Europe do not report detection of an increasing trend. No significant trend in the extreme rainfall intensity was found, e.g. in Italy [6.4], Cyprus [6.1], and Uccle (Belgium), where Vaes et al. [6.31] examined a 100-year daily record.

Longer series of observation records make it possible to place recent events in a broader context. Indeed, some recent rainfall events in Europe have exceeded all-time records. According to Czech sources, several records in the category of 1-, 2-, and 3-day precipitation observed in July 1997 and August 2002 at a number of gauges entered the lists of the 10 highest observations ever made in the Czech territory. In Zinnwald-Georgenfeld (Saxony, Germany), 312 mm of 1-day precipitation was also recorded from 12 August 2002, 6 a.m. to 13 August 2002, 6 a.m. (usual time interval for measuring one-day precipitation), matching the observation at a nearby Czech station in Cinovec and beating the all-time German national record. The maximum 24-h precipitation recorded in Zinnwald-Georgenfeld from 12 August 2002, 3 a.m. to 13 August 2002, 3 a.m. was even higher and reached 352.7 mm. The list of all-time extreme precipitation totals, in several rainfall duration classes, observed in Germany contains several entries from the last 10 years.

From the viewpoint of areal extent of precipitation, its intensity and duration (a wave of two wet spells), the August 2002 event was exceptional in the instrumental data.

6.3 Observations of River Flow

There have been a number of recent flood events when river flow or stage records have been broken. For instance, the 2002 flood peak level of the Vltava in Prague (Czech Republic) exceeded all the events recorded in the last 175 years. This was the only time when the flow rate of 5 000 m^3/s was exceeded, while between 1941 and 2001 Vltava flow never reached 2 500 m^3/s. The water level of the Elbe in the profile Dresden on 17 August 2002, i.e. 940 cm, has considerably (by 63 cm) exceeded the former historical record. In the past, stages in excess of 800 cm were observed in Dresden five times during less than a century – from 1785 to 1879 (four out of five times – in February or March), but this level was never reached in 1880–2001. However, such a long period of lower annual maxima has not been uncommon in the historical data. For instance, between 1502 and 1783, i.e. 281 years, the level of 800 cm was exceeded only once. Apart from the record set in Dresden, the all-times historical flood records on the river Mulde in Döbeln (1897) and Grimma (1573) were exceeded in August 2002 by 126 cm and 136 cm, respectively.

In view of recent dramatic floods, search for trends in long time series of flood data is of scientific interest and practical importance. It is essential for planning of future water resources and flood protection systems, where system design is traditionally based on the assumption of stationarity in hydrological processes such as river stage or discharge. Most studies (except [6.23]) have only analysed a few rivers, or at most a region or a single country.

A study on change detection in worldwide hydrological time series of annual maximum river flow was carried out by Kundzewicz et al. (2005a) [6.14]. The annual maximum flow was used as a surrogate for floods, recognizing that it does not always represent an out-of-bank flow. There are advantages to this approach: it is a straightforward and well-established concept. However, time series of annual maximum flow conveys information on some extremes only. In years with more than one high-flow episode only one flood event per year is selected, while in years with no large flows at all, a non-flood (medium or even low) flow, in absolute terms, is extracted as the annual maximum. An alternative is a peak-over-threshold (POT) approach, where all independent floods above a certain threshold (i.e. possibly several events in one year and none in another year) are considered. Results of a complementary study of POT are reported in Svensson et al. [6.30] or with regard to a combination of POT and point processes in Chap. 8.

The study by Kundzewicz et al. [6.14] is limited to a subset of discharge time series held at the Global Runoff Data Centre (GRDC) in Koblenz, Germany. Out of more than a 1000 long time series made available by GRDC, a data set consisting of 195 long series of daily mean flow records was selected.

The choice of stations used was based on the following criteria:

- A record length of at least 40 years (with exceptions in regions with scarce data). The longest series, from the River Main at Würzburg, spans 178 years from 1824 to 2001)
- Currency of records, preferably extending through at least the late 1990s, with only a few exceptions, in regions having scarce data
- No gaps in the records that could misrepresent the annual maxima series. Missing values and gaps in data are always complicating factors. Whether or not to fill them, and if so, in what way, is a complex issue. In the reported study no gaps were filled. If gaps exist, the data are only conditionally useful for studies of annual maxima, e.g. where flooding is clearly seasonal and gaps occur in a non-flood season (e.g. gaps in autumn in a catchment subject to snowmelt flooding), they were ignored
- Representative geographic coverage and avoiding the use of many neighbouring (and possibly correlated) stations
- Catchment size, with priority given to smaller catchments which are more likely to have less strong anthropogenic influence, especially in developed countries

Selection of stations to use in a study is very important (cf. [6.16]). River flow is the integrated result of both natural factors, such as precipitation, catchment storage, and evaporation loss, and management practices that alter the river basin and the river conveyance system over time. This complicates the problem of detecting a climate change signature in river flow data.

Hence, particular care is needed in selecting data and sites for use in studying change. In order to assess climatically forced hydrological changes, data should be taken from pristine drainage basins; that is, those minimally affected by such human activities as deforestation, urbanization, reservoirs, drainage systems, water abstraction, river engineering, etc. Catchments featuring significant land-use and land-cover changes or river regulation are not appropriate. The data should be of high quality and extend over a long period. Detailed suggestions on how to select a network of stations for climate change detection are given in Pilon [6.27].

However, there are no metadata available for the flow series used in the study by Kundzewicz et al. [6.14], which means that any observed trends may be the result of both climatic and non-climatic influences. The GRDC metadata only include basic characteristics of the gauging stations, which severely constrained selection of optimal streamgauge records. A metadata needed to adequately screen and select appropriate streamgauge records should include changes in the catchment (land-use change); data collection methods; data from nearby sites; related variables, etc. The data used in the analysis have been subject to quality control both at the GRDC and within this study. Unfortunately, the coverage is not globally uniform with many stations in three regions (North America, Australia and the Pacific, and Europe) and only a few stations in Africa, Asia, and South America.

Whether it is appropriate to use the classical test procedure for trend detection depends on the assumptions that can be made about the data (cf. [6.16]). In the study by Kundzewicz et al. [6.14], the series of annual maxima was subject to two tests for independence. Both tests indicated non-randomness in very few of the annual maxima series. Given that the number of non-random series was small, it was decided to omit them.

Since the data analysed in the present study [6.14] are non-normal, but are independent and non-seasonal, hence, any of the basic distribution-free tests are suitable. The seasonal variation inflow is removed by the use of annual maxima rather than a continuous daily series. Extremes, such as series of annual maxima, generally have a positively skewed distribution. Due to the global coverage of the study, calendar years were used, since hydrological years start in different months in different regions.

Kundzewicz et al. [6.17] used the Mann–Kendall test, a non-parametric distribution-free method which is frequently used to detect trends (cf. [6.16]). This approach allows the investigator to make minimal assumptions (constancy of distribution and independence) about the data. It makes it possible to avoid assumptions about the form of the distribution that the data derive from, e.g. there is no need to assume that data are normally distributed. However, observations must be independent for the computed significance levels to be correct (cf. [6.13, 6.16]).

The analysis of annual maximum flows reported by Kundzewicz et al. [6.17] does not support the hypothesis of ubiquitous growth of high flows worldwide. Although 27 cases of strong, statistically significant increase were identified by the Mann–Kendall test, there are 31 decreases as well, and most (137) time series do not show any significant changes. Caution is advised in interpreting these results as flooding is a complex phenomenon, caused by a number of factors that can be associated with local, regional, and hemispheric climatic processes. Moreover, river flow has strong natural variability and exhibits long-term persistence which can confound the results of trend and significance tests.

In several cases, the highest flow was observed after 1990. In some series (from Chao Phraya in Thailand, and three rivers in Norway), generally decreasing trend was observed over decades, but the highest flows occurred in 1990s.

In a number of cases, the highest value in a long time series of annual maximum flow was more than twice as high as the second highest record. This can be due to special, and really extreme, conditions related to the absolute maximum. For example, the highest annual maximum flow in Chena River, Fairbanks (US) of the amplitude 1 809 m^3/s occurred in 1967, while the second highest annual maximum flow was 456 m^3/s (in 1960). In Salcha River, near Salchaket (US), the highest annual maximum flow was 2 635 m^3/s in 1967, while the second highest annual maximum flow was 879 m^3/s (in 1986).

Since all European series considered in the study started no later than in 1961, one can take the year 1961 as the starting point for a 40-year common period and then divide this common period into two 20-year subperiods. It was found [6.17] that the overall 1961–2000 maxima occurred considerably more frequently (46 times) in the later subperiod, 1981–2000 than in the earlier subperiod, 1961–1980 (24 times). This was despite the fact that not all-time series lasted until the year 2000 (series at 15 stations end in 1999 and at 6 end in 1998). Hence, it may well be that even more maxima fall into the subperiod 1981–2000.

A closer look into particularities of individual stations concerned would be needed to discriminate the driving factors. Since this information is not available in the GRDC holdings, there is a need to augment the collected data by accommodating more detailed metadata with more information about a station, including history of river development for navigation and energy generation. It would be useful to attempt to describe deterministically the reasons for atypical behaviour of some series (as compared to their spatial neighbourhood). Here, influence of a local event (e.g. flood resulting from a very high-intensity local storm), reservoirs, polders, flood control could play an important role.

Analysis should also differentiate the flood generation mechanisms (snowmelt vs. rainfall). In the present study, all floods (summer, winter, spring) were treated as one category. In a seasonal analysis, Mudelsee et al. [6.24] demonstrated a significant decrease in winter floods at the Elbe and the Odra.

A regional change in timing of floods has been observed in many areas, with increasing late autumn and winter floods, cf. McCarthy et al. [6.21]. Mudelsee et al. [6.24] and Bronstert [6.3] present evidence for decreasing ice-jam floods on rivers in central Europe.

Apart from the inherent complexity involved in detecting a greenhouse (climate change) signal in flow records, there are serious problems with the data (gaps, errors) and with the methodology to detect changes as well. Even if the data are perfect, however, it is worthwhile to re-state a tautology: extreme (hence rare) events are rare [6.16]. They do not happen frequently, so even with a very long time series of instrumental records, there is a small sample of truly extreme and destructive floods.

As stated by Radziejewski and Kundzewicz [6.28]

- failure to detect a significant trend should not be considered as a proof of the absence of change (while a detection of a trend is not a proof of presence of change);
- examination of detectability of artificially introduced, hence fully controlled, trends in time series leads to the following common sense results:

  If a change is weak and lasts for a short time, it is not likely to be detected.
  If a change is stronger and lasts longer, the likelihood of detection grows.

6.4 Projections of Intense Precipitation and High River Flow

Changes in precipitation extremes are projected by climate model simulations in many areas of the globe. The extremes in precipitation are likely to be impacted more than the means in the future. As the water holding capacity of the atmosphere, and thus its absolute potential water content, increases with temperature, the possibility of intense precipitation also increases.

Recent works on changes in precipitation extremes in Europe (e.g. [6.8]) project that the intensity of daily precipitation events will predominantly increase. The number of wet days is projected to decrease [6.8], which leads to longer dry periods except in the winter of West and Central Europe.

Palmer and Räisänen [6.26] analysed the modelled differences between the control run and an ensemble with transient increase in CO_2 and calculated around the time of CO_2 doubling. A considerable increase of the risk of a very wet winter in Europe and a very wet monsoon season in Asian monsoon region was found. This would have consequences to flood hazard. The modelling results indicate that the probability of total boreal winter precipitation exceeding two standard deviations above normal will considerably increase over large areas of Europe. For example, an over fivefold increase is projected over parts of British isles and much of the Baltic Sea basin and even over sevenfold increase for parts of Russia.

The "Modelling the Impact of Climate Extremes" (MICE) Project within the 5th Framework Programme of the European Union aimed to demonstrate how the occurrence of extremes over Europe may be expected to change due to anthropogenic warming. Among other things, the project examined changes in intense precipitation.

The data used in the MICE project are results of simulations from the Hadley Centre Regional Climate Model (HadRM3-P) for the whole European region, with the spatial resolution $0.44° \times 0.44°$. In the MICE project, the daily precipitation totals data from the Hadley Centre HadRM3-P regional model simulations over the European continent, were compared for two 30-year periods (1961–1990 and 2070–2099). Analyses of extremes were carried out both on the continental European scale and on a local scale.

Results obtained in the MICE project for the European region show that even if, according to climate model results (HadRM3-P), the mean summer precipitation over much of Europe is likely to decrease from the control period, 1961–1990, to the period of interest in the twenty-first century, 2070–2099, the behaviour of precipitation extremes is considerably different. The highest quantiles of daily precipitation amounts and annual maximum daily precipitation increase over many areas, also where decrease in means is projected. This result is in agreement with the findings of Christensen and Christensen [6.5], who examined another climate model. The results of comparison of the number of days with intense precipitation, exceeding an arbitrarily selected threshold (10 mm), simulated by HadRM3-P for the periods 1961–1990 and 2070–2099 show increase of the number of days with intense precipitation over most of Europe [6.17]. A larger increase in the frequency than in the magnitude of precipitation extremes was found.

As stated by Arnell and Liu Chunzhen [6.2], changes in future flood frequency in Europe are complex, depending on the generating mechanism, e.g. increasing flood magnitudes where floods result from heavy rainfall and decreasing magnitudes where spring floods are generated by snowmelt. Climate change is likely to cause an increase of the risk of riverine flooding across much of Europe.

In many parts of Europe, significant changes in flood risk are expected under IPCC IS92a scenario (similar to SRES A1) for the 2020s and the 2070s [6.18]. The regions most prone to a rise in flood frequencies are northern to north-eastern Europe, for climate change projected by both the ECHAM4 and the HadCM3 model.

Milly et al. [6.22] demonstrated that for 15 out of 16 large basins (over 200 000 km^2) they analysed worldwide, the control 100-year flood (in monthly time-scale) is projected to be exceeded more frequently as a result of CO_2 quadrupling. In some areas, what is given as a 100-year flood now (in the control run) is projected to occur much more frequently, even every 2–5 years. A particularly strong increases in frequency are projected in Northern Asia, albeit a large uncertainty in these projections is recognized.

Impacts of extremes on human welfare are likely to occur disproportionately in countries with low adaptation capacity [6.20]. Up to 20% of the world population live in river basins that are likely to be affected by increased flood hazard in the course of global warming [6.12].

Summarizing, it can be stated that increase in intense precipitation leads to increase of flood hazard in areas, where inundations are typically triggered by intense summer rain (cf. [6.17]). Also, during wetter and warmer winters, with increasingly more frequent rain and less frequent snow in many areas, flood hazard may increase. On the other hand, ice-jam floods are likely to become less frequent and less severe over much of the warming Europe (robust, temperature-related statement). Since snowmelt is earlier and less abundant, the risk of spring floods decreases, yet such floods happened in spring of 2006 in Central Europe.

There is still a great deal of uncertainty in findings about climate change impacts on water resources. Part of it is due to the differences between estimates of precipitation in different climate models and for different emission scenarios. Only in some, but not all, areas, the direction of projected change of hydrological processes is consistent across different models and scenarios. The scale mismatch between coarse-resolution climate models and the hydrological (catchment) scale is another source of uncertainty. The uncertainty level is particularly strong for extremes.

6.5 Conclusion

The weight of observational evidence indicates an ongoing intensification of the water cycle [6.10]. Hence, the question of detection and attribution of changes in various flood-related indices attracts considerable interest. Among the mechanisms that can impact flood risk are changes in socio-economic systems, which influence terrestrial systems, and changes in the climatic system. The atmosphere's water-holding capacity (and hence potential for intense precipitation) increases with temperature and more intense precipitation has been documented in many areas in the warming world. In contrast, globally, no ubiquitous, general, and coherent increase in high river flows has been detected worldwide [6.14]. However, in presence of multiple contributing (and partly compensating) factors, and a weak but persistent trend, a series must be sufficiently long in order for the change to be detected.

Based on the results of the climate models, it is projected that the water cycle will further intensify [6.32], with possible consequences to extremes. Some model-based projections for the future (e.g. [6.22]) show a clear increase in rain-caused flood hazard.

However, there is a discontinuity between some observations made so far, where increase in flood maxima is not evident (e.g. [6.19, 6.24]) and the increasing flood hazard in future projections.

Flood hazard tends to increase over many areas, due to a range of climatic and non-climatic impacts, whose relative importance is site specific. Apart from changes in climatic system discussed above, changes of relevance to flood hazard have also occurred in economic and social systems and in terrestrial systems (hydrological systems and ecosystems). Land-use changes, which induce land-cover changes, control the rainfall–runoff relations. Deforestation, urbanization, and reduction of wetlands diminish the available water storage capacity and increase the runoff coefficient, leading to growth in the flow amplitude and reduction of the time to peak of a flood triggered by a "typical" intense precipitation. However, the nature of a "typical" intense precipitation event has also been changing due to climatic reasons, becoming more intense in the warmer climate.

Furthermore, humans have been encroaching into unsafe areas thereby increasing the damage potential. Societies become more exposed, developing flood-prone areas (maladaptation). However, even an over-dimensioned and perfectly maintained dike does not guarantee complete protection, as it may not withstand an extreme flood (much higher than the design flood). When a dike is overtopped or when it breaks, the damage in the inundated areas is likely to be higher than it would have been in a levee-free case.

Climate-related changes in flood frequency are complex and dependent on the flood-generating mechanism (e.g. heavy rainfall vs. spring snowmelt), affected in different ways by climate change. Changes in future flood frequency in Europe depend on the generating mechanism. Increase in intense precipitation in the warming Europe is plausible, as shown in projections in this chapter, and is likely to lead to a rise of flood hazard. On the other hand, snowmelt and ice-jam floods are decreasing in many areas with growing temperature, so in the regions where floods can be caused by several possible mechanisms, the net effect of climate change on flood risk is not immediately clear. Flood risk is controlled by a number of non-climatic factors such as population growth, increase in wealth per capita, increasing vulnerability of property, risk ignoring behaviour, and compensation culture. It is clear that flat-rate statements on the direction of change in flood risk cannot be formulated in a general and ubiquitously valid way.

The inherent uncertainty in analysis of any set of extreme flood flows stems from the fact that accuracy of measurements of high floods is problematic. Rating curves are not available for the extreme flow range, hence extrapolation has to be made. Gauges are destroyed by the flood wave, observers evacuated, yet – indirect determination of the highest stage is often possible. Since, as stated in McCarthy et al. [6.21], the analysis of observations of extreme events is underdeveloped, there is a considerable merit in continuing change detection studies in order to identify a change sufficiently early and to react adequately.

Acknowledgments The present study is a background activity of the author within the WATCH (Water and Global Change) Project of the 6th Framework Programme of the EU.

References

6.1. P. Alpert, T. Ben-Gai, A. Baharad, Y. Benjamini, D. Yekutieli, M. Colacino, L. Diodato, C. Ramis, V. Homar, R. Romero, S. Michaelides, S. Manes, The paradoxical increase of Mediterranean extreme daily rainfall in spite of decrease in total values. Geoph. Res. Lett. **29**(11), 1536 (2002)

6.2. N. Arnell, L. Chunzhen, Hydrology and water resources. In *Climate Change 2001: Impacts, Adaptation, and Vulnerability*, ed. by J.J. McCarthy, O.F. Canziani, N.A. Leary, D.J. Dokken, K.S. White (Cambridge University Press, Cambridge, Intergovernmental Panel on Climate Change (IPCC), 2001)

6.3. A. Bronstert, Floods and climate change: interactions and impacts. Risk Anal. **23**, 545–557 (2003)

6.4. M. Brunetti, M. Maugeri, T. Nanni, A. Navarra, Droughts and extreme events in regional daily italian precipitation series. Int. J. Climatol. **22**, 543–558 (2002)

6.5. J.H. Christensen, O.B. Christensen, Severe summertime flooding in Europe. Nature **421**, 805 (2003)

6.6. H.J. Fowler, C.G. Kilsby, A regional frequency analysis of United Kingdom extreme rainfall from 1961 to 2000. Int. J. Climatol. **23**, 1313–1334 (2003)

6.7. C. Frei, C. Schär, Detection probability of trends in rare events: Theory and application to heavy precipitation in the Alpine region. J. Clim. **14**, 1568–1584 (2001)

6.8. F. Giorgi, X. Bi, J. Pal. Mean, interannual variability and trend in a regional climate change experiment over Europe. ii: climate change scenarios (2071–2100). Clim. Dyn. **23**, 839–858 (2004)

6.9. P.Y. Groisman, R.W. Knight, D.A. Easterling, T.R. Karl, V.N. Razuvaev, Trends in intense precipitation in the climate record. J. Clim. **18**, 1326–1350 (2005)

6.10. T.G. Huntington. Evidence for intensification of the global water cycle: Review and synthesis. J. Hydrol. **319**, 83–95 (2006)

6.11. A.M.G. Klein Tank, G.P. Koennen, Trends in indices of daily temperature and precipitation extremes in Europe 1946–1999. J. Clim. **16**, 3665–3680 (2003)

6.12. T. Kleinen, G. Petschel-Held, Integrated assessment of changes in flooding probabilities due to climate change. Clim. Change **81**, 283–312 (2006)

6.13. Z.W. Kundzewicz, A. Robson (eds.), Detecting trend and other changes in hydrological data. Wcdmp-45, wmo/td – no. 1013 (World Climate Programme Data and Monitoring, Geneva, 2000)

6.14. Z.W. Kundzewicz, D. Graczyk, T. Maurer, I. Piñskwar, M. Radziejewski, C. Svensson, M. Szwed, Trend detection in river flow time series: 1. annual maximum flow. Hydrol. Sci. J. **50**(5), 797–810 (2005)

6.15. Z.W. Kundzewicz, M. Radziejewski, I. Piñskwar, Too much – too little: Precipitation extremes in the changing climate of Europe. Clim. Res. **31**, 51–58 (2006)

6.16. Z.W. Kundzewicz, H.J. Schellnhuber, Floods in the IPCC TAR perspective. Nat. Hazards **31**, 111–128 (2004)

6.17. Z.W. Kundzewicz, U. Ulbrich, T. Brücher, D. Graczyk, A. Krüger, G. Leckebusch, L. Menzel, I. Piñskwar, M. Radziejewski, M. Szwed, Summer floods in central Europe—climate change track? Nat. Hazards. **36**(1/2), 165–189 (2005)

6.18. B. Lehner, P. Döll, J. Alcamo, H. Henrichs, F. Kaspar, Estimating the impact of global change on flood and drought risks in Europe: a continental, integrated assessment. Clim. Change. **75**, 273–299 (2006)

6.19. H.F. Lins, J. R. Slack, Streamflow trends in the United States. Geoph. Res. Lett. **26**(2), 227–230 (1999)

6.20. S. Manabe, P. C. D. Milly, R. Wetherald, Simulated long term changes in river discharge and soil moisture due to global warming. Hydrol. Sci. J. **49**(4), 625–642 (2004)

6.21. J.J. McCarthy, O.F. Canziani, N.A. Leary, D.J. Dokken, K.S. White (eds.), *Climate Change 2001: Impacts, Adaptation, and Vulnerability* (Cambridge University Press, Cambridge, Intergovernmental Panel on Climate Change (IPCC), 2001)

6.22. P.C.D. Milly, R.T. Wetherald, K.A. Dunne, T.L. Delworth, Increasing risk of great floods in a changing climate. Nature **415**, 514–517 (2002)
6.23. H.T. Mitosek, Climate variability and change within the discharge time series: A statistical approach. Clim. Change. **29**, 101–116 (1995)
6.24. M. Mudelsee, M. Börngen, G. Tetzlaff, U. Grünewald, No upward trends in the occurrence of extreme floods in central Europe. Nature **421**, 166–169 (2003)
6.25. T.J. Osborn, M. Hulme, Evidence for trends in heavy rainfall events over the U.K. Phil. Trans. Roy. Soc. A. **360**, 1313–1325 (2002)
6.26. T.N. Palmer, J. Räisänen, Quantifying the risk of extreme seasonal precipitation events in a changing climate. Nature **415**, 512–514 (2002)
6.27. P. Pilon, Detecting trend and other changes in hydrological data. WCDMP-45, WMO/TD No. 1013 (World Climate Programme Data and Monitoring, Geneva, 2000)
6.28. M. Radziejewski, Z.W. Kundzewicz, Detectability of changes in hydrological records. Hydrol. Sci. J. **49**(1), 39–51 (2004)
6.29. J. Schmidli, C. Frei, Trends of heavy precipitation and wet and dry spells in Switzerland during the 20th century. Int. J. Climatol. **25**(6), 753–771 (2005)
6.30. C. Svensson, Z. W. Kundzewicz, T. Maurer, Trend detection in river flow series: 2. flood and low-flow index series. Hydrol. Sci. J. **50**(5), 811–824 (2005)
6.31. G. Vaes, P. Willems, J. Berlamont, 100 years of rainfall registration: are there trends? Water Sci Technol. **45**(2), 55–61 (2002)
6.32. R.T. Wetherald, S. Manabe, Simulation of hydrologic changes associated with global warming. J. Geoph. Res. **107**(D19), 4379 (2002)

Warta river, near Wronki. Photo by Robson at obiezyswiat.org (haveseen.org)

Chapter 7
About Trend Detection in River Floods

Maciej Radziejewski

This chapter reviews some methods for studying changes in the occurrence of extreme river flows. Some general background information on change detection is given. Changes in various characteristics of high flows are investigated using a variety of approaches. Although many cases of statistically significant changes are found, they go in different directions and are of different character in different regions. There does not seem to be a clear, uniform signal of change, except for, possibly, the occurrence of most extreme flows that seems to be increasing in the recent decades.

7.1 Introduction

Reliable detection of changes in natural processes is important because major changes often threaten our further survival or quality of life, and because major anthropogenic changes bear the moral aspect of human responsibility even if one is not directly affected by them. Changes in extremes are the most important ones because of their potential consequences. They are also the most difficult ones to detect, because extreme events are rare by definition.

Flood risk and flood management are among the major concerns related to river flow. While it is usually possible to quantify flood risk based on historical records, the risk may change, e.g., in response to climate change. Large temperature rise has already been observed worldwide [7.7, p. 5]. Increase in the frequency of heavy precipitation events, projected over many areas in the 21st century [7.7, p. 8], may contribute to more frequent flood events [7.8, p. 11]. However, no global and uniform increasing trend in flood occurrence has been found so far [7.12, 7.16, 7.20, 7.27]. The common impression of floods becoming more frequent is attributed to increased media coverage whereas the increase in flood losses is partly explained by economic

M. Radziejewski (✉)
Faculty of Mathematics and Computer Science, Adam Mickiewicz University, 61-614 Poznań, Poland; Institute for Agricultural and Forest Environment Polish Academy of Sciences, 60-809 Poznań, Poland
e-mail: maciejr@amu.edu.pl

J.P. Kropp, H.-J. Schellnhuber (eds.), *In Extremis*,
DOI 10.1007/978-3-642-14863-7_7, © Springer-Verlag Berlin Heidelberg 2011

growth, population growth, and human encroachment into flood plains [7.12]. Temperature change influences river flows indirectly and in a complex way. Increasing precipitation and increasing evapotranspiration have opposite effects on the flow process. The net results may differ between regions and they need not be visible, or detectable, until sufficiently long records of data are collected [7.12, 7.25]. Warming leads to less severe spring snowmelt and ice-jam floods contributing to downward trends for floods in the winter season [7.1, 7.11, 7.20].

7.2 Methods

7.2.1 Testing of Significance

The question whether or not an observed natural process exhibits changes presents some theoretical difficulties. The concept of change in such a process builds upon the assumption that some kind of constantness or repeatability is possible in the phenomenon that we study. Change is a negation of such constantness. The usual assumption is that in the absence of changes the natural process in question may be described as a (possibly complex) stochastic process that admits some randomness, but some of its properties do not change over time. This is already a simplification, although it is generally accepted without question. The main problems considered in change detection are as follows: what kind of stochastic process matches the natural process studied and whether or not observations of the natural process support the conclusion that the "underlying" stochastic process is subject to changes.

Most natural processes that evolve in time exhibit some changes in observed statistical properties estimated from samples. Such changes, inherent from the internal structure of a system, are termed natural variability. In change detection one is interested in changes in the process, as opposed to natural variability. This distinction can never be sharp, because for a stochastic process with enough randomness apparent changes of arbitrary magnitude are possible, only very unlikely, under natural variability alone.

The method of significance testing (cf. [7.15] and also [7.13, 7.14]) offers an unsharp distinction between "real changes" and natural variability. Instead of answering the question of the presence of changes one states how unlikely a change this large would be under natural variability alone. Thus one considers the null hypothesis of no change (only natural variability) and an alternative hypothesis of a specific kind of change. The null hypothesis, denoted H_0, should reflect the properties of the process in question, i.e., the statistical description of the process adopted by the researcher. Mathematically it defines a family of random processes. The alternative hypothesis, H_1, also defined as a family of processes, should be plausible and supported by initial exploratory analysis of the available data. The strength and possibly the direction of changes in a data set D are quantitatively measured by a test statistic $S(D)$. In a two-sided (two-tailed) test large values of $S(D)$ usually indicate an increase, and low ones a decrease of whatever $S(D)$ measures. In a one-sided (one-tailed) test only large (say) values of $S(D)$ indicate changes and low values lack of changes.

For a one-sided test with large values indicating change we say that the null hypothesis is rejected, and the alternative hypothesis supported, on a significance level α, where $0 < \alpha < 1$, if the probability of this large or larger change occurring for any process defined by the null hypothesis is smaller than α. It means that the null hypothesis is rejected if

$$P(S(D') \geq S(D)) \leq \alpha, \tag{7.1}$$

for all random processes in H_0, where D' denotes a random realization of the process. For a two-sided test one would like to detect the presence of not only changes but also their direction, i.e., the alternative hypothesis is split into two. One of the possible approaches is to reject the null hypothesis in favor of that of increase if

$$P(S(D') \geq S(D)) \leq \frac{\alpha}{2} \tag{7.2}$$

for all processes in H_0, with D' as before. Accordingly, the null hypothesis is rejected in favor of that of decrease if

$$P(S(D') \leq S(D)) \leq \frac{\alpha}{2} \tag{7.3}$$

for all processes in H_0, with D' as before. A result of $S(D)$ leading to the rejection of H_0 is called significant. The significance level is selected arbitrarily (usually 5 or 1%) and guarantees that the probability of rejecting H_0 when it is true (type I error) is at most α. In other words, the probability of obtaining a false positive in the test is at most α, provided that the null hypothesis is stated correctly. One can also consider an observed significance, or p-value, i.e., the smallest significance level on which the result is significant (equal to 1 if the result is not significant at any level). A statistical test consists of a defined null hypothesis, alternative hypothesis, a test statistic, and a specified significance level.

The choice of the class of stochastic processes adopted by the researcher may lead to underestimation or overestimation of the likely magnitude of natural variability, and consequently to overdetection or underdetection of changes (cf., e.g., [7.5] for a discussion of this issue). In most cases one does not have a perfect mathematical model to completely describe the process in question, including all possible states of the system of which the process is part (e.g., the climatic system). Hence there is not enough a priori knowledge to formulate perfect null and alternative hypotheses (H_0 and H_1). It is possible to construct an adequate H_0 based on the analysis of data in question, although it leads to a paradox: detecting changes means precisely that the data violates H_0, so, logically, it cannot completely support it at the same time. In practice it is up to the researcher to judge which of the known properties of the data to use for constructing H_0 or how to select a subset of data unaffected by changes.

Estimation of the probability of type II error (accepting H_0 when it is false) and the complementary probability of correctly rejecting H_0 (called the power of the test) are possible if the alternative hypothesis defines a parametric family of stochastic processes. The power is a function of the parameter(s) of H_1 and it is defined for given H_0, H_1, and test statistic. In practice one often cannot give an exact description of how the process might change, so H_1 cannot be stated explicitly. Instead one simply looks for changes of several types using appropriate test statistics and indices or other transformations of data. Without knowing all possibilities of how the process might evolve one also cannot claim that a detected trend will continue in the future. Detecting the direction of changes opens a possibility for a third type of error: detecting increase in a decreasing process or vice versa. Intuitively, consistency between the notions of change measured by $S(D)$ and reflected in H_1 should ensure that the probability of such an error never exceeds $\frac{\alpha}{2}$ and this is implicitly assumed.

The result of a test may be conveniently expressed by the "trend index (TI)" (cf. [7.12] and also [7.25]) that encompasses the p-value and (in case of two-sided tests) the apparent direction of changes. For one-sided tests the trend index is simply

$$\mathrm{TI} = 1 - p, \tag{7.4}$$

where p is the p-value. For two-sided tests the sign of TI corresponds to the direction of change (positive indicates an increase and negative a decrease) and the absolute value is as above:

$$|\mathrm{TI}| = 1 - p. \tag{7.5}$$

If the distribution of $S(D')$ is the same for every random process in H_0 then this definition simplifies to $\mathrm{TI} = P(S(D') < S(D))$ for one-sided tests. For two-sided tests in that case one has $\mathrm{TI} = 2P(S(D') < S(D)) + P(S(D') = S(D)) - 1$, with possible correction toward 0 by at most $P(S(D') = S(D))$ if the latter probability is non-zero due to a discrete nature of $S(D')$. For two-sided tests the value of TI is always between -100 and $+100\%$, with negative values indicating a decrease and positive values an increase. For one-sided tests TI is in the range from 0 to $+100\%$. This measure puts results of different tests on a common scale. The p-value always equals $1 - |\mathrm{TI}|$, so one can readily see if the result is significant on any desired level. For example, for a two-sided test, a negative trend is significant at the 5% level precisely if $\mathrm{TI} < -95\%$. A value of $\mathrm{TI} = +99.14\%$ implies a strong increase, significant on the 1% level.

7.2.2 Resampling

Resampling means generating random data from given data. It is a flexible method of assessing the significance of changes and computing the trend index. Given a statistic $S_0(D)$ that quantitatively measures change, and a resampling method, the procedure for computing the p-value and TI is straightforward: reference data D' are generated randomly from given data D, using the selected method, a large number of times, e.g., 10 000 times. Then TI is estimated using the formulas

$$\mathrm{TI} = P(S_0(D') < S_0(D)) \tag{7.6}$$

or

$$\mathrm{TI} = 2P(S_0(D') < S_0(D)) + P(S_0(D') = S_0(D)) - 1, \tag{7.7}$$

respectively, for one-sided and two-sided S_0.

This procedure fits quite well into the general testing framework outlined at the beginning. The choice of the resampling method directly corresponds to the choice of H_0, namely, each process in H_0 must be invariant upon resampling. TI estimated above will play the role of the test statistic $S(D)$ and, since it is approximately uniformly distributed for each process in H_0, it will indeed be equal to the trend index in the resampling test. For example, the simplest resampling methods for use with time series are permutation testing and bootstrapping (sampling with replacement), cf. [7.6, 7.14]. They correspond (roughly) to the null hypothesis of series values being independent and identically distributed. The "block" versions of these methods involve permuting or resampling the data in blocks to account for seasonality and correlations within the block size.

In [7.23] an alternative method was developed, based on randomization of phases in the Fourier transform of the series. It can be applied to long time series of river flows of annual, monthly, and even daily resolution. It involves the following steps:

1. Standardization:
   (a) Normal (or probit) scores are computed for the entire time series, i.e., for each term in the time series its percentile rank is computed and the term is replaced with a corresponding quantile of the standard normal distribution. This way the relative ranks of the data are preserved and the sample distribution becomes normal.
   (b) Deseasonalization of the series is performed by subtracting the seasonal mean and dividing by seasonal standard deviations. This step is omitted in case of annual series.
2. Phase randomization:
   (a) Fourier transform of the time series is computed.
   (b) The phase spectrum is replaced by independent random phases. The power spectrum is kept unchanged.
   (c) Inverse Fourier transform is computed.

3. Reversing standardization:

   (a) Data are multiplied by original seasonal standard deviations (from 1b) and then the original seasonal means (again, from 1b) are added to it. This step is omitted in case of annual series.
   (b) For each term in the randomized time series its percentile rank is computed and the term is replaced with a corresponding value from the original series having the same percentile rank.

As a result we obtain a time series with the same sample distribution as the original, almost the same seasonal means (regime) and seasonal standard deviations and almost the same autocorrelation function. However, any systematic trend in the data will dissipate, because introducing random phases makes the probabilistic random series obtained in step 2 invariant under circular shifts (dividing the series into two parts and swapping them) and reversing direction. The choice and order of standardization procedures was worked out in a case study reported in [7.21]. In particular the standardization step 1a is necessary, because phase randomization tends to make the distribution normal, so it should be applied to a normally distributed series. Some studies [7.9, 7.17] (cf. also Chap. 13) also note that other information is lost in phase randomization without step 1a (and 1.3b). It is unknown to the author if this problem affects the procedure described above, but in any case it does not seem to be directly related to trend detection.

Phase randomization is a convenient alternative to methods like block permutation and block bootstrapping, as it requires no additional parameters (like block size); it uses all the available information in the time series; and it can deal with very strong and very long-term autocorrelation. It is particularly useful with daily time series of river flow, as these time series may possess strong autocorrelation. If a trend or step change is present in the data, it will cause increase of low-frequency components in the power spectrum; hence, the probability of obtaining a large increase/decrease in the randomized series will be higher than it would have been without a trend. In other words the presence of a trend may cause overestimation of natural variability and underestimation of TI. In contrast, the use of methods like permutation or bootstrapping, and their block versions with small block size, is likely to result in greater power (smaller probability of type II error), but underestimated probability of type I error.

7.2.3 Tests for Changes

Tests may be divided into parametric, where H_0 and H_1 are parametric families with finite-dimensional parameter spaces (e.g., with finitely many numeric parameters), and non-parametric, where this is not the case. A large family of parametric tests for changes in time series includes the null hypothesis of series values being independent and identically distributed, with a specific form of distribution, whose parameters need not be known. The alternative hypothesis then assumes that the series is affected by a specific type of trend, whose parameters (strength, time of occurrence, some shape parameters, etc.), again, need not be known. As long as the hypotheses are adequate for the problem the test may be performed and, in addition, trend parameters may be estimated if a trend is detected. In many cases a specific form of distribution cannot be assumed. The easiest solution is to define the test statistic based on ranks, normal scores, or the standardized series (see Sect. 7.2.2) and thus make it independent of the underlying distribution. This gives rise to another, non-parametric, family of tests. A test is called distribution dependent or distribution free, depending on whether or not it relies on distributional assumptions. The notions of parametric and distribution-dependent tests (respectively, non-parametric and distribution-free tests) are often considered equivalent, precisely because of the above-mentioned two families of tests.

Resampling (see Sect. 7.2.2) is another method of relaxing test assumptions. Given a statistic it allows for the construction of a test, which is, for most resampling methods, non-parametric and distribution free. The use of resampling alleviates the need for normalizing factors that, in case of parametric tests, are needed to make the statistic independent of distribution parameters. Computation of probabilities in resampling plays essentially the same role. This makes it easier to develop new kinds of tests suited to the kind of change one expects. This is also the reason why different statistics may turn out equivalent when resampling is used. Resampling was used in all the studies presented here. The statistics are based on well-known tests or least-squares fitting of a trend:

1. Linear regression: The statistic is the correlation coefficient (also called Pearson's r) of the time variable and the observations, cf. [7.14].
2. Normal scores linear regression or standard scores linear regression: Linear regression is applied to the series of normal scores or, in case of daily data, to the standardized series, cf. [7.14].
3. Spearman's rank correlation: Linear regression is applied to the series of ranks, cf. [7.14].
4. Mann–Kendall's test: This is a robust non-parametric test based on the "tau" statistic introduced by Kendall and later adapted by Mann to time series analysis [7.10]. The test statistic is Kendall's sum:

$$S = \sum_{1 \le i < j \le n} \operatorname{sgn}(a_j - a_i), \tag{7.8}$$

   where n is the length of the time series and $a_1, a_2, \ldots, a_n$ denote the time series values.
5. Jump fitting by the least-squares method: A step-shaped change (abrupt change in mean) is fitted to the time series using the least-squares method. The statistic used is the difference in means between the two periods. This statistic is related to Worsley's likelihood ratio [7.2, 7.4], except it includes the information on the direction of changes.

6. Jump fitting to normal or standard scores: Jump fitting is applied to the series of normal scores or, in case of daily data, to the standardized series.
7. Jump fitting to ranks: Jump fitting is applied to the series of ranks.

The statistics listed here measure a change in the mean, but by applying them to a series of appropriate indices, or to a weighted series, or to other kinds of derived series, one obtains a new statistic that measures changes of different kinds or changes occurring in a specific season. Both gradual changes (linear or other trend) and step-shaped changes in time series may be detected by any of these tests; however, the test nos. 1–4 are better suited (in terms of power) to detect a gradual change, while test nos. 5–7 are more sensitive to step-shaped changes [7.24, 7.25]. Therefore, the test nos. 1–4 may be called "trend tests," and test nos. 5–7 "step change tests."

There are numerous statistical tests for changes developed, and there is no clear "winner" among them, as every test is sensitive to a slightly different kind of changes. While it is not feasible, nor desirable, to run all known tests, it is good practice to use several tests when one does not know exactly what kind of change to expect. In the studies described here, when several related tests were applied to a number of data sets, the number of significant trends is summarized in the following way:

1. Changes detected by a minority of tests (e.g., by one of three), not confirmed by other tests even on a weaker significance level, are discarded altogether.
2. Number of significant trends is counted for each test.
3. The range of numbers of significant trends are reported.

For example, 10–12 significant trends detected by 3 tests on significance level 5% may mean that 9 trends were detected by all 3 tests, 1 trend detected by 1 test (with 2 others showing changes in the same direction on a greater significance level), and 2 more were detected by 2 other tests. Such an approach provides a good summary of trends found excluding unconfirmed results.

7.3 Trends in Time Series of River Flows

In this section some results regarding trends in time series of river flows without regard to seasons are presented. Svensson et al. [7.27] studied changes in numerous annual indices and peaks over threshold (POT) series. They used numerous daily flow series (not standardized) from GRDC data holdings [7.3]. They have found no evidence of overall increase in floods/dry spells and attributed the lack of such evidence possibly due to the increasing number of operational reservoirs, as "low flows are augmented and flood flows are mitigated downstream of the reservoir," cf. [7.28]. They also note that there were more significant trends found in annual maxima series than in POT series.

Kundzewicz et al. [7.12] studied 195 long time series of worldwide annual maximum flows stemming from GRDC data holdings [7.3]. The time series were selected based on data quality (no missing values in critical periods), sufficient record length, coverage of the recent decades, catchment area (including small catchments), and geographic distribution. Trends were found (significant on the 10% level) in 58 time series. In the European time series (70) trends were found in 20 and in North American (70) trends were found in 26. The number of significant results is much larger than what could be expected, but the trends are more-or-less symmetrically divided into increasing and decreasing (Fig. 7.1). In an earlier study of flows of

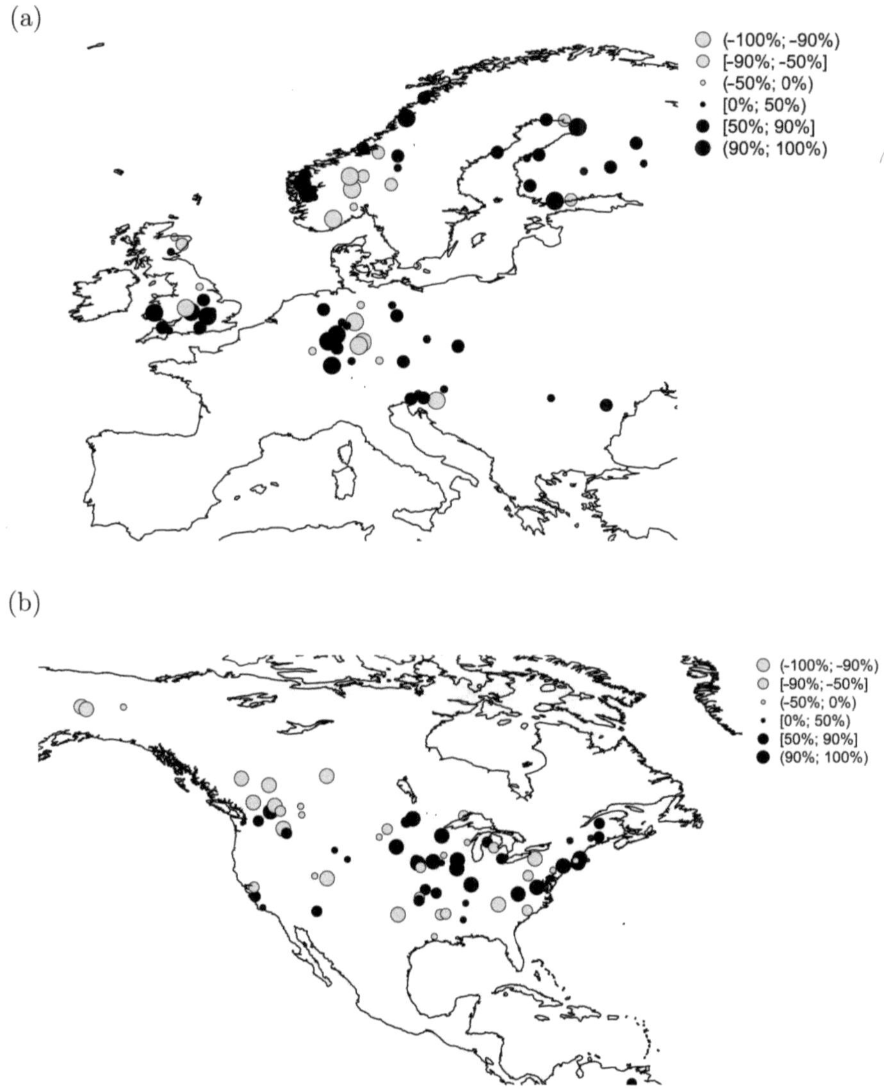

Fig. 7.1 Changes in annual maximum flows over Europe (**a**) and North America (**b**) according to Mann–Kendall test. *Circle sizes* correspond to the trend index. *Black circles* indicate increase, *gray* – decrease. *Large circles* denote changes significant at 10% level. Reproduced from [7.12]

the river Warta at Poznań gauge [7.24] high flow events (POT) in the standardized series of daily flows were considered. One of the objects of the study was the structure of sets of days with high flows. The "box-counting" fractal dimension, or packing dimension [7.18], when applied to subsets of the real line (e.g., time) measures clustering of points on all scales. Although not directly applicable to the discrete set of days, the concept may be used to describe fractal-like clustering of events on a certain range of scales. The study has shown that the sets of days with value over a given high threshold in the standardized series of Warta flows exhibit fractal-like clustering in time scales from 10–15 days to 2–4 years, depending on the threshold. The fractal dimension corresponding to such scaling depends on the selected high flow threshold and, generally, the higher the threshold the lower the dimension. An attempt to analyze nonstationarity of this characteristic of POT was made by computing POT dimensions for a number of thresholds for different 30-year windows along the standardized time series. In addition, the Hurst coefficient was computed for each window based on the rescaled adjusted range [7.19]. It was found that changes in the dimension correspond to changes in the Hurst coefficient, as shown in Fig. 7.2. However, these changes look rather like oscillations, with no clear trend or change point. In [7.23] 202 daily flow time series of US rivers from the HCDN database [7.26] were tested for changes in the mean. The data were chosen based on the criteria of quality and length of continuous daily record (minimum 60 years). Changes in daily flow were studied using phase randomization and the seven tests described in the previous section. Tests based on non-parametric tests (test nos. 1–3 and 5–7) were applied to standardized data (obtained by computation of normal scores and deseasonalization). Tests based on parametric tests were applied to original data. Significant increases on the 5% level were found in 12–14 time series (depending on the test) in the continental United States using three tests: Mann–Kendall's test, normal scores linear regression, and Spearman's rank correlation; see Fig. 7.3. There were also three significant decreases detected by these tests,

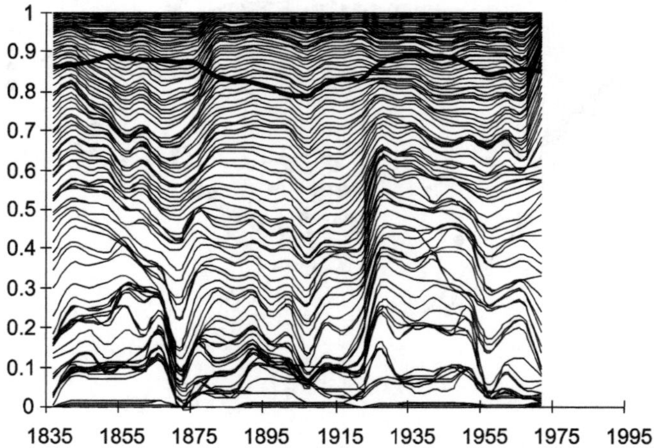

Fig. 7.2 Nonstationarity of clustering of high flows of the river Warta at Poznań gauge and the Hurst coefficient. *Thin lines*: Fractal dimensions corresponding to different thresholds. *Thick line*: Hurst coefficient. The x-axis denotes the center of a 30-year window. Reproduced from [7.24]

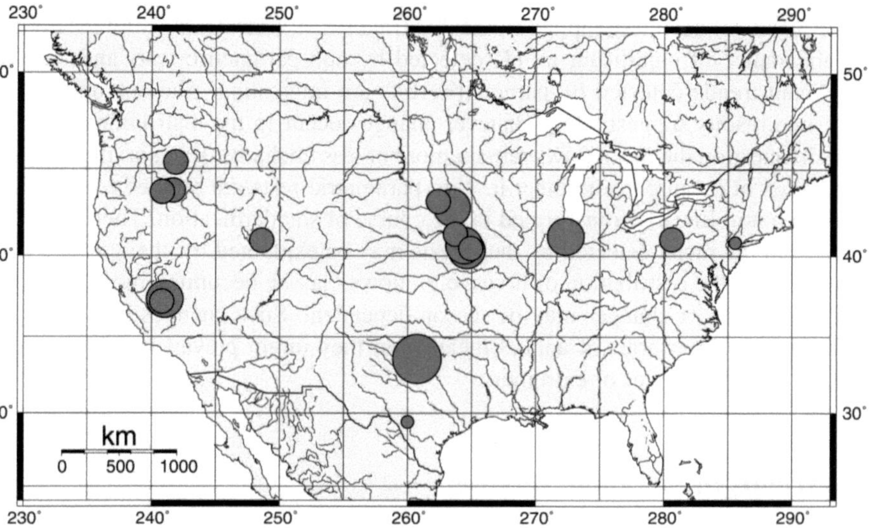

Fig. 7.3 Significant changes in US streamflow, detected by at least one of three tests: Mann–Kendall's test, normal scores linear regression, and Spearman's rank correlation. *Circle size* (from smallest to largest) corresponds to TI $\geq +90, +95, +98, +99\%$

in the flows of rivers located in Hawaii, cf. also Table 3 in the paper cited for the complete account of these results. In comparison to two parametric tests (applied to original data) the non-parametric ones uniformly show a greater than expected number of increases while the two parametric tests give a less than expected number of significant results. This can be readily seen on Fig. 7.4 that shows the distribution

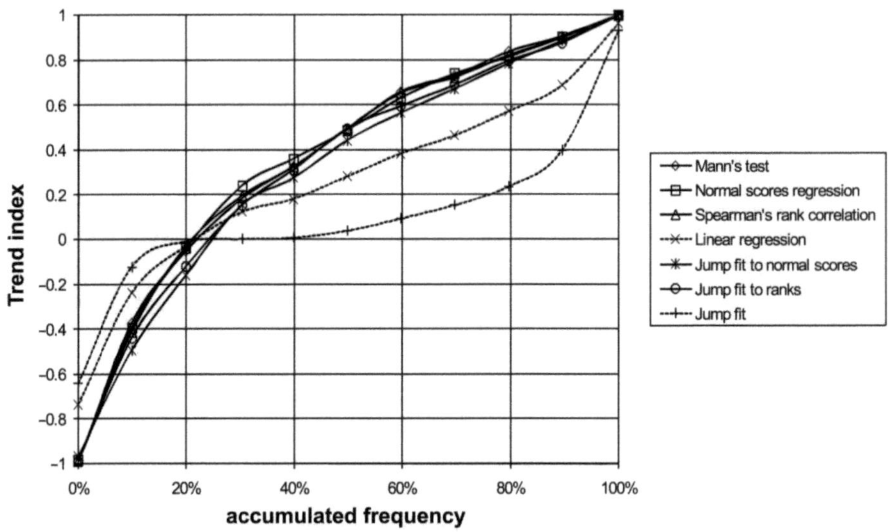

Fig. 7.4 Quantiles of test results (TI) of 7 tests applied to 202 flow series selected from HCDN. The TI was computed using phase randomization. *Dotted/dashed lines* were used where the statistics were taken from parametric tests, for easy distinction. Reproduced from [7.23]

of the results of each test applied to the 202 time series considered. Both the apparent abundance of significant trends detected by non-parametric tests and lack of significant trends detected by parametric tests are interesting. The conjunction of these two results suggests that whatever changes occur in the distribution of river flows in question the mean flows remain more or less unchanged, so parametric tests for change in mean cannot detect it. Non-parametric tests are applied to data that have been non-linearly transformed in the course of standardization. Therefore, the changes in the mean detected by these tests may be explained by changes of a different kind in the distribution of flows, e.g., flows may be becoming more extreme, or less extreme, or changes may be season dependent. Such changes would not be visible in the time series of annual means, but they might prevail in other annual indices, e.g., the indices of extremes.

7.4 Trends in Seasonal Maxima

It has been observed [7.5, 7.12] that regional change in timing of floods occurs in many areas. For example, there are increasing late autumn and winter floods worldwide, and less ice-jam-related floods in Europe. The results described in this section and in the following section were also reported on in [7.22]. In this section an attempt is made to separate different types of floods by considering high flows in different 3-month seasons:

OND (Oct–Nov–Dec), corresponding to floods caused by rain in autumn/winter,
DJF (Dec–Jan–Feb), corresponding to ice-jam floods,
FMA (Feb–Mar–Apr), corresponding to snowmelt floods, and
JJA (Jun–Jul–Aug), corresponding to rain-caused floods.

Such selection of periods is suitable for studying floods specifically in the European continent.

Time series of daily flows were tested for trends or changes in seasonal maxima. The study was based on the set of 70 European stations used in [7.12]. Only the data in the period 1961–2000 were considered. The seasonal maxima were computed from original (not standardized) daily records. Three different tests for changes were used: Mann–Kendall's test, Spearman's rank correlation, and linear regression. The significance of changes was computed using permutation testing (2 500 permutations). All the test results are quoted using the trend index.

Derived time series of seasonal maxima of daily flow (OND, DJF, FMA, resp., JJA) were studied for each time series selected, i.e., 4 records of 40 values for each series. Out of 17 time series from Germany, increasing trends in DJF maxima were found in 6 time series. Table 7.1 lists the test results and characteristics of these six time series. There was also a fair number of increases in maximum DJF and FMA flows in Norway: trends were found in 6 time series (of the 15 selected) by all tests (in 7 by at least some tests). All the results for DJF and FMA flow in Norway

Table 7.1 Significant changes in DJF maxima, Germany. Seasonal maxima (Dec–Jan–Feb) were computed from daily data and tested for change in mean. Test results are given as the values of trend index, between −1 (strong decrease) and 1 (strong increase)

Station	Catchment area (km^2)	Mann–Kendall's test	Spearman's rank correlation	Linear regression
Danube, Hofkirchen	47 496	0.990	0.991	0.993
Danube, Ingolstadt	20 001	0.996	0.997	0.998
Diemel, Helminghausen	103	0.986	0.978	0.893
Kinzig, Schwaibach	954	0.957	0.951	0.950
Rhine, Kaub	103 488	0.991	0.990	0.993
Rhine, Worms	68 827	0.998	0.998	0.998

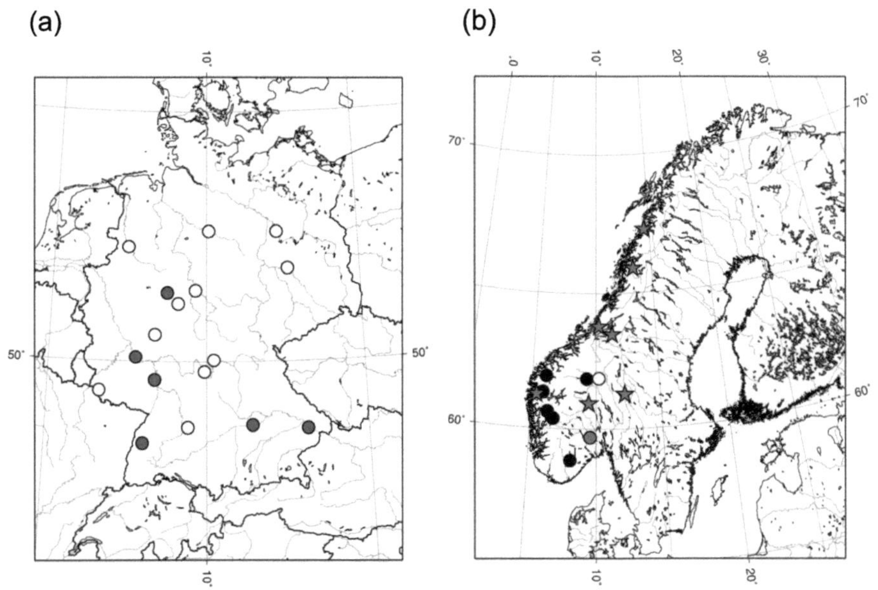

Fig. 7.5 Trends in seasonal maxima: (**a**) Germany, *dark circles* – increase, *white* – no increase; (**b**) Norway, *black circles* – increase in DJF and FMA, *stars* – no significant changes in DJF or FMA, *gray circles* – increase in DJF, *white circles* – decrease in DJF and increase (according to linear regression) in FMA

are given in Table 7.2. The geographical distribution of stations referenced in both tables is shown in Fig. 7.5.

Decreases in maximum FMA flows were found in three (out of five) time series from Slovenia. The number of series from other countries where trends were detected is within the margin of error, although it must be noted that some of the trend indices were quite large. To conclude, the number of trends found is greater than expected to occur by chance, and the number of trends is particularly large in Germany and Norway. However, again, there is no clear and uniform pattern. It must be noted that the records used in the study lack detailed metadata, so it was

Table 7.2 Significant changes in DJF and FMA maxima, Norway. Seasonal maxima (Dec–Jan–Feb and Feb–Mar–Apr) were computed from daily data and tested for change in mean. Test results are given as the values of trend index, between −1 (strong decrease) and 1 (strong increase)

Station	Catchment area (km^2)	DJF Mann–Kendall's test	DJF Spearman's ρ	DJF Linear regression	FMA Mann–Kendall's test	FMA Spearman's ρ	FMA Linear regression
Austena, Austena	286	0.997	0.994	0.990	0.990	0.995	0.998
Eidselv, Hornindalsvatn	387	0.950	0.950	0.896	0.980	0.992	0.995
Gaular, Viksvatn	505	0.989	0.990	0.979	0.995	0.997	0.999
Kinso, Holen	229	0.960	0.978	0.945	0.998	0.999	0.999
Otta, Lalm	3 982	0.959	0.946	0.974	0.999	0.999	0.998
Vosso, Bulken	1 102	0.997	0.997	0.998	0.997	0.998	0.999
Jondalselv, Jondal	150	0.988	0.995	0.994	0.078	0.069	0.314
Atna, Atnasjo	465	−0.961	−0.967	−0.935	0.687	0.744	0.984

not possible to discriminate between the effects of changes in land use, effects of reservoirs operation, and impacts of climate change on the results.

7.5 Seasonal Peaks Over Threshold

Seasonal occurrence of extreme flows was analyzed for the same time series and the same seasons as in the previous section using the POT approach. For each of the time series included in the study (daily series, without standardization) peaks over threshold were found in the following way:

- Independent peaks were identified as flows larger than any other flows within a given time horizon, based on catchment area, as in [7.27].
- A threshold was chosen in order to obtain a required average number of independent peaks per year. Four classes of "severity" were used: three peaks per year, one per year, one per 3 years, one per 9 years.

As a result, for each time series and threshold a sequence of "peaks" was constructed that included dates and values of the selected peaks. In order to visualize seasonal occurrence of extreme flows the time of year of each peak was plotted against the year of occurrence. Taking several time series from a given region together it is possible to observe a shift in seasonal occurrence of floods over the decades. All the series shown on such a plot had to cover the same period of time, otherwise a change in the number of floods might be falsely perceived. The graph obtained for Germany (Fig. 7.6a) does seem to have some structure, but no obvious changes in the occurrence of floods are visible. It may be possible to notice the increase in DJF floods, but it is unclear. The graph for Norway (Fig. 7.6b) clearly shows two "wet seasons" in May–June and September–November. However, the largest peaks do not seem at all confined to these seasons. It may also be noticed that over the last decade the wet seasons are not as clearly marked and the frequency of middle-severity peaks outside of the wet seasons (in particular December–April) seems to be greater than before. This is another manifestation of increasing DJF and FMA flows in Norway.

For each of the time series included in the study derived series of seasonal peaks over threshold (one series for each original time series, season, and threshold) were constructed by including only the peaks within a given season. Frequency and size of seasonal peaks were subject to trend analysis with the methods described in the previous section. Frequency was studied by counting the peaks in each season and size by computing the height of the largest peak in a season (missing value was taken if there were no peaks in a season). It must be noted that the usual practice is to consider peaks of frequency "three per year" or sometimes "one per year." More severe peaks are rare by definition and it is almost impossible to obtain strong results based on them. Nevertheless, it seems essential to pay special attention to the "most extreme extremes" in a flood hazard analysis and to study them in concert with larger sets of extremes. It may be necessary to develop dedicated methods,

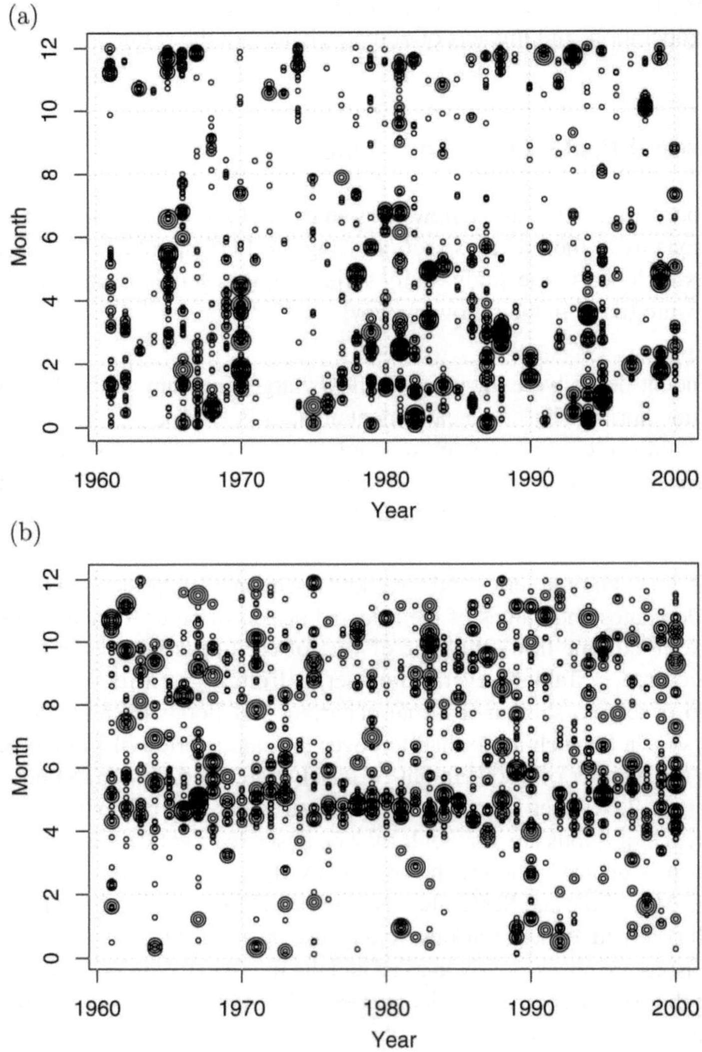

Fig. 7.6 Years and seasons of occurrence of extreme flows: (**a**) Germany; (**b**) Norway; *y-axis values* correspond to the month and day of month, i.e., the interval 0–1 is January, 1–2 February, etc. *Sizes of circles* depend on the relative size of the peak, i.e., whether it is included in one per 9-year peaks, 1/3 years, 1/year, or 3/year peaks

more powerful than those presented here, to afford the study of this kind of extreme events.

Similar to [7.27], the number of significant results obtained for POT series was considerably smaller than for seasonal maxima. There were several increases (3–4, 1, 3–4, 3) in the number of OND peaks of different severity classes (resp., three per year, one per year, one per 3 years, one per 9 years) in Great Britain. There were also four (resp. three) increases in the number of three-per-year (resp. one-per-year) FMA peaks in Norway. Other results are scarce and scattered and do not seem to form any pattern. In general the results (not all quoted here) suggest that trends in maxima are to some extent reflected in the frequency of peaks, but not always. An interesting example is Danube river at Hofkirchen gauge. It shows a strong increasing trend in DJF maxima (see Table 7.1 and Fig. 7.7a) but no significant trends in the number or frequency of DJF peaks of any severity (Fig. 7.7b shows the peaks).

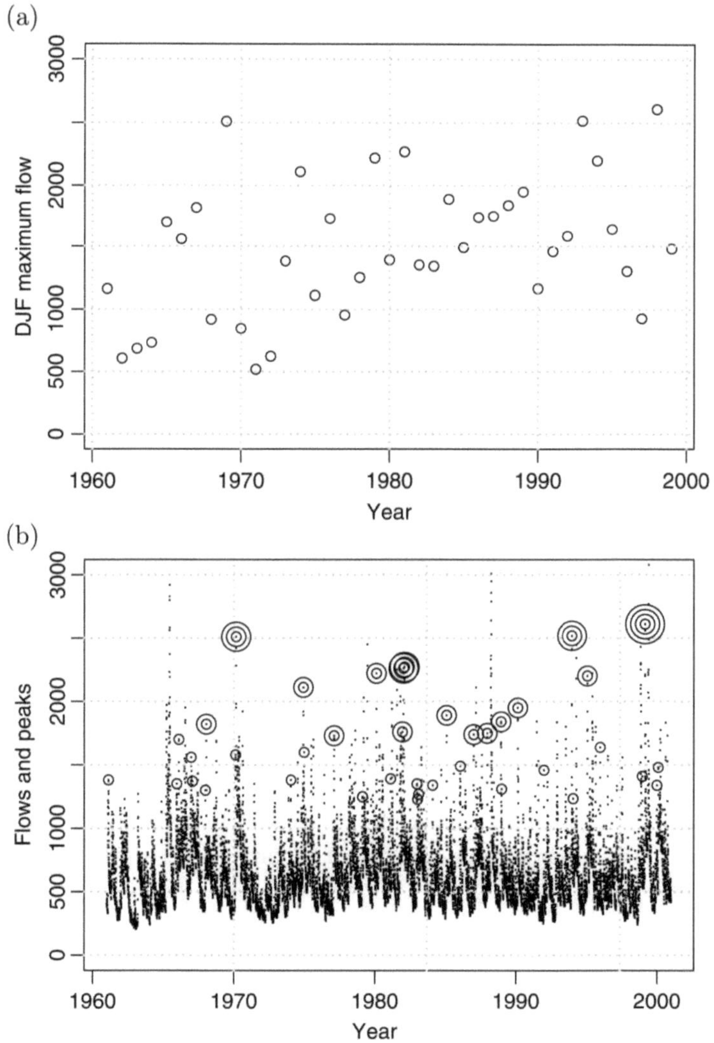

Fig. 7.7 Danube river, Hofkirchen, 47 496 km^2: (**a**) DJF maxima, (**b**) the daily series and DJF peaks. *Sizes of circles* depend on the relative size of the peak, i.e., whether it is included in one per 9-year peaks, 1/3 years, 1/year, or 3/year peaks

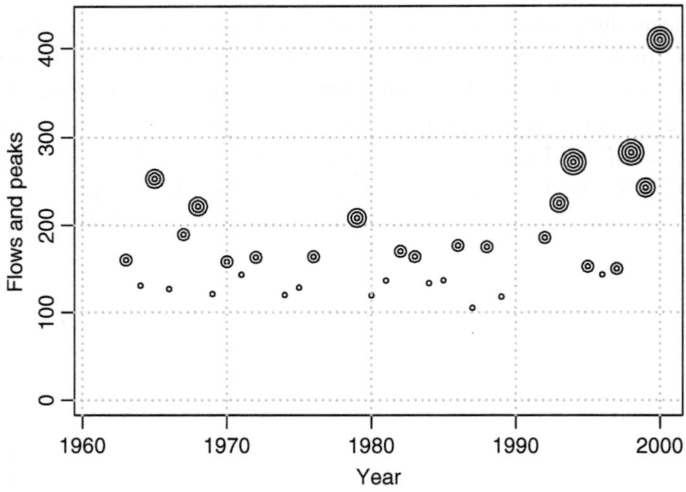

Fig. 7.8 Taw river, Umberleigh, UK, 826 km^2, annual maximum size of OND peaks. Only one test (*linear regression*) gave a significant result, for 3/year peaks. Other results are up to 0.91. *Sizes of circles* depend on the relative size of the peak, i.e., whether it is included in 1/9-year peaks, 1/3 years, 1/year, or 3/year peaks

Very few significant results were obtained of trends in peak size. The method used for this purpose is complementary to studying the changes in frequency and almost ignores changes in peak frequency, so it is not very powerful. Still, the example of Taw river, Umberleigh, UK (Fig. 7.8), shows that a visible trend in peaks size may escape statistical testing, as the issues of data scarcity and short duration of change come into play. It should be noted that one has to be very careful about interpreting any statistical results for a small number of values, as in the case of one-per-3-year and one-per-9-year peaks. Even three-per-year peaks may be very scarce in the seasonal approach, where the threshold is chosen based on the entire series, not adjusted for different seasons.

7.6 Most Extreme Flows

Kundzewicz [7.12] noted that out of the 70 daily flow series of European rivers, described in the previous sections, the global maximum over 1961–2000 occurs:

- in the period 1961–1980 for no more than 24 series and
- in the period 1981–2000 for at least 46 series.

The exact number could not be given there because only the data until 1998 were available for five of the series. A question arises: striking as it may be, is this result statistically significant? The simplest choice for the test statistic $S(D)$ here is the difference in the number of absolute maxima between the second period and the first period. The data set D to be resampled consists of 70 time series of 40 years each.

The data may be resampled 10 000 times according to a selected null hypothesis. Comparisons of the test statistic for the original and resampled data let one compute the trend index. The only remaining step is to choose the null hypothesis and the resampling method. As noted in Sect. 7.2, the choice of the null hypothesis plays the key role in determining the significance of changes and the TI, and this is also demonstrated here.

For the first case the null hypothesis of no spatial and no inter-annual dependence is considered. Then resampling may involve random permutations of each series independently, with block size of 1 year. For a random resampled data set D' the probability of having exactly n maxima in the second half of the 40-year period in D' (hence $m = 70 - n$ maxima in the first half) is, by elementary combinatorics

$$p_n = \frac{70!}{m!n!2^{70}}. \tag{7.9}$$

Therefore the accumulated probability corresponds to the binomial distribution and equals

$$\begin{aligned} P &= P(S(D') < S(D)) + \tfrac{1}{2}P(S(D') = S(D)) \\ &= p_0 + p_1 + \cdots + p_{45} + \tfrac{1}{2}p_{46} \\ &= 0.9957. \end{aligned}$$

Accordingly we have

$$\mathrm{TI} = 2P - 1 = +99.14\%. \tag{7.10}$$

Hence the result would be significant on the 1% level.

Another approach would be to assume a very high degree of spatial dependence, but still no inter-annual dependence. Then resampling would involve permuting each time series in blocks of 1 year, but the same permutation would have to be used for each time series in the resampled data set D'. It is easy to see that this approach is equivalent to a slightly simpler procedure. First, the absolute maxima in the original data set D are counted year by year, in order to obtain an annual time series C whose each value corresponds to the number of 40-year maxima that occurred in a given year among the 70 original time series. Now $S(D)$ equals the difference of the sums of values of time series C in the second half and in the first half. Resampling D to obtain D' is equivalent to permuting the time series C. The results of permutation testing using 10 000 permutations turn out to be not significant on the 5% level, with TI = +92%.

It may be argued that the first approach underestimated the spatial and possibly temporal dependences in the data and overestimated the trend index. The second approach certainly overestimated the spatial correlations and probably underestimated the trend index. An adequate resampling method that preserves spatial correlations in a similar way that phase randomization preserves temporal correlations would be desirable in this case.

7.7 Conclusion

Changes in various characteristics of river flow are visible in many time series of flow throughout the world. Changes in frequency, size, seasonal occurrence, and clustering of high flows may be investigated using a variety of approaches. There does not seem to be a clear signal of an overall global trend in the occurrence of floods in any of the characteristics studied. In many cases the number of significant results is greater than expected; however, it is problematic to assess the overall significance to be attributed to a multi-site study, because of possible dependence between different time series. Seasonal timing of high flows has changed most notably in Germany (increase in DJF maximum flows) and Norway (increase in DJF and FMA maximum flows). The changes affect only some series in a region. Changes are much better visible in the series of seasonal maxima than in POT series. Possibly the methods applied to POT could be improved. Changes are clearly visible in the very rare, most extreme events, and these require dedicated methods.

Acknowledgments A part of this research was performed within the ENSEMBLES (Ensembles-Based Predictions of Climate Changes and their Impacts) project financed by the European Commission under the 6th Framework Program. The author thanks the Foundation for Polish Science for the financial support of his work. The daily mean river flow data were provided under a project-specific license by the Global Runoff Data Centre [7.3], Federal Institute of Hydrology, Koblenz, Germany. Constructive remarks of the anonymous referee are gratefully acknowledged.

References

7.1. N. Arnell, C. Liu, Hydrology and water resources. In *Climate Change 2001: Impacts, Adaptation, and Vulnerability*, ed. by J.J. McCarthy, O.F. Canziani, N.A. Leary, D.J. Dokken, K.S. White (Cambridge University Press, Cambridge, Intergovernmental Panel on Climate Change (IPCC), 2001)
7.2. T.A. Buishand, Some methods for testing the homogeneity of rainfall records. J. Hydrol. **58**, 11–27 (1982)
7.3. GRDC (Global Runoff Data Centre), *Dataset of 195 river discharge time series from the Global Runoff Data Centre*. (Federal Institute of Hydrology, Koblenz, 2003)
7.4. F. H. S. Chiew, T.A. McMahon, Detection of trend or change in annual flow of Australian rivers. Int. J. Climatol. **13**, 643–653 (1993)
7.5. T.A. Cohn, H.F. Lins, Natures style: Naturally trendy. Geoph. Res. Lett. **32**, L23402 (2005)
7.6. P. Good. *Permutation Tests: A Practical Guide to Resampling Methods for Testing Hypotheses* (Springer, Berlin, 1993)
7.7. IPCC, Summary for policymakers. In *Climate Change 2007: The Physical Science Basis. Contribution of Working Group I to the Fourth Assessment Report of the Intergovernmental Panel on Climate Change*, ed. by S. Solomon, D. Qin, M. Manning, Z. Chen, M. Marquis, K.B. Averyt, M.Tignor, H.L. Miller (Cambridge University Press, Cambridge, United Kingdom and New York, NY, 2007)
7.8. IPCC, Summary for policymakers. In *Climate Change 2007: Impacts, Adaptation and Vulnerability. Contribution of Working Group II to the Fourth Assessment Report of the Intergovernmental Panel on Climate Change*, ed. by M. L. Parry, O. F. Canziani, J. P.

Palutikof, P. J. van der Linden, C. E. Hanson (Cambridge University Press, Cambridge, 2007)
7.9. T. Kalisky, Y. Ashkenazy, S. Havlin, Volatility of linear and nonlinear time series. Phys. Rev. E. **72**, 011913 (2005)
7.10. M.G. Kendall, A new measure of rank correlation. Biometrika. **30**, 81–93 (1938)
7.11. Z.W. Kundzewicz, Floods in the context of climate change and variability. In *Climatic Change: Implications for the Hydrological Cycle and for Water Management*, ed. by M. Beniston, (Kluwer, Dordrecht, The Netherlands, 2002), pp. 225–247
7.12. Z.W. Kundzewicz, D. Graczyk, T. Maurer, I. Pi nskwar, M. Radziejewski, C. Svensson, M. Szwed, Trend detection in river flow time series: 1. annual maximum flow. Hydrol. Sci. J. **50**(5), 797–810 (2005)
7.13. Z.W. Kundzewicz, M. Radziejewski, Methodologies for trend detection. In *Climate Variability and Change—Hydrological Impacts (Proceedings of the Fifth FRIEND World Conference)*, IAHS Publ. 308, 2006, pp. 538–549, Havana, Cuba, November 2006
7.14. Z.W. Kundzewicz, A. J. Robson, Change detection in hydrological records—a review of the methodology. Hydrol. Sci. J. **49**(1), 7–19 (2004)
7.15. E.L. Lehmann, J.P. Romano. *Testing Statistical Hypotheses*, 3rd edn. (Springer Texts in Statistics. Springer, Berlin, 2005)
7.16. H.F. Lins, J. R. Slack, Streamflow trends in the United States. Geoph. Res. Lett. **26**(2), 227–230 (1999)
7.17. V. Livina, Y. Ashkenazy, P. Braun, R. Monetti, A. Bunde, S. Havlin, Nonlinear volatility of river flux fluctuations. Phys. Rev. E. **67**, 042101 (2003)
7.18. B. Mandelbrot. *The Fractal Geometry of Nature*. (W. H. Freeman, New York, 1982)
7.19. B. Mandelbrot, M. Taqqu, Robust R/S analysis of long-serial correlation. Bull. Int. Stat. Inst. **48**(Book 2), 59–104 (1979)
7.20. M. Mudelsee, M. Börngen, G. Tetzlaff, U. Grünewald, No upward trends in the occurrence of extreme floods in central Europe. Nature **425**, 166–169 (2003)
7.21. M. Radziejewski, Fractal and non-fractal analysis of hydrological time series. Unpublished report of the work during tempus mobility to wageningen agricultural university (Wageningen Agricultural University, The Netherlands, 1994)
7.22. M. Radziejewski, Detection of change in world-wide time series of river flow records. In *Anwendungen der weltweiten Sammlung von Abflussdaten des Global Runoff Data Centre (Kolloquium am 8./9. März 2005 in Koblenz)*, Bundesanstalt für Gewässerkunde Veranstaltungen 3/2005, (Koblenz, Germany, 2005), pp. 23–30
7.23. M. Radziejewski, A. Bardossy, Z. W. Kundzewicz, Detection of change in river flow using phase randomization. Hydrolo. Sci. J. **45**(4), 547–558 (2000)
7.24. M. Radziejewski, Z.W. Kundzewicz, Fractal analysis of flow of the river Warta. J. Hydrolo. **200**(1), 280–294 (1997)
7.25. M. Radziejewski, Z.W. Kundzewicz, Detectability of changes in hydrological records. Hydrolo. Sci. J. **49**(1), 39–51 (2004)
7.26. J.R. Slack, A.M. Lumb, J.M. Landwehr, *Hydro-Climatic Data Network (HCDN): Streamflow Data Set, 1874–1988*. US Geological Survey, 1993. CR-ROM and Manual
7.27. C. Svensson, Z.W. Kundzewicz, T. Maurer, Trend detection in river flow series: 2. flood and low-flow index series. Hydrol. Sci. J. **50**(5), 811–824 (2005)
7.28. C.J. Vörösmarty, K.P. Sharma, B.M. Fekete, A.H. Copeland, J. Holden, J. Marble, J.A. Lough, The storage and aging of continental runoff in large reservoir systems of the world. Ambio. **26**, 210–219 (1997)

River discharge. Already Heraclitus stated that “you cannot step twice into the same river” to allegorise change, so can you? (Source: www.cambridge2000.com)

Chapter 8
Extreme Value Analysis Considering Trends: Application to Discharge Data of the Danube River Basin

Malaak Kallache, Henning W. Rust, Holger Lange, and Jürgen P. Kropp

This chapter proposes and applies an extreme value assessment framework, which allows for auto-correlation and non-stationarity in the extremes. This is, e.g., useful to assess the anticipated intensification of the hydrological cycle due to climate change. The costs related to more frequent or more severe floods are enormous. Therefore, an adequate estimation of these hazards and the related uncertainties is of major concern. Exceedances over a threshold are assumed to be distributed according to a generalised Pareto distribution and we use a point process to approximate the data. In order to eliminate auto-correlation, the data are thinned out. Contrary to ordinary extreme value statistics, potential non-stationarity is included by allowing the model parameters to vary with time. By this, changes in frequency and magnitude of the extremes can be tracked. The model which best suits the data is selected out of a set of models which comprises the stationary model and models with a variety of polynomial and exponential trend assumptions. Analysing winter discharge data of about 50 gauges within the Danube River basin, we find trends in the extremes in about one-third of the gauges examined. The spatial pattern of the trends is not immediately interpretable. We observe neighbouring gauges often to display distinct behaviour, possibly due to non-climatic factors such as changes in land use or soil conditions. Importantly, assuming stationary models for non-stationary extremes results in biased assessment measures. The magnitude of the bias depends on the trend strength and we find up to 100% increase for the 100-year return level. The results obtained are a basis for process-oriented, physical interpretation of the trends. Moreover, common practice of water management authorities can be improved by applying the proposed methods, and costs for flood protection buildings can be calculated with higher accuracy.

M. Kallache (✉)
Climpact, 75009, Paris, France
e-mail: mk@climpact.com

J.P. Kropp, H.-J. Schellnhuber (eds.), *In Extremis*,
DOI 10.1007/978-3-642-14863-7_8, © Springer-Verlag Berlin Heidelberg 2011

8.1 Introduction

In Europe floods have been a known natural hazard for centuries (cf. [8.4, 8.11, 8.31]). Recently occurred extreme river floods have had severe effects in central Europe. The Elbe flood in August 2002, for example, caused 36 deaths and over 15 billion USD damages and the Oder flood in July 1997 caused 114 deaths and around 5 billion USD damages [8.27]. An appropriate assessment of extreme events is an important tool for water management authorities to develop mitigation and adaptation strategies for severe effects of floods on society.

The occurrence of extreme flood events seems to have grown considerably over recent decades. References [8.19, 8.24], for example, show a higher occurrence of flood events worldwide in the time span of 1990–1998 than in the nearly four times longer period of 1950–1985. River discharge data may exhibit trends because of a variety of reasons, thereby climate change is an anticipated factor. This is because the capacity of the atmosphere to hold water grows with increasing temperature. In this way the potential for intense precipitation and thus floods also increases. Moreover, climate change is likely to change atmospheric circulation patterns. The coupling between atmospheric circulation and the water cycle takes place on many levels [8.17] and flood-producing atmospheric circulation patterns have already been identified [8.1].

Here river discharge records of the Danube River basin in Southern Germany are examined to track the question whether there are trends in river discharge extremes observable. To do so we set up an extreme value assessment framework which allows for non-stationary extremes. Standard extreme value theory assumes independent and identically distributed extremes, which is often not fulfiled by hydrological data. Discharge is influenced by complex dynamical processes such as precipitation, topography, land use changes, or climate change; therefore, a univariate distribution with two or three parameters is not necessarily a good approximation of its multi-faced nature [8.14]. An unjustified assumption of stationarity may lead to considerable underestimation of the probability of a disastrous extreme event (cf. [8.7]). The framework presented assesses trends in extremes, despite of the scarcity of available data. There exist approaches to conclude from changing properties of mean values of a series to the change in the extremes. However, this goes along with a specification of the relationship of mean values and extremes. In general, the function between a trend in the extremes and a trend in the mean values is non-linear and does not show an obvious relation [8.29]. The frequency of extreme events might change dramatically as a result of even a small change in the mean of the sample [8.37].

We look at about 50 gauges in the Danube River basin, all of them cover the time period between 1940 and 2003. The precipitation pattern in this area has undergone some significant changes. Heavy winter precipitation events become more frequent in Bavaria and Baden-Württemberg [8.12] and are anticipated to become even more frequent in the future [8.35]. Local water management authorities expect an increase in heavy floods due to climate change. As a consequence, a *climate change factor* of 1.15 was introduced in Bavaria in 2006. Due to this factor, every design value for river discharge is expanded about 15% [8.3]. This chapter illustrates an alternative calculation of design values by incorporating the assumption of a trend already in the extreme value assessment. By doing so, a trend can be tested for and the strength of the trend is provided for each gauge separately. Universal trend patterns have yet not been found for whole river basins [8.5, 8.23]. Therefore, design values are calculated more adequately by proceeding gauge by gauge which might reduce costs of mitigation or costs of severe effects. Moreover the uncertainty of the assessment measures is provided by using the methodology proposed in this chapter. We proceed as follows: in Sect. 8.2 the preparation of the data and details of extreme value analysis using point processes are provided. The methodology proposed is applied to the Danube River basin in southern Germany, which is outlined in Sect. 8.3. Finally, in Sect. 8.4 a discussion of the results and a conclusion are given.

8.2 Method

Here point processes (PP) are used to represent extreme events. In this way it is possible to model the frequency of occurrence in time as well as the magnitude of the extremes. The PP approach is based on the Fisher–Tippett theorem [8.16, 8.18], which determines the limiting distribution of block maxima of a sequence of independent and identically distributed (i.i.d.) random variables $\{X_i\}$ with $i = 1, \ldots, n$. This theorem holds for a wide range of the common distribution function $F(\cdot)$ of the $\{X_i\}$; therefore, our approach is widely applicable.

8.2.1 Choice of the Extreme Values

To extract the set of extreme values some preparations are necessary. Hydrological data underly the annual cycle. This periodic non-stationarity has either to be modelled or eliminated by classifying the data [8.6]. We chose the latter option and examined the different seasons. The winter season is of special interest because of the observed and predicted precipitation changes (cf. Sect. 8.1). We found the intra-seasonal variability of this season (December to February) to be sufficiently low and therefore regard it as negligible in the following.

To be able to use a point process model we specify the set of extreme values as excesses above a threshold u. Thus an appropriate threshold u has to be found. On the one hand, a high threshold is preferred, such that it is reliable to assume a limiting distribution for the excesses. For threshold excesses, this limiting distribution is the generalised Pareto distribution (GPD) ([8.13]). On the other hand, a sufficient number of excesses must be available to estimate the model with a reasonable amount of uncertainty. Therefore, a tradeoff between bias and variance has to be found. To operationalise this choice, we apply the non-parametric mean residual life (MRL) plot and a parametric approach, i.e. the comparison of a PP fit over a range of thresholds [8.6]. The MRL plot depicts the sample mean of the threshold excesses for different u. In case the extreme values are distributed according to the limiting distribution, the sample mean is an estimator for the expected value of the extremes. This expected value changes linearly with u. Thus we select the lowest u_0 such that the MRL plot changes linearly for $u > u_0$. In the parametric approach a GPD distribution is fitted to the extreme values for a range of thresholds. In case the asymptotics is reached, we expect the parameters of the GPD distribution to stay stable when the threshold is varied. Therefore the lowest u_0 is selected, such that parameter estimates for $u > u_0$ stay approximately constant. For each individual river gauge we compare the results of both methods to cross-check and to derive reliable results.

Subsequently, we take care of the auto-correlation structure in the data. The Fisher–Tippett theorem holds for iid or dependent but stationary data [8.13]. Thus, to introduce non-stationarity we remove the auto-correlation structure. To do so the threshold excesses are grouped into clusters and only the maxima of these clusters are kept. For the empirical data of the Danube River basin a cluster size of 6–12 days has shown to be sufficient. The extreme value theory applied holds also for the thinned-out series [8.15].

The choice of u and the cluster size are verified by examining the GPD model which suits best to the resulting set of extreme values. The probability and quantile plots are examined and the goodness of fit of the model is tested by applying a Kolmogorov–Smirnov test (see [8.20]). We find the GPD to be a suitable model for all but three of the stations examined; therefore, these three records are excluded from the further analysis. Moreover the set of threshold excesses is compared to the set of block maxima which are drawn from intervals of the length of 1 year. Block maxima are an alternative representation of the extreme events. For all stations the set of declustered threshold excesses contains the more extreme events in the series. This effect is due to the auto-correlation being present in discharge data. We therefore regard the threshold approach as more suitable for the empirical data analysed here.

8.2.2 *Point Processes*

We approximate threshold excesses by a point process, which may be seen as a stochastic rule on a set $\mathcal{A}$ for the occurrence of extreme events. Regions of $\mathcal{A}$ have the form $A := (t_1, t_2) \times [u, \infty)$ with $[t_1, t_2]$ being the observation period which is mapped to [0, 1]. The point process can be represented as two independent Poisson processes $\text{Poi}(\Lambda_1([t_1, t_2]))\times \text{Poi}\Lambda_2([x, \infty))$, with intensities

$$\Lambda_1([t_1, t_2]) = (t_2 - t_1) \tag{8.1}$$

and

$$\Lambda_2([x, \infty)) = \left[1 + \xi\left(\frac{x - \mu}{\sigma}\right)\right]^{-1/\xi}, \tag{8.2}$$

where the parameters μ and σ control the location and scale of the distribution and ξ controls its shape. In this way $\mathrm{Poi}(\Lambda_1([t_1, t_2]))$ is a homogenous Poisson process with constant intensity λ and gives the random times at which $X_i > u$. Their sizes $Y_i = X_i - u$ are represented by $\mathrm{Poi}(\Lambda_2([x, \infty]))$. Conditional on the number of exceedances $N = n_u$ over u, $Y_1 = X_1 - u, \ldots, Y_{n_u} = X_{n_u} - u$ are a random sample of the GPD and we get

$$P\{(X_i - b_n)/a_n > x | (X_i - b_n)/a_n > u\} = \left[1 + \xi\left(\frac{x-u}{\psi}\right)\right]^{-1/\xi}, \tag{8.3}$$

with ψ being the scale parameter and ξ being the shape parameter of the GPD and a_n and b_n being normalizing constants. Here inferences for λ, ψ and ξ are based on the point process likelihood.

8.2.2.1 Relation of Moments to Point Process Parameters

The parameters $\Theta = (\lambda, \psi, \xi)$ of the point process are related to the moments of the distribution of the extremes (see, e.g. [8.26]). Let Y_i be excesses over a threshold u which are distributed as $\mathrm{GPD}(y) = 1 - \exp(-y/\psi)$ for $y > 0$, i.e. $\xi = 0$. Then we get for $\gamma = 0.57722$ being Euler's constant

$$\begin{aligned} E(Y) &= u + \psi\gamma, \\ \mathrm{var}(Y) &= \psi^2\frac{\pi^2}{6} \approx 1.645\psi^2. \end{aligned} \tag{8.4}$$

For $\{Y_i\}$ having a GPD distribution with shape parameter $\xi > 0$ or $\xi < 0$ we derive

$$E(Y) = u + \frac{\psi}{\xi}[1 - \lambda^{\xi}\Gamma(1+\xi)]$$

and for $\xi > -0.5$

$$\mathrm{var}(Y) = \left(\frac{\lambda^{\xi}\psi}{\xi}\right)^2 \{\Gamma(1+2\xi) - [\Gamma(1+\xi)]^2\}, \tag{8.5}$$

with $\Gamma(\cdot)$ being the gamma function.

8.2.3 *Test for Trend*

Non-stationarity is incorporated in the extreme value assessment framework by allowing the model parameters to vary with time, thereby we proceed fully parametric and assume a predetermined trend shape for each of the parameters. By doing so, standard methods like the delta method can be provided to obtain uncertainty bands for the parameter estimates and the trend can be extrapolated in the future.

For the rate parameter λ_i we assume a constant $\lambda_i \equiv \lambda = c_0$ and exponential trend shapes

$$\lambda_i = \exp\left\{\sum_{i=0}^{k} c_i t^i\right\}, \qquad k = 1, \ldots, 4, \tag{8.6}$$

with c_i being constants. These shapes assure a slowly varying λ_i. Furthermore, the exponential relation is regarded as the "most natural" simple relationship for rates of occurrence (cf. [8.9]). It is chosen by various authors (see [8.25, 8.27]).

With respect to the scale parameter we consider the following shapes:

$$\text{(i)} \quad \psi_i = \sum_{i=0}^{k} d_i t^i, \qquad k = 0, \ldots, 4,$$

$$\text{(ii)} \quad \psi_i = \exp\left\{\sum_{i=0}^{k} d_i t^i\right\}, \qquad k = 1, \ldots, 4, \tag{8.7}$$

with d_i being constants. Polynomials are capable to represent a variety of trend shapes which might occur empirically. Tests with simulated data showed that even maxima with an S-curve-shaped trend are well approximated by using models with polynomial-shaped trend guesses. Furthermore, we utilise the exponential trend guesses described in (8.7) (ii). This trend form does not allow for a shape parameter $\psi_i < 0$. This is a necessary condition, because the scale parameter is related to the variance of the extreme values.

The framework presented here is capable to allow for the non-stationarity of the shape parameter ξ. However, the variation of ξ around 0 lacks a theoretical interpretation. According to the Fisher–Tippett theorem, different marginal distributions of the random variables $X_1, \ldots, X_n$ have different domains of attraction of the limiting extreme value distribution and they are separated by ($\xi < 0$), ($\lim \xi \to 0$) or ($\xi > 0$). In case the best suiting point process model includes a ξ_i which changes sign, this may just indicate that the model class we assume is not appropriate to represent the set of extreme values. Therefore we chose to only use models with a constant ξ (see also [8.6, 8.8, 8.30]). Small time variations of the shape parameter can be emulated by time variation of the other parameters [8.21, 8.38]. Thus, even if the assumption of a constant ξ actually is not fulfilled for a specific empirical time series, non-stationary models with a fixed ξ might still be adequate. The goodness-of-fit tests applied confirm the feasibility of our reduction.

To test for a trend, we create a class of extreme value models by using all possible combinations of (8.6) and (8.7) and the stationary PP model. Then the deviance statistic (see [8.10, 8.28]), i.e. the likelihood ratio test, is applied to choose the best suiting model. This test weights the complexity of the model against the degree of explanation of the variance of the empirical time series. The best suiting model might be a non-stationary one, thus the trend test is transferred to a model selection problem. Further details are provided in [8.20].

The deviance statistic is based on the maximum likelihood. To derive the likelihood of a non-stationary point process let $X_1, \dots, X_n$ be iid random variables and let δ_i be an indicator for X_i lying above some given threshold u. λ is the rate of threshold exceedances and we get $P(X > u) = \lambda/n$ for n being the length of the time series. The likelihood contribution for an X_i falling below u is thus $1 - \lambda/n$, and for an X_i above u it is λ/n times the GPD approximation with parameters (ψ, ξ). This is because we actually model excesses, i.e. the conditional probabilities $P(X > y + u|X > u)$.

The density GPD$(\cdot)$ of GPD(y) is given by

$$\frac{\mathrm{dGPD}(y)}{\mathrm{d}y} = \psi_i^{-1}\left[1 + \xi_i\left(\frac{x_i - u}{\psi_i}\right)\right]^{-\frac{1}{\xi_i}-1}. \tag{8.8}$$

So the likelihood function for sufficiently large u and n is, apart from a scaling factor,

$$\begin{aligned}\mathcal{L}(\lambda, \psi, \xi; x_1, \dots, x_n) &= \{P(X_i < u)\}^{1-\delta_i}\{P(X_i > u)P(X_i = x|X_i > u)\}^{\delta_i} \\ &\propto \prod_{i=1}^{n}\left[\left\{1 - \frac{\lambda_i}{n}\right\}^{1-\delta_i}\left\{\frac{\lambda_i}{n}\psi_i^{-1}\left[1 + \xi_i\left(\frac{x_i - u}{\psi_i}\right)\right]^{-\frac{1}{\xi_i}-1}\right\}^{\delta_i}\right].\end{aligned} \tag{8.9}$$

The log likelihood then can be written as

$$l(\lambda, \psi, \xi) = l_N(\lambda) + l_Y(\psi, \xi), \tag{8.10}$$

where l_N and l_Y are the log-density of the number of exceedances N and the log-conditional density of $Y_1, \dots, Y_{n_u}$ given that $N = n_u$. Thus inferences on λ can be separated from those on the other two parameters.

8.2.3.1 Power of the Trend Test

For $n \to \infty$ the deviance statistic is asymptotically χ^2 distributed. We study the power of the trend test (based on the deviance statistic) by means of a simulation study. Thereby the length of the time series and the magnitude of trends resemble the situation given in our empirical analyses: The parameters are taken from a stationary PP fit to the Danube River at Beuron, i.e. $u = 65\,\mathrm{m}^3/\mathrm{s}$, $\psi_0 = 43.9$ and $\xi = 0.076$. We generate time series with 5 415 (18 050) data points, which equals 60 (200) years of daily measurements evaluated at a season of 3 months. The occurrence rates λ_0 of 0.01 and 0.04 are chosen; they cover the range found in the empirical series. Thus 60–800 extrema are provided.

A trend in the occurrence rate is tested for by generating artificial data from a Poisson process with a linear trend in the exponential function of λ_t. We get $\lambda_t = \exp(\log(\lambda_0) + s \cdot t)$ for $t = 1, \ldots, n$ where s is a constant slope, which is varied between 0 and 0.0005. Then two models, namely a stationary Poisson process and one assuming an exponential trend in λ (the "right" model), are fitted to the artificial Poisson process data and the better suiting model is chosen according to the deviance statistic with a significance level of 5%. For each value of the slope s this study is repeated 1 000 times. In case no trend is present in the artificial data, we therefore expect to falsely detect a trend in 50 of the cases. Furthermore we would like the test to detect even weak trends in the data. If the test detects a trend in all cases it has a power of 1. A trend in the scale parameter ψ of the GPD distribution is tested for by simulating artificial data from a GPD model with $\psi_t = d_0 + s \cdot t$ with the slope s ranging from 0 to 0.05 and we proceed as described for the Poisson process.

For every simulation setting we obtain as result that a trend is not falsely detected too often in case no trend is present in the data. In case a trend is present in the data we detect it as given in Table 8.1. The smallest change of the frequency of occurrence which is detected with a sufficient power is from once a year to 2.7 times a year when looking at series with a length of 60 years (winter season only) and about 60 extrema at hand. In case 200 winter seasons and about 800 extrema are available much weaker trends are detected, namely a change from once a year to 1.6 times a year within 200 years or equivalently from once a year to 1.2 times a year within 60 years. This reveals quite good trend detection qualities of the methodology regarding the frequency of occurrence. Changes in the ψ parameter are detected with sufficient power up from a change of 20% of the mean within 60 years and 60 extrema at hand and up from a change of 2% of the mean for 800 extrema being available. The power of the deviance statistic to discriminate between types of trends, i.e. linear and quadratic trends, is also assessed and comparable results are obtained. We therefore do not regard it as necessary to split up the trend detection into two steps, i.e. first detecting a possible trend and then determining its type. We rather choose out of the complete class of models using the deviance statistic. Further details are given in [8.20].

Table 8.1 Power of the trend test based on the deviance statistic. Smallest changes which are detected by the trend test with sufficient power (i.e. percentage of the cases) for given simulation settings. The settings differ by length of the time series (60 years winter season and 200 years winter season) and the initial rate of occurrence ($\lambda = 0.01$ and 0.04)

	60 years (winter season) $\lambda = 0.01$ $\rightarrow$ 60 extrema available	200 years (winter season) $\lambda = 0.04$ $\rightarrow$ 800 extrema available
Trend in λ	1 per year $\rightarrow$ 2.7 per year is detected in 95% of the cases	1 per year $\rightarrow$ 1.6 per year is detected in 98% of the cases
Trend in ψ	$\Delta E(Y) = 20\%$ is detected in 80% of the cases	$\Delta E(Y) = 2\%$ is detected in 100% of the cases

8.2.4 Return-Level Estimation

Return levels are quantiles of the distribution of the extreme values. They are of crucial importance in water management since they are the basis for design flood probability calculations. Let p be the probability that the return level z_p is exceeded once in a year. Then the return period is given by $1/p$. p depends on the occurrence of an extreme event which is determined by λ, i.e.

$$p = P\left\{\frac{(X_i - b_n)}{a_n} > x \middle| \frac{(X_i - b_n)}{a_n} > u\right\} P\left\{\frac{(X_i - b_n)}{a_n} > u\right\}. \tag{8.11}$$

By using (8.3) and setting $x = z_p$ we get

$$z_p = \begin{cases} u + \frac{\psi}{\xi}[(\lambda/p)^{\xi} - 1] & \text{for} \quad \xi \neq 0 \\ u + \psi \log(\lambda/p) & \text{for} \quad \xi = 0 \end{cases}. \tag{8.12}$$

In the non-stationary case all three parameters of $\Theta = (\lambda, \psi, \xi)$ may be time dependent. Therefore we have to determine a fixed time point t_0.[1] The return level is then calculated such that it is valid from t_0, i.e. $\Theta(t_0)$ is inserted in (8.12). By doing so an estimated trend in the extreme values is considered but not extrapolated into the future. We proceed in this manner (see also [8.2]).

By using the fully parametric approach presented here it is possible to extrapolate the estimated trend into the future. In this case we have to define a *prediction period*. Then all return levels $(z_p^{t_0}, \ldots, z_p^{t_k})$ within this prediction period $[t_0, t_k]$ can be calculated, according to a specified return period $1/p$. Then, e.g. the highest return level out of this range of return levels can be chosen to get an estimate for the return level within this prediction period, which is crossed with probability p once in a year. This choice follows the precautionary principle.

8.2.4.1 Uncertainty

Confidence intervals for the estimates can be obtained by using the asymptotic normality property of maximum likelihood estimators. These consistency and efficiency properties of maximum likelihood parameter estimates of the skewed GPD distribution are given for $\hat{\xi} > -0.5$ [8.36]. By using the *delta method* we then can approximate confidence intervals for the return-level estimates (see (8.12)). This method expands the function $\hat{z}_p$ in a neighbourhood of the maximum likelihood estimates $\hat{\Theta}$ by using a one-step Taylor approximation and then takes the variance (cf. [8.6]).

---

[1] Any time point t_0 may be chosen as reference point. The occurrence of extreme events may vary from time series to time series when using threshold excesses, but the non-stationary parameters of the extreme value distribution are at hand for each i in $1, \ldots, n$ through interpolation of the assumed trend shape with its estimated parameter values.

The delta method therefore gives

$$\mathrm{var}(\hat{z}_p) \approx \nabla z_p^{\mathrm{T}} \Sigma \, \nabla z_p, \tag{8.13}$$

where Σ is the variance–covariance matrix of $(\hat{\lambda}, \hat{\psi}, \hat{\xi})$. $\mathrm{cov}(\hat{\lambda},\hat{\psi})$ and $\mathrm{cov}(\hat{\lambda},\hat{\xi})$ are zero. For $\xi \neq 0$ we get

$$\begin{aligned} \nabla z_p^{\mathrm{T}} &= \left[\frac{\partial z_p}{\partial \lambda}, \frac{\partial z_p}{\partial \psi}, \frac{\partial z_p}{\partial \xi}\right] \\ &= \left[\frac{\psi}{p^{\xi}}\lambda^{\xi-1}, \xi^{-1}\left\{\left(\frac{\lambda}{p}\right)^{\xi} - 1\right\}, \frac{\psi}{\xi}\left(\frac{\lambda}{p}\right)^{\xi} \ln\left(\frac{\lambda}{p}\right) - \frac{\psi}{\xi^2}\left\{\left(\frac{\lambda}{p}\right)^{\xi} - 1\right\}\right], \end{aligned} \tag{8.14}$$

whereas for $\xi = 0$, z_p depends only on ψ and λ and

$$\nabla z_p^{\mathrm{T}} = \left[\frac{\partial z_p}{\partial \lambda}, \frac{\partial z_p}{\partial \psi}\right] = \left[\frac{\psi}{\lambda}, \ln\left(\frac{\lambda}{p}\right)\right] \tag{8.15}$$

is obtained. In the non-stationary case, $\Theta_i = (\lambda_i, \psi_i, \xi)$ itself is time dependent. The functional form of this dependence is given; therefore, derivatives of z_p in direction of λ, ψ and ξ can be calculated by using the chain rule.

This standard methodology assumes independence of the extreme values. Auto-correlation in the data causes a loss of information. To avoid an underestimation of the uncertainty we therefore either have to consider the auto-correlation structure in the data as presented in Sect. 8.2.1 or uncertainty measures can be provided by using the bootstrap (cf. [8.34]).

8.3 Results

Daily discharge observations of 50 gauges in the Danube River basin in southern Germany are investigated. The examined time period is restricted to a jointly covered period of 60 years, i.e. from 1 January 1941 to 31 December 2000. We investigate data of the winter season (December to February). For each time series, a threshold u is chosen and the data are declustered as described in Sect. 8.2.1. About 5 000 data points are available for analysis purposes, which results in approximately 30–120 extrema per time series.

To assess the extreme values a stationary and a range of non-stationary point processes are fitted to them. From this class of models the best one is chosen using the deviance statistic and assessment measures, such as the return level, are derived from this model. We use the results only in case more than 30 extreme values are left after declustering and in case the best model passes all goodness-of-fit tests, which is the case for 47 of the 50 stations.

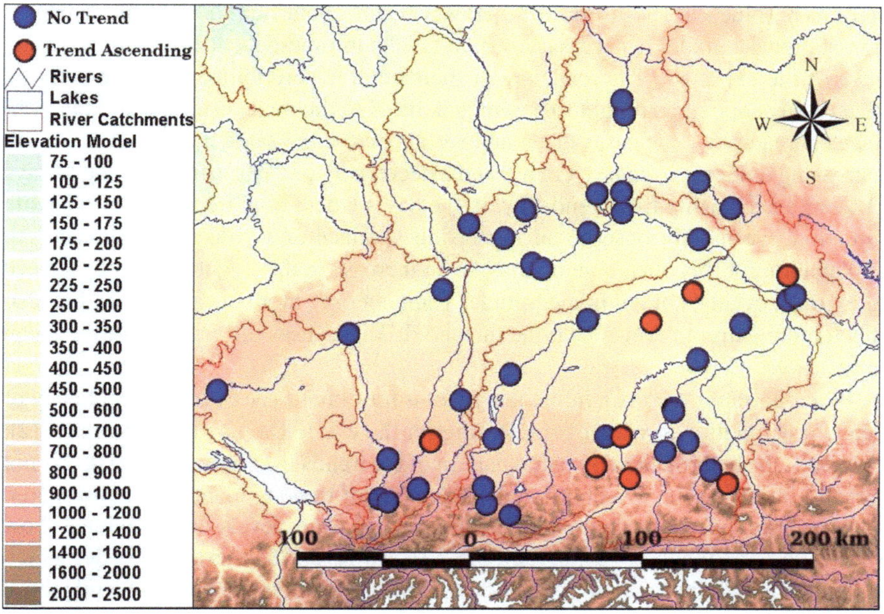

Fig. 8.1 Trend in frequency of occurrence of extreme events. Trend tendency of the possibly non-stationary rate λ of the best point process model fitted to extremes of discharge data at 47 gauges of the Danube River catchment. A model with a constant rate λ is marked in *blue*, whereas *red* indicates an increasing frequency

The change in the frequency of occurrence of extreme events can be measured by examining the possibly non-stationary rate λ of the point process. In Fig. 8.1 the trend tendency of the rate λ is depicted. In most of the cases a model with a constant λ represents the data best (blue circles). In eight cases the frequency of occurrence increases. These gauges are the Geltnach River at Hörmannshofen, the Vils River at Rottersdorf, the Große Vils River at Vilsbiburg, the Ramsauer Ache River at Ilsank, the Inn River at Oberaudorf and Rosenheim, the Leitzach River at Stauden and the Ilz River at Kalteneck. A possible cause may be the observed increase in frequency of heavy winter precipitation events in Bavaria and Baden-Württemberg (cf. [8.12, 8.22]). These events are anticipated to become even more frequent in the future [8.35].

In Fig. 8.2 the trend tendencies of the estimated mean and variance of the excesses are depicted. They are calculated using the parameter estimates of the best point process model. As apparent in (8.4) and (8.5), the rate λ is needed for the calculation which results in trends at the same sites where a non-stationary rate λ is indicated in Fig. 8.1. Furthermore, in Fig. 8.2 the influence of a non-stationary GPD model, i.e. a time-varying scale parameter ψ, becomes visible. In our study the discharge extremes never require an inhomogeneous Poisson process and a non-stationary GPD at the same time. Throughout, the best suiting model possesses at most one non-stationary component. In most of the cases the extreme events are stationary (blue circles in Fig. 8.2). However, for about one-third of the gauges a non-stationary model is necessary, which results in time-dependent mean and variance of the extremes. The tendency of their trend is determined by evaluating the sign of the slope of a straight line, which is fitted to the mean estimate and the variance estimate. In some cases mean and variance have the same trend tendencies (an increasing tendency is marked in red and a decreasing one in green), then we always observe a decreasing trend tendency. In the cases where both trends are of opposite direction, we mostly observe an increasing variance and a decreasing mean. An explanation for this phenomenon may be an increase in the frequency of occurrence of extreme events, but not in magnitude. Then not only more outliers occur, but also much more extreme events near the threshold. This causes the estimate of the mean to be lower.

Precipitation is the main influencing factor for floods. The observed frequency of heavy precipitation in winter is increasing for the whole Danube River basin [8.22]. However, the heterogeneous spatial pattern of the trends in extremes in Fig. 8.2 is not directly interpretable. We observe all three sorts of trend tendencies and no spatial accumulation becomes apparent. Maximum daily water levels during winter show spatial patterns related to topography [8.32]. However, the elevation model which is depicted at the back of Fig. 8.2 does not suffice to interpret the trend tendencies. Further influencing factors might be land use changes, river regulation measures or changes of the hydraulic conditions within the river system. Land use change, for

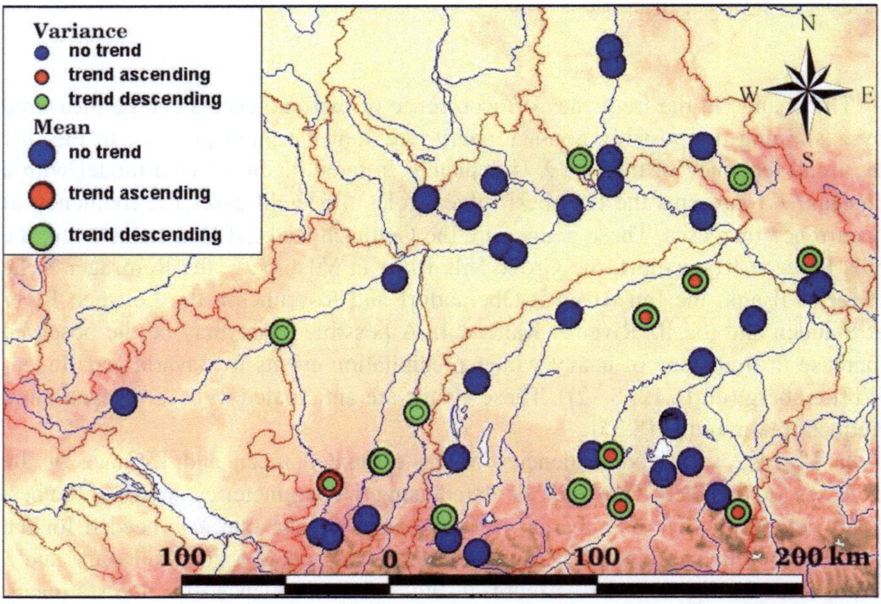

Fig. 8.2 Trend in mean and variance of extreme events. A point process model is fitted to excesses of discharge data of 47 gauges in the Danube River basin. Mean (*small circles*) and variance (*big circles*) are calculated from the parameter estimates. Stationary mean and variance are depicted in *blue*, whereas *green* indicates a decreasing trend tendency and *red* an increasing trend tendency

example, has a strong impact on changes of the rainfall–runoff relationship and land use has undergone significant changes over the last few centuries in all European countries (cf. [8.32]).

To assess the impact of incorporating non-stationary models in the extreme value analysis, we examine the influence on the return-level estimation. For this purpose the 100-year return level, which is crossed once a year with probability $p = 0.01$, is calculated for each station using a stationary point process model. Then we calculate the probability of exceedance p^* of the corresponding return level z_p for the best suiting, possibly non-stationary, model. In Fig. 8.3 the difference between p and p^* is depicted for stations in the Danube River basin. White circles indicate no change implying that the stationary model suits best. A larger probability of exceedance is marked with red circles and $p^* < p$ is indicated by green circles. In this context a change of 100% denotes $p^* = 0.02$, that is z_p is expected to be exceeded twice every 100 years when using the best suiting model. As outlined in Sect. 8.2.4, a return level has to be calculated for a certain time point t_0 in case parameter estimates of a non-stationary model are used. We here choose $t_0 = 01.01.1996$.

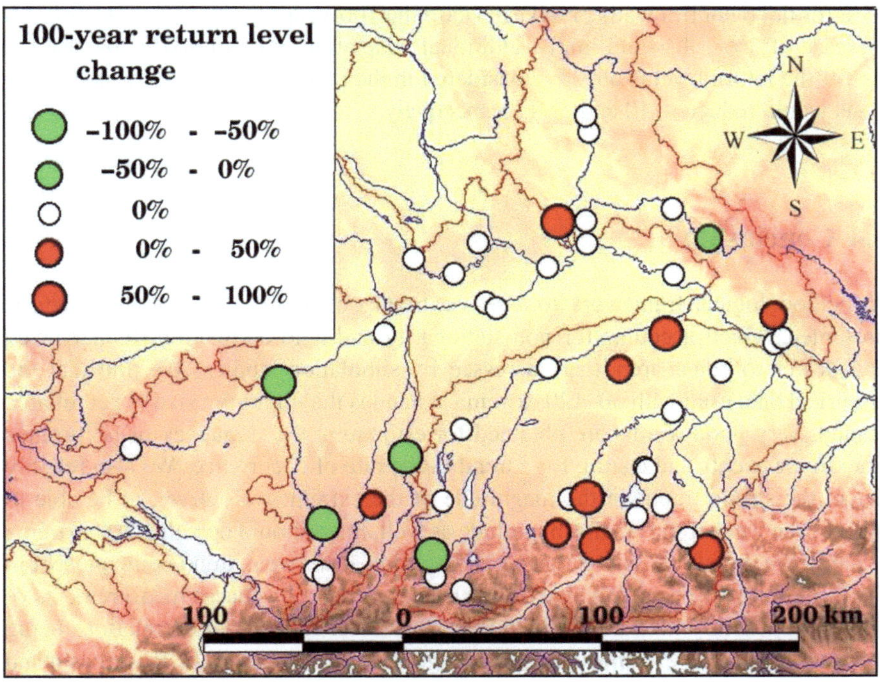

Fig. 8.3 Change of probability of exceeding a 100-year return level. The 100-year return level is calculated according to a stationary point process model which is fitted to the extremes. Then we calculate p^*, the probability of exceeding this 100-year return level, using the parameter estimates of the best fitting, possibly non-stationary point process model at time point $t_0 = 01.01.1996$. The corresponding changes of probability are depicted. *White* indicates no change, i.e. the best fitting model is a stationary one. A decreasing probability is marked in *green* and an increasing one in *red*

As shown in Fig. 8.3, we observe severe changes in the probability of exceedance in case non-stationarity in the extremes is considered. The occurrence of non-stationary river discharge extremes in the Danube River basin and the corresponding change of flood risk is already recognised by water management authorities [8.22], but up to now the common methodological canon does not incorporate the systematic evaluation of this kind of change. Our results show that the probability of exceedance gets lower for some stations; most of them are located in the southeastern part of the Danube River basin. The stations with a decreasing probability of exceedance are the Iller River at Kempten and Wiblingen, the Lech River at Landsberg, the Schwarzer Regen River at Teisnach and the Ammer River at Oberammergau. For 9 of the 47 assessed stations the probability of exceedance gets higher, in some cases more than 50%. Those gauges are the Geltnach River at Hörmannshofen, the Naab River at Heitzenhofen, the Vils River at Rottersdorf, the Große Vils River at Vilsbiburg, the Ramsauer Ache River at Ilsank, the Inn River at Oberaudorf and Rosenheim, the Leitzach River at Stauden and the Ilz River at Kalteneck.

The adaptation costs which correspond to an increase in flood risk do not change linearly with an increase in flood risk (cf. [8.22]). In an exemplary study for Baden-Württemberg cost calculations suggest, for example, additional adaptation costs of 55% for flood control walls or dykes in case the 100-year return level rises 20%. An increase of 30% of would imply additional adaptation costs of 157%. In case the rise of the return level is already considered in the planning phase, those additional costs would reduce to 10 and 13%, respectively.

8.4 Conclusion

A methodological framework to assess extremes of hydro-meteorological data is presented, which accounts for non-stationarity and auto-correlation. Its adequacy and power to detect trends are assessed by simulation studies. We find that the empirical data used with 30–120 extrema at hand is the minimum size to get reliable results. Very good results are obtained for estimating the frequency of occurrence of extreme events and testing for a trend in the rate of occurrence. We also reliably detect the strong trends in the magnitude of the extremes. However, to be able to detect even weak trends in the magnitude as well, there should be rather 200 extrema at hand. We therefore assume that we did not find every weak trend being present in the Danube River basin. We analysed extremes of daily discharge measurements of about 50 stations within the Danube River basin. We found auto-correlations being present in the set of extremes of all discharge records of the Danube River basin and had to decluster them. One-third of the stations exhibit also non-stationary extremes. Importantly, the estimates of the extreme value distribution parameters may be biased in case a stationary model is used to represent non-stationary extremes. The magnitude of this bias depends on the trend strength. To demonstrate the relevance of the changes which arise when allowing for non-stationary extreme value models, we exemplarily assess return levels. They are an important assessment measure for water management authorities. We identified changes up to 100% for the probability of exceedance of the 100-year return level. This implies a potential doubling of the damage costs.

Regarding the trend tendencies, the frequency of occurrence of extreme events always increases when it is detected to be non-stationary. In case a change in the magnitude of the extreme events is found, we observe decreasing and increasing tendencies. The spatial pattern of the trend in extremes is not immediately interpretable. Our results suggest that we have to proceed gauge by gauge. We found specific, site-related structures for all statistical characteristics assessed and often neighbouring gauges to behave distinctively. These results show that the impact of climate change is not directly detectable in river discharge yet, but is blurred by other factors. Evidence for trends in precipitation, for example, is globally stronger than for changes in flooding [8.23, 8.33]. Possible factors, which are not necessarily in tune with gross climate-related drivers, may be land use, soil conditions, weather regimes, elevation, moisture or catchment size.

Our methodology provides uncertainty intervals along with each parameter estimate and assessment measure. The resulting confidence intervals show that the examination of uncertainty is a crucial prerequisite for the evaluation and interpretation of the results of extreme value analysis.

We conclude that the application of an extended extreme value analysis framework, as presented here, is necessary to adequately assess non-stationary extreme values. Trends in frequency and magnitude of extremes of river discharge are anticipated because of climate change and we already find a noteworthy fraction of the empirical records analysed exhibiting non-stationary extremes. The incorporation of non-stationarity in extreme value analysis leads to different results of important assessment measures, such as the return level. The trend shape determines whether its influence lowers or augments the return level. The results obtained may serve as a basis for a physical interpretation of the trends. Besides, these statistical characteristics of river discharge may be used to validate the output of, e.g. hydrological models. Furthermore, common practice of water management authorities can be improved and costs for flood protection buildings can be calculated with higher accuracy. Thus, e.g. construction costs are reduced in case return-level estimates are overestimated and possible future damage costs are dampened in case it is shown that those estimates are too low.

Acknowledgments The study presented was partly carried out within the framework of the project *Skalenanalyse*. We thank the German Federal Ministry of Science for financial support (under grant no. 0330271) and the Bavarian Environmental Protection Agency for discharge data.

References

8.1. A. Bárdossy, F. Fili z, Identification of flood producing atmospheric circulation patterns. J. Hydrol. **313** 48–57 (2005)
8.2. A. Bárdossy, S. Pakosch, Wahrscheinlichkeiten extremer Hochwasser unter sich ändernden Klimaverhältnissen. Wasserwirtschaft, **7–8**, 58–62 (2005)

8.3. BlfW. Bayrisches Landesamt für Wasserwirtschaft: Klimaänderungsfaktoren bei Planungen für den Hochwasserschutz. Water Management Directive, 2006
8.4. R. Brázdil, C. Pfister, H. Wanner, H. von Storch, J. Luterbacher. Historical climatology in Europe – the state of the art. Clim. Change, **70**, 363–430 (2005)
8.5. H.J. Caspary, A. Bárdossy, Markieren die Winterhochwasser 1990 und 1993 das Ende der Stationarität in der Hochwasserhydrologie infolge von Klimaänderungen? Wasser Boden, **47**(3), 18–24 (1995)
8.6. S. Coles, *An Introduction to Statistical Modeling of Extreme Values*. (Springer, Berlin, 2001)
8.7. S. Coles, L.R. Pericchi, S. Sisson, A fully probabilistic approach to extreme rainfall modeling. J. Hydrol. **273**, 35–50 (2002)
8.8. D.R. Cox, V.S. Isham, P.J. Northrop, Floods: Some probabilistic and statistical approaches. Philos. Trans. R. Soc. A: Math., Phys. Eng. Sci., **360**, 1389–1408 (2002)
8.9. D.R. Cox, P.A.W. Lewis, *The Statistical Analysis of Time Series Events* (Methuen, London, 1966)
8.10. A.C. Davison, D.V. Hinkley, *Bootstrap Methods and Their Application* (Cambridge University Press, Cambridge, 1997)
8.11. DFO. Dartmouth Flood Observatory. Technical report, 2004. Data available at: `http://floodobservatory.colorado.edu/`, last access: 2010
8.12. B. Dietzer, Th. Günther, A. Klämt, H. Matthäus, T. Reich, Langzeitverhalten hydrometeorologischer Größen. Technical report, DWD, 2001. Klimastatusbericht
8.13. P. Embrechts, C. Klüppelberg, T. Mikosch, *Modelling Extremal Events* (Springer, Berlin, 1997)
8.14. K. Engeland, H. Hisdal, A. Frigessi, Practical extreme value modelling of hydrological floods and droughts: A case study. Extremes, **7**, 5–30 (2004)
8.15. Ch. A.T. Ferro, *Statistical Methods for Clusters of Extreme Values*. PhD thesis, University of Lancaster, Great Britain, 2003
8.16. R.A. Fisher, L.H.C. Tippett, Limiting forms of the frequency distribution of the largest or smallest member of a sample. Proc. Cambridge Phil. Soc., **24**, 180–190 (1928)
8.17. C. Frei, H.C. Davies, J. Gurtz, C. Schär, Climate dynamics and extreme precipitation and flood events in Central Europe. Integr. Assess., **1**, 281–299 (2000)
8.18. B.V. Gnedenko, Sur la distribution limite du terme maximum d'une série aléatoire. Ann. Math., **44**, 423–453 (1943)
8.19. IPCC. *Climate Change 2001: The Scientific Basis. Contribution of Working Group I to the Third Assessment Report of the Intergovernmental Panel on Climate Change* (Cambridge University Press, Cambridge, 2001)
8.20. M. Kallache, *Trends and Extreme Values of River Discharge Time Series*. PhD thesis, University of Bayreuth, Bayreuth, 2007
8.21. R.W. Katz, M.B. Parlange, P. Naveau, Statistics of extremes in hydrology. Adv. Water Resourc., **25**, 1287–1304 (2002)
8.22. KLIWA, Klimaveränderung und Konsequenzen für die Wasserwirtschaft. 2. Symposium. Technical report (Arbeitskreis KLIWA, München, 2004)
8.23. Z.W. Kundzewicz, D. Graczyk, I. Przymusinska, T. Maurer, M. Radziejewski, C. Svensson, and M. Szwed, Detection of change in world-wide hydrological time series of maximum annual flow. Technical report, Global Runoff Data Centre (GRCD) 2004
8.24. Z.W. Kundzewicz, S.P. Simonovic, Non-structural flood protection and sustainability. Water Int., **27**, 3–13 (2002)
8.25. C.J. MacLean, Estimation and testing of exponential polynomial rate function within the nonstationary Poisson process. Biometrika, **61**(1), 81–85 (1974)
8.26. D.R. Maidment, *Handbook of Hydrology* (McGraw-Hill, New York, 1993)
8.27. M. Mudelsee, M. Börngen, G. Tetzlaff, U. Grünewald. Extreme floods in central Europe over the past 500 years: Role of cyclone pathway "Zugstrasse Vb". J. Geophy. Res., **109**, D23101 (2004)

8.28. J.A. Nelder, R.W.M. Wedderburn, Generalized linear models. J. R. Stat. Soc. A, **135**: 370–384 (1972)
8.29. M. Nogaj, P. Yiou, S. Parey, F. Malek, P. Naveau, Amplitude and frequency of temperature extremes over the North Atlantic region. Geophy. Res. Lett., **33**, L10801, (2006)
8.30. S. Pakosch, Statistische Methoden zur stationären und instationären Auswertung von gemessenen Maximalabflüssen mit Hilfe theoretischer Verteilungsfunktionen. Master's thesis, Universität Stuttgart, 2004
8.31. C. Pfister, R. Brázdil, R. Glaser, M. Barriendos, D. Camuffo, M. Deutsch, P. Dobrovolný, S. Enzi, E. Guidoboni, O. Kotyza, S. Militzer, L. Raczii, F.S. Rodrigo. Documentary evidence on climate in sixteenth-century Europe. Clim. Change, **43**, 55–110 (1999)
8.32. L. Pfister, J. Kwadijk, A. Musy, A. Bronstert, and L. Hoffmann. Climate change, land use change and runoff prediction in the Rhine-Meuse basins. River Res. Appl., **20**, 229–241 (2004)
8.33. A.J. Robson, Evidence for trends in UK flooding. Philos. Trans. R. Soc. A, **360**, 1327–1343 (2002)
8.34. H.W. Rust, *Detection of Long-Range Dependence – Applications in Climatology and Hydrology*. PhD thesis, Potsdam University, Potsdam, 2007
8.35. D. Schröter, M. Zebisch, T Grothmann, Climate change in Germany – vulnerability and adaptation of climate-sensitive sectors. Technical report, DWD, 2005. Klimastatusbericht.
8.36. R.L. Smith, Maximum likelihood estimation in a class of non-regular cases. *Biometrika*, **72**, 67–90 (1985)
8.37. T.M.L. Wigley, Climatology: Impact of extreme events. *Nature*, **316**, 106–107 (1985)
8.38. X. Zhang, F.W. Zwiers, G. Li, Monte Carlo experiments on the detection of trends in extreme values. J. Clim., **17**(10), 1945–1952 (2004)

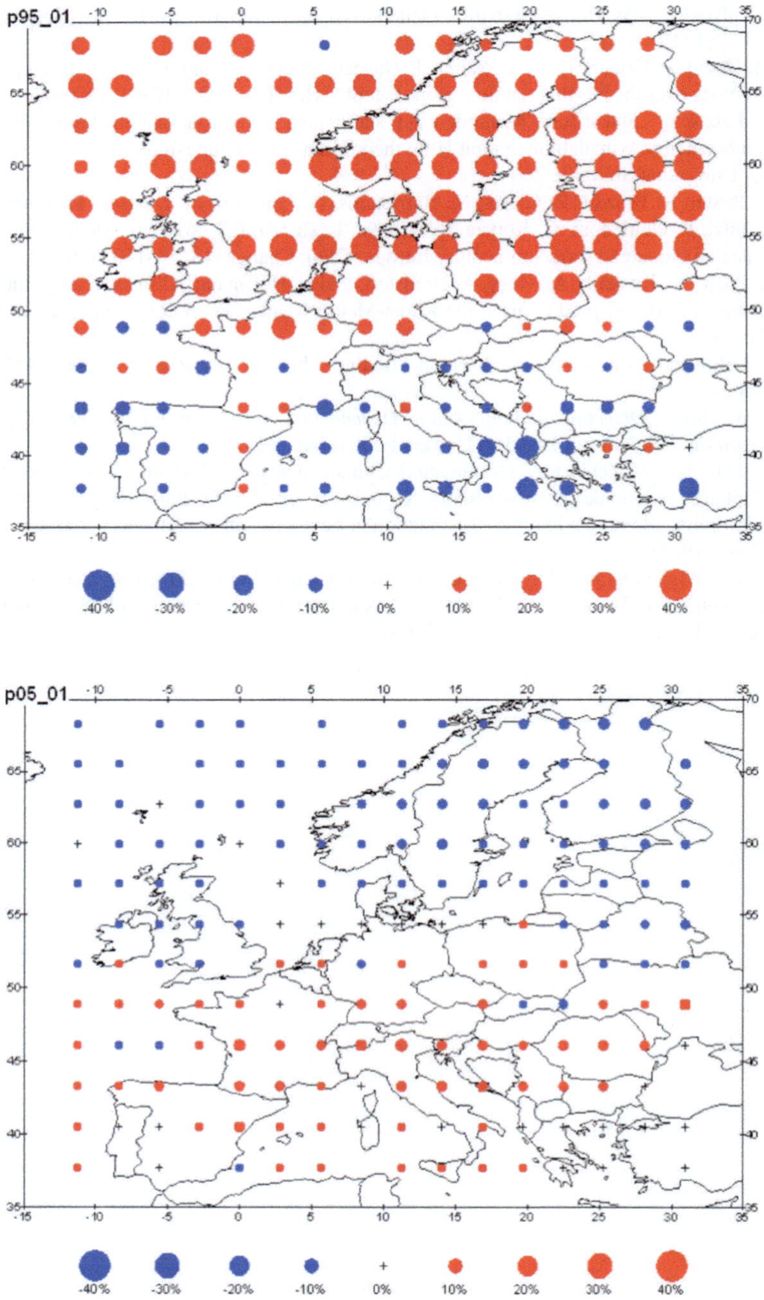

Application of the generalized time series decomposition technique on modelled time series of the coupled global model ECHAM4/OPYC3 [9.12], 1990–2100, under SRES A2 scenario shows in January in the northern part of Europe strong increases in the probability for exceeding the 95th percentile (*top*) and less pronounced decreases in the probability for falling under the 5th percentile (*bottom*). In central Europe an area with increases in both kinds of extremes can be seen

Chapter 9
Extreme Value and Trend Analysis Based on Statistical Modelling of Precipitation Time Series

Silke Trömel and Christian-D. Schönwiese

Application of a generalized time series decomposition technique shows that observed German monthly precipitation time series can be interpreted as a realization of a Gumbel-distributed random variable with time-dependent location parameter and time-dependent scale parameter. The achieved complete analytical description of the series, that is, the probability density function (PDF) for every time step of the observation period, allows probability assessments of extreme values for any threshold at any time. So, we found in the western part of Germany that climate is getting more extreme in winter. Both the probability for exceeding the 95th percentile and the probability for falling under the 5th percentile are increasing. Contrary results are found in summer. The spread of the distribution is shrinking. But in the south, relatively high precipitation sums become more likely and relatively low precipitation sums become more unlikely in turn of the twentieth century.

Furthermore, the decomposition technique provides the mean value of the Gumbel-distributed random variable for every time step, too. So, an alternative approach for estimating trends in observational precipitation time series is achieved. On that way, the non-Gaussian characteristics can be taken into account and robust estimates can be provided. In contrast, application of the least-squares estimator to non-Gaussian climate time series often leads to everestimated trends in the expected value.

9.1 Introduction

The analysis of climate variability as reflected in observational records is an important challenge in statistical climatology. In particular, it is important to estimate reliable trends in the mean value as well as changes in the probability of extreme values. In case of monthly or annual temperature data, often Gaussian distributions are adequate for their description. If any linear or non-linear trend of the time

---

S. Trömel (✉)
Institut für Meteorologie Universität Bonn, Bonn, 53121, Germany
e-mail: silke.troemel@uni-bonn.de

J.P. Kropp, H.-J. Schellnhuber (eds.), *In Extremis*,
DOI 10.1007/978-3-642-14863-7_9, © Springer-Verlag Berlin Heidelberg 2011

series average occurs, this may be described by a shift of a Gaussian probability density function (PDF). Additionally, some other time series components like the annual cycle, for example, may also vary and cause a shift to higher and lower values again during the observation period as discussed below[9.2]. Grieser et al. (2002)[9.2] consider temperature time series as a superposition of trends, annual cycle, episodic components, extreme events and noise. Thereby, Gaussian assumptions are used, which implies that residuals cannot be distinguished from a realization of a Gaussian-distributed random variable. However, this simple model of a shifting Gaussian distribution with constant variance over the observation period is not suitable to describe the variability of precipitation time series. We observe a skewed distribution as well as seasonal and sometimes long-term changes in the shape of the distribution. Moreover, changes in the spread of the distribution have to be considered, too. In summary, in addition to the expectation, also the variance and/or further moments may vary with time.

Evidently, wrong assumptions concerning the PDF lead also to biased estimators of trends or probabilities of extreme values. In consequence, we present a generalized time series decomposition technique which allows any PDF and any related parameter change in time, that is, changes in the location, the scale and the shape parameter of a PDF are allowed. In case of Gaussian distributions, for example, the location parameter is realized by the average and the scale parameter is realized by the variance.

Here, we apply this technique on precipitation time series from a German station network focussed on extreme value and trend analysis. The detection of structured time series components like trends, called signals, is based on several parameters of a Gumbel or Weibull distribution, respectively, as described in the following. Sect. 9.2 provides the definition of the components for the analysis of monthly climate time series and Sect. 9.3 presents a brief overview of the detection of the analytical functions reflecting the time dependence of the different distribution parameters. For all details, see [9.10, 9.11].

9.2 Components

The equation

$$S_{j,k}(t) = d_{j,k} t^k \cos\left(2\pi \frac{j}{12} t\right) + e_{j,k} t^k \sin\left(2\pi \frac{j}{12} t\right) \tag{9.1}$$

with wave number $j = 1, \ldots, 6$ per year and $k = 0, 1, 2$ gives the basis functions to describe the seasonal component. Evidently, the maximum wave number 6 is tuned to the analysis of climate data with a monthly sampling rate. Besides fixed annual cycles, changes in amplitude and phase are allowed. For the amplitude linear and quadratic time dependence is considered. Superposition of two or three harmonics of the annual cycle with the same wave number j but different time dependence k in one time series yields

$$S_j(t) = A_j(t)\cos\left(2\pi\frac{\mathrm{j}}{12}\left(t - t_j\right)\right), \tag{9.2}$$

with amplitude

$$A_j(t) = \sqrt{\sum_{k=0}^{2}\left(d_{j,k}^2 + e_{j,k}^2\right)t^{2k}} \tag{9.3}$$

and phase

$$t_j(t) = \frac{12}{2\pi}\arctan\left(\frac{\sum_{k=0}^{2} d_{j,k}t^k}{\sum_{k=0}^{2} e_{j,k}t^k}\right). \tag{9.4}$$

In this way, the detection of linear, progressive and degressive shaped changes in phase and amplitude of the annual cycle is possible.

In addition, trends up to the order 5 are considered:

$$T_i(t) = g_i + h_i t^i, \quad \text{with } i = 1, \ldots, 5. \tag{9.5}$$

To detect all significant structures but neglecting, on the other hand, all unsignificant structures, in a first step the detection of the seasonal component and the trend component of the parameters is performed simultaneously within the modified stepwise regression procedure (see Sect. 9.3 and [9.9]).

In a second step we observe sometimes relatively low-frequency variations superposed on the components mentioned above. So we offer also polynomial equations up to the order 5:

$$V_l(t) = c_o + \sum_{i=1}^{l} c_i t^i. \tag{9.6}$$

In a third step a search for extreme events, which are independent from changes in the parameters of the distribution, is performed. According to Grieser et al. [9.2], we define extreme events as a relatively small number of extreme values which are unexpected within the scope of the fitted statistical model. In contrast to extreme events, extreme values are relatively high or low values which occur by chance. A positive trend in the mean value, for example, increases the probability of occurrence of relatively high extreme values.

Within the strategy introduced detected extreme events are extracted and replaced by a random value distributed conform with the PDF and the two parameters at the given time (see [9.2] for further details). The iterative procedure for the detection of trends, seasonal component and low-frequency variations is applied until no further extreme events are found. In the application of Gaussian assumptions and the least-squares method, the quadratic function which has to be minimized in order to fit any analytical functions is called distance function of the Gaussian distribution. Generally speaking, it depends on the robustness [9.3] of the applied distance function whether structured components are more or less influenced by extreme events.

Using an iterative procedure to find within these functions the best model equations of two parameters of a PDF leads to a very extensive procedure. Note that the model equation of one parameter influences the equation for the second parameter and vice versa. To reduce this effort, an additional restriction is introduced. For one of the parameters all functions mentioned above are offered:

$$P_g(t) = \sum S_{j,k}(t) + \sum T_i(t) + V_l(t). \tag{9.7}$$

The second parameter is assumed to be of minor relevance to describe the time series. Only the one cycle per year harmonic and one trend function can be chosen:

$$P_r(t) = S_{1,0} + T_i. \tag{9.8}$$

At the end of the time series decomposition procedure, a priori assumed residual distribution is tested. In case of a chosen Gumbel distribution, residuals should follow a Gumbel distribution $G(0, 1)$ with a location parameter equals 0 and a location parameter equals 1 after elimination of detected structured components in the parameters $P_g(t)$ and $P_r(t)$. Additional stationarity of the distribution points to a complete description of the time series within the PDF and its time-dependent parameters.

It depends on the characteristics of the time series under consideration, which PDF has to be chosen and which PDF parameter implies a larger number of degrees of freedom in comparison to others. Not before the end of the procedure, including the analysis of the statistical properties of the residuals, it can be decided whether the chosen model provides an adequate description of the time series.

9.3 The Distance Function and the Model Selection Criterion

The basis of any time series decomposition technique is the distance function and a model selection criterion. The least-squares estimator broadly used in trend analyses is the maximum likelihood estimator under the assumption of Gaussian-distributed residuals with constant variance. This quadratic function which has to be minimized is called distance function of the Gaussian distribution. So, consistently with the maximum likelihood principle another distance function, defined as the negative logarithm of the PDF, replaces the function of squared errors to be minimized, if we choose another distribution as basis of the decomposition procedure, for example, the Gumbel or the Weibull distribution. Then time dependence for different distribution parameters can be allowed. With the exchange of the distance function, fitted basis functions describe changes in the location, scale or shape parameter of an appropriate PDF. As an example, the corresponding distance function of the Gumbel distribution is

$$\rho(x,t) = \ln(b(t)) + \exp\left(-\frac{x-a(t)}{b(t)}\right) + \frac{x-a(t)}{b(t)}. \tag{9.9}$$

Under the assumption of statistically independent random variables the coefficients in the model equations for estimating $a(t)$ and $b(t)$ are chosen by minimizing

$$\sum_t \rho\,(x,t) = \min., \tag{9.10}$$

equivalent to the maximum of the loglikelihood function. In this work, Powells method [9.6], p. 406, is used to minimize ρ in the multidimensional space. Figure 9.1 shows the PDF of the Gaussian distribution and on the right-hand side its distance function, the quadratic function. The more deviant the points, the greater the weight. The influence increases very fast because the probability of occurrence of relatively high or relatively low values is very small in the Gaussian case. For comparison the Gumbel distribution for two different location and scale parameters and the associated distance functions are shown in Fig. 9.2. The tails are more prominent and consequently, if we take a look at the distance functions, the influence increases less rapidly, and one value in a given distance from the location estimator has more weight the smaller the scale parameter. So, structured components can be detected in different parameters and estimators of different parameters compete with each other. Finally, Fig. 9.3 shows the Weibull distribution with two different scale and shape parameters. And on the right-hand side the distance functions clearly show the dependence on the shape of the distribution. One important point is that the PDF chosen and the basis functions used to describe the signals are complementary. The other point is that the basis used to describe one parameter of the PDF influences the basis necessary to describe the second parameter. That is why a dynamic procedure is necessary to estimate the coefficients of the basis functions

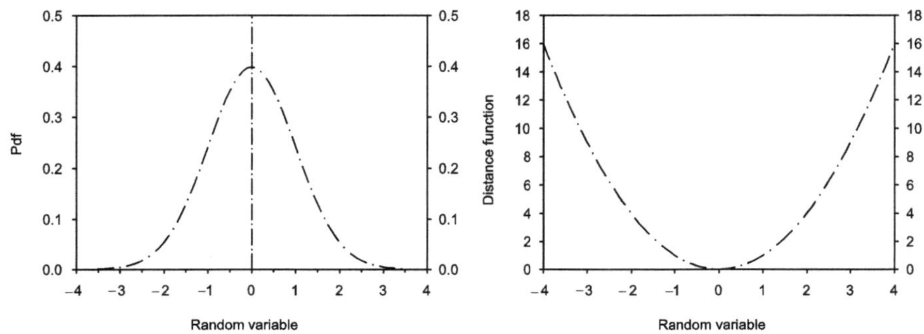

Fig. 9.1 The PDF (*left*) of the Gaussian distribution and its distance function (*right*) under the assumption of constant variance

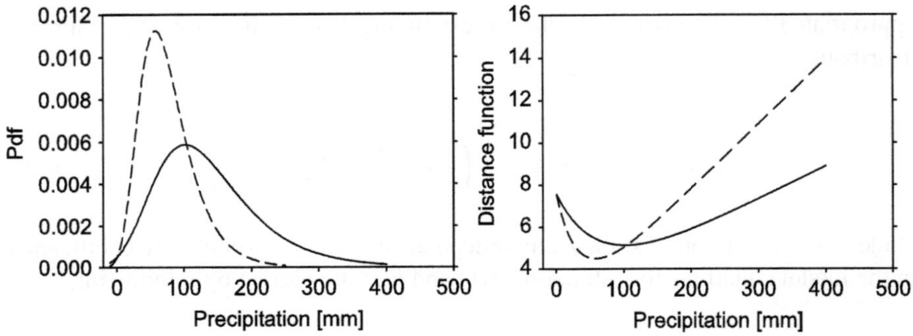

Fig. 9.2 The PDF (*left*) of the Gumbel distribution with two different location and scale parameters and the associated distance functions (*right*)

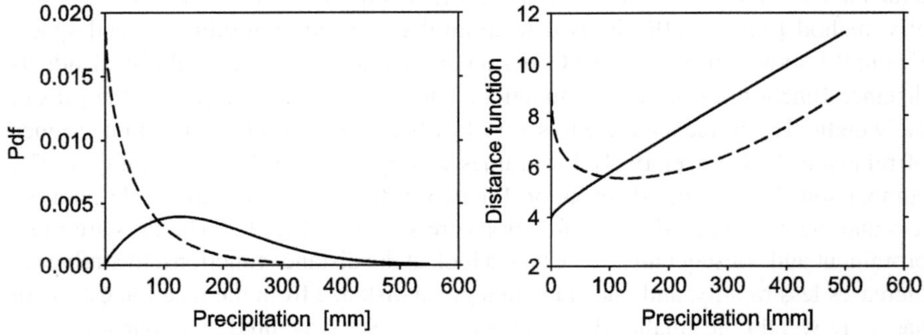

Fig. 9.3 The PDF (*left*) of the Weibull distribution with two different scale and shape parameters and the associated distance functions (*right*)

of the different components of both parameters simultaneously. A flexible model selection criterion often used is the stepwise regression. Stepwise regression [9.9, p. 166], represents a dynamic model selection criterion in order to find the optimal regression equation. Within the generalized time series decomposition a modification to handle two distribution parameters is required.

The common iterative application of forward selection and backward elimination is used to determine the model equation of a first distribution parameter as, for example, the scale parameter $b(t)$. But the distance function used now, also depends on the selected model equation describing the second distribution parameter (see (9.9)). Between these alternate parts redefinition of the second parameter ($a(t)$) is inserted, now taking into account the selected model of the first parameter ($b(t)$). In the modified version, the flexible strategy of stepwise regression is used twice. Now the parameters of the model equations influence themselves mutually. A regressor of the first parameter selected at an earlier stage can be superfluous because of a new entry candidate in the model equation of the first parameter or because of the actualization of the second parameter and vice versa.

The common F-test statistic broadly used in regression analysis to decide whether a specific regressor contributes significantly to explained variance is sensitive to departures from the Gaussian distribution and, therefore, has to be replaced. A test statistic based on a likelihood ratio test seems to be more applicable, because likelihood values are computed anyhow minimizing the distance function (9.10). Define $D(p)$ as the minimum value of (9.10) subject to the model containing p regressors and $D(q)$ as the minimum value of (9.10) subject to the model containing q regressors, with $p > q$, a direct generalization of the common F-test may be based upon the statistic

$$F_{\mathrm{M}} = (D(p) - D(q))\,/(p - q) \tag{9.11}$$

[9.8] with $p - q$ and $N - p$ degrees of freedom. Here N represents the number of data. If the density function of the residuals has not the form $\exp(-\rho)$, the model assumptions are not fulfilled and a correction term is necessary. For small sample sizes Schrader and Hettmansperger propose to compare F_{M} with a critical value from a central F distribution.

The modified stepwise regression including a modified F-statistic (9.11) represents the basis of a generalized time series decomposition technique. Inserting the corresponding distance function it can be used for time series analysis based on any data distribution and any statistical model. Thereby, the initial choice of the probability density function and the corresponding distance function should depend on the characteristics of the climate variable under consideration. The Gumbel distribution, the lognormal distribution as well as the Gamma distribution are promising distributions already by reason of the skewness. However, the Weibull distribution seems to be a promising choice when changes in the shape of the distribution occur. Not until the final residual analysis at the end of the decomposition tests whether a priori assumed statistical model can be confirmed. If so, a complete analytical description of the time series is achieved.

Actually, the generalized time series decomposition technique in a deterministic and a statistical part is applied to four different models: Observed time series can be interpreted as

- a realization of a Gaussian-distributed random variable with time-dependent location (mean value) and additional time-dependent scale parameter (standard deviation);
- a realization of a Gumbel-distributed random variable with time-dependent location and scale parameter, it depends on the choice to offer the location or the scale parameter the greater pool of regressors if we talk about
  - the Gumbel model with special emphasize on location or
  - the Gumbel model with special emphasize on scale;
- a realization of a Weibull-distributed random variable with time-dependent scale and shape parameter.

9.4 Application to a German Station Network

9.4.1 General Remarks

The generalized time series decomposition technique is applied now to monthly precipitation sums from a German station network of 132 time series covering 1901–2000. At least for the most part of the sample the decomposition technique shows that observed time series can be interpreted as a realization of a Gumbel-distributed random variable with time-dependent location parameter $a(t)$ and time-dependent scale parameter $b(t)$. So the decomposition is based on the probability density function (PDF) of the Gumbel distribution

$$f(x,t) = \frac{1}{b(t)}\left\{\exp\left(-\frac{x-a(t)}{b(t)}\right)\exp\left[-\mathrm{e}^{-(x-a(t))/b(t)}\right]\right\}. \quad (9.12)$$

Since the Gumbel model with emphasis on scale leads to a better description of the time series, we define the location parameter $a(t) = P_r(t)$ and the scale parameter $b(t) = P_g(t)$ (see (9.7) and (9.8)). However, 7 out of 158 400 (132 × 1200) monthly precipitation sums are extracted as extreme events which are unexpected within the scope of the fitted Gumbel model. As aforementioned the analysis of the remaining residuals represents an important part of the analysis procedure. After elimination of the detected structures in the location and in the scale parameter, residuals should fulfill the condition of the a priori assumed statistical model. Concerning the German station network of 132 monthly precipitation sums, the residuals should be undistinguishable from the realization of a Gumbel-distributed random variable $G(0, 1)$ with the location parameter 0 and the scale parameter 1. In fact the Kolmogorov–Smirnov test [9.6] rejects in 7 out of 132 cases this hypothesis with a probability larger than 90%. This is less (<10%) than may be expected by chance. Furthermore, again a Kolmogorov–Smirnov test statistic is used in order to check the stationarity of the residuals. Now in 6 out of 132 cases stationarity is rejected with a probability larger than 90%. Consequently, the residuals confirm the model applied and a complete description of the observed time series on the basis of the Gumbel model with emphasis on the scale could be achieved.

The provided PDF for every time step of the observation period allows probability assessments of extreme values, i.e. the probability for exceeding a given precipitation sum (threshold) at any time. In this context it should be reemphasized that in contrast to analyses dealing with the familiar generalized extreme value distribution [9.1, 9.4], the generalized time series decomposition technique presented in this issue requests the analytical description of the whole times series instead of addressing exclusively the maxima of the observations. Furthermore, with the PDF the mean value of the Gumbel-distributed random variable can be given for every time step, too.

In Sect. 9.4.2 the time series decomposition of an exemplary time series is discussed in more detail for a better understanding of the method introduced.

Subsequently, results concerning changes in the probability of extreme values and the expected value of the entire station network are presented in Sects. 9.4.3 and 9.4.4.

9.4.2 Example: Eisenbach–Bubenbach

In case of the Gumbel model with emphasis on scale the greater pool of regressors is offered to the scale parameter $b(t)$ of the Gumbel distribution and the smaller one to the location parameter $a(t)$. Table 9.1 shows the significant functions detected in the location parameter $a(t)$ and the scale parameter $b(t)$ of the Gumbel distribution to describe the time series observed in Eisenbach–Bubenbach (47.97°N, 8.3°E). Phase angles are defined with reference to December 15. Consequently, we observe a superposition of two harmonics with wave number one per year. The function $S_{1,0}$ has a constant amplitude with maximum in September but $S_{1,1}$ reveals linear time dependence in the amplitude and has its maximum in January. So, the winter becomes more variable and the summer becomes less variable during the observation period. The seasonal component in the scale parameters shows a phase shift of 109 days from September to January. The resulting effects can also be seen in Figs. 9.4 and 9.5. The former figure illustrates the decrease in the scale parameter in

Table 9.1 Significant functions in the location parameter $a(t)$ and scale parameter $b(t)$ for the description of the monthly precipitation time series observed in Eisenbach–Bubenbach

Parameter	Function	Amplitude (mm)	Phase (days)
$b(t)$	$S_{1,0}$	$7.04 \times t^k$	−82.73
	$S_{1,1}$	$0.148 \times 10^{-1} \times t^k$	44.87
	T_1	5.78	–
$a(t)$	$S_{1,0}$	$-8.89 \times t^k$	24.21

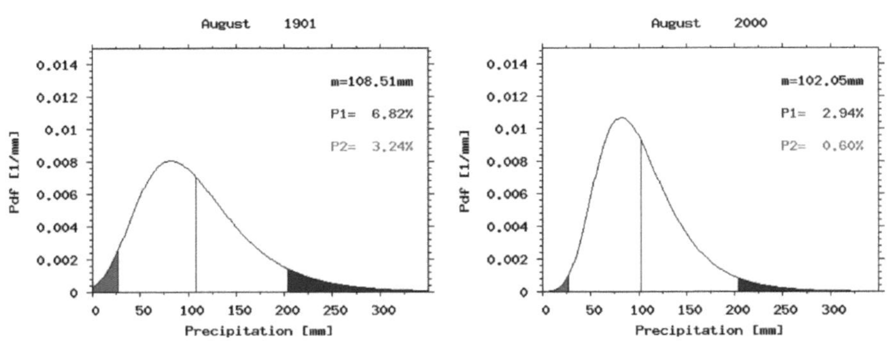

Fig. 9.4 On the basis of the entire time series estimated PDFs for two time steps: August 1901 (*left*) and August 2000 (*right*). Areas below the 5th percentile are marked in *light grey*, the areas above the 95th percentile in *dark grey* and the *vertical line* marks the expected value m. The respective value m, the probability $P1$ for exceeding the 95th percentile and the probability $P2$ for falling under the 5th percentile are given in the *upper right corner*, respectively

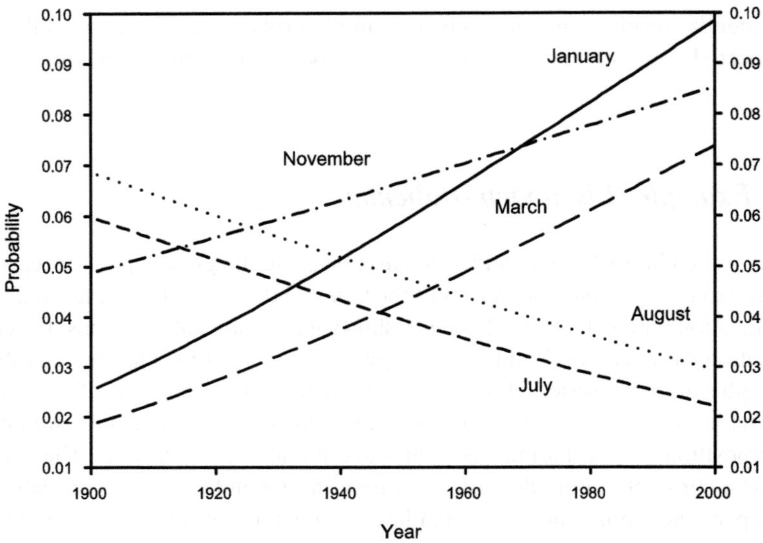

Fig. 9.5 Estimated probabilities for exceeding the 95th percentile between 1901 and 2000 in January, March, July, August and November at the station Eisenbach–Bubenbach

August. The functions given in Table 9.1 define the scale and the location parameter of the Gumbel distribution for every time step. No low-frequency variations ($V_1(t)$) and no unexpected values defined as extreme events (see Sect. 9.2) within the time series decomposition technique can be detected in the time series. However, probability assessments of extreme values can be based on the analytical description provided. Figure 9.4 shows the estimated PDFs for the first and the last August in the observation period. The probability $P1$ for exceeding the 95th percentile, in this case 204 mm, and the probability $P2$ for falling under the 5th percentile, in the present case 27 mm, are also given for the two time steps. The corresponding areas within the PDF are marked in dark grey and light grey, respectively. First of all, we see is a decrease in the scale parameter, going along with a decrease in the probability of exceeding the 95th percentile from 6.8 to 2.9% and we see a smaller decrease in the probability of falling under the 5th percentile from 3.2 to 0.6%. The probability of both kinds of extremes has decreased and in contrast to the Gaussian model the expected value is also affected. The expected value decreases from 109 to 102 mm in the observation period. It is worth to mention that the least-squares estimator is not able to describe those decreases or increases in variability and not the influence on the expected value either. In Fig. 9.5 integration over the dark grey area for every time step and different months has been done. Contrary tendencies in the probability of exceeding the 95th percentile can be seen. At the beginning the relatively high precipitation sums occured with higher probability in summer but at the end of the observation period the probability for exceeding the threshold is highest in January. Both the phase shift in the seasonal component and the positive linear trend detected in the scale parameter describe an increasing variability in winter. The distribution is widening. The tendencies in summer are contrary. Both kinds of extremes become more unlikely between 1901 and 2000.

9.4.3 Probability Assessment of Extreme Values

If the time series decomposition in a statistical and a deterministic part succeeded, the time-dependent PDF $f(x,\ P_g(t),\ P_r(t))$ as a complete analytical description is provided (see again Fig. 9.4).

In this model

$$
\begin{aligned}
p_{\mathrm{G}}(x \geq x_s, t) &= 1 - \int_{-\infty}^{x_s} \frac{1}{P_g(t)} \left\{ \exp\left(-\frac{x - P_r(t)}{P_g(t)} \right) \exp\left[-\mathrm{e}^{-(x - P_r(t))/P_g(t)} \right] \right\} \mathrm{d}x \\
&= 1 - \exp\left\{ - \exp\left(-\frac{x_s - P_r(t)}{P_g(t)} \right) \right\}
\end{aligned}
\tag{9.13}
$$

gives the probability for exceeding the threshold x_s at time t. Significant structured components in the location and the scale parameter cause changes in the probability of occurrence of these high precipitation sums. The probability for falling under x_s is given by

$$
\begin{aligned}
p_{\mathrm{G}}(x \leq x_s, t) &= \int_{-\infty}^{x_s} \frac{1}{P_g(t)} \left\{ \exp\left(-\frac{x - P_r(t)}{P_g(t)} \right) \exp\left[-\mathrm{e}^{-(x - P_r(t))/P_g(t)} \right] \right\} \mathrm{d}x \\
&= \exp\left\{ - \exp\left(-\frac{x_s - P_r(t)}{P_g(t)} \right) \right\}.
\end{aligned}
\tag{9.14}
$$

In Fig. 9.6 tendencies in the probability of a monthly precipitation total greater than the 95% percentile (left map) and tendencies in the probability of a monthly precipitation total smaller than the 5% percentile (right map) are shown, that is, for every single time series the threshold is selected so that 95% (5%) of all monthly rainfall totals of the series are smaller than the respective threshold. The result is in January, an increase in the probability of exceedance in the overwhelming majority of stations. In the right map the tendencies in the probability of a monthly precipitation total less than the 5th percentile are seen. In January we see in the northern part of Germany increases in the probability of occurrence for these small precipitation totals, too. So relatively high and relatively small precipitation totals become more likely during the twentieth century. The distribution is widening. However, in the south the distribution is shifting to higher values.

Figure 9.7 shows the respective results in August. In the northern part of Germany decreases in the probability of relatively high precipitation sums are detected during the twentieth century. But in the south we see several increases in August as well as in January. In the right map it can be seen that decreases in the probability of

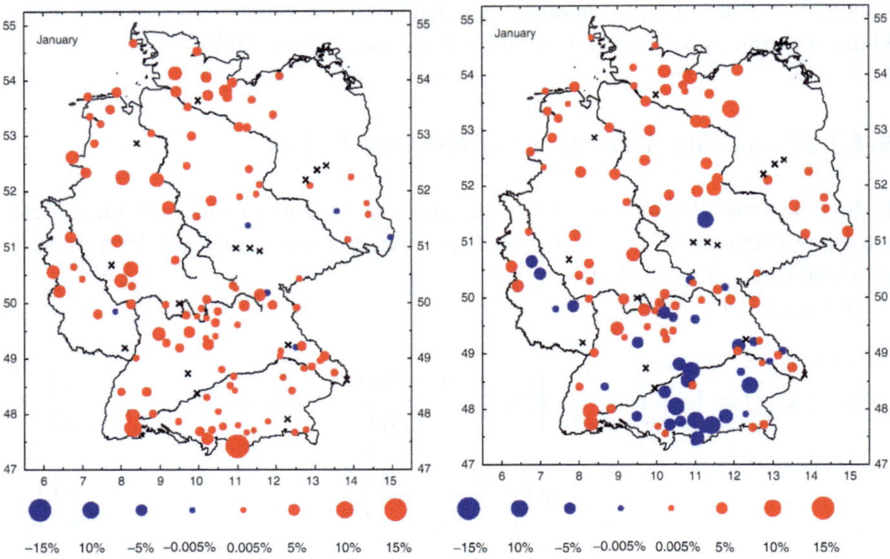

Fig. 9.6 Changes in the probability of occurrence of a monthly precipitation sum greater than the 95th percentile (*left*) and in the probability of occurrence of a monthly precipitation sum smaller than the 5th percentile (*right*). Results are given for January in the observation period 1901–2000. *Red dots* indicate an increase and *blue dots* indicate a decrease in the probability of occurrence. The size of the dots is proportional to the magnitude in probability change

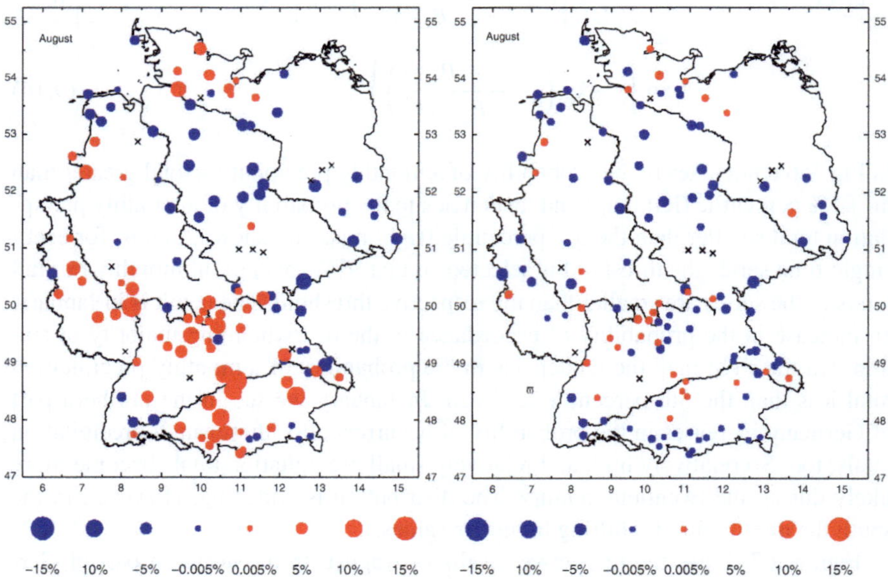

Fig. 9.7 Analogous maps to Fig. 9.6 with results for August

occurrence of relatively small precipitation totals are detected in the overwhelming majority of stations. Combined with the tendencies in the probability of exceedance of the 95th percentile we come to the conclusion that in summer in the northern part of Germany the extreme high and extreme low precipitation totals are getting more unlikely. In the south we observe again at several stations a shift of the distribution to higher values but results are not as uniform in that region in summer.

9.4.4 Changes in the Expected Value

As could already be seen in Fig. 9.4 the statistical modelling provides the expected value for every time step of the observation period. Consequently, the method introduced represents an alternative approach to estimate trends. The difference in the expected value between January 2000 and January 1901 is defined as the trend in the expected value at every station in that month. With the use of an appropriate distance function relatively high precipitation sums do not get more weight than can be justified from a statistical point of view. Additionally, changes in different parameters of the distribution can be taken into account, because the expected value of a Gumbel-distributed random variable is defined as

$$\mu(t) = a_{\mathrm{G}}(t) + b_{\mathrm{G}}(t)\gamma \approx a_{\mathrm{G}}(t) + 0.57722 \cdot b_{\mathrm{G}}(t) \qquad (9.15)$$

with the location parameter $a_{\mathrm{G}}(t)$, the scale parameter $b_{\mathrm{G}}(t)$, and Euler's constant γ. Because of the skewness of the distribution a change in the scale parameter causes changes in the expected value, too. And the expected value of a Weibull-distributed random variable depends on the location parameter a_{W}, the scale parameter b_{W} and the shape parameter c_{W}:

$$\mu(t) = a_{\mathrm{W}}(t) + b_{\mathrm{W}}(t)\Gamma\left(1 + \frac{1}{c_{\mathrm{W}}}\right). \qquad (9.16)$$

The gamma function Γ is defined as

$$\Gamma(z) = \int_0^\infty t^{z-1}\mathrm{e}^{-t}\mathrm{d}t. \qquad (9.17)$$

A familiar nonparametric trend test would be the Mann–Kendall trend test [9.5]. Even though this test contains information about a increase or decrease in the time series considered, no information about the temporal evolution nor the amplitude of the trend is provided. In most cases, changes in the expected values of non-Gaussian climate time series, e.g. ordinary trend maps, are estimated using the least-squares method, too. Comparison of the trend map on the basis of the least-squares method with the trend map estimated on the basis of the method introduced (Fig. 9.8) shows in January similar spatial structures. We observe increases in the western and south-

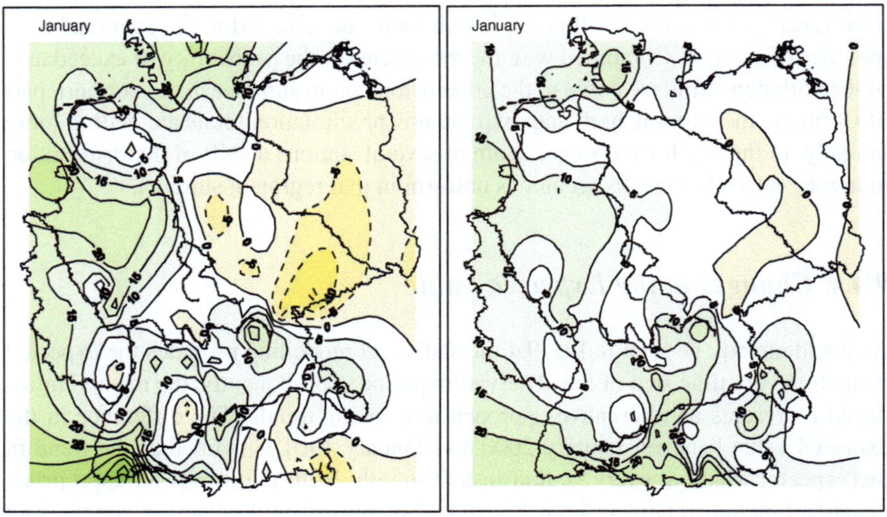

Fig. 9.8 Observed trends 1901–2000 in January estimated with the least-squares method (*left*) and on the basis of the statistical modelling of the time series using the distance function of the Gumbel distribution (*right*)

ern part of Germany, but decreases in the eastern part. However, the amplitudes are smaller using the distance function of the Gumbel distribution.

Another advantage of the method introduced is that trends of a particular month are estimated on the basis of the whole sample size. In order to demonstrate the large effect of the sample size on trend analysis results, Monte Carlo simulations have been performed. A total of 100 time series with a length of 100 or 1 000 time steps, respectively, of Gumbel-distributed variables with trends $\Delta\mu$ have been generated. Subsequently, we applied the least-squares estimator and tried to find the prescribed trend. In Table 9.2 the mean least-squares estimator $\overline{\Delta\mu}_{KQ}$, its standard deviation $\sigma_{\Delta\mu}$, as well as the smallest and the greatest estimator, $\Delta\mu^-_{KQ}$ and $\Delta\mu^+_{KQ}$, are shown for eight different experiments using the longer generated time series (1 000

Table 9.2 Trends of 100 generated Gumbel-distributed time series with 1 000 time steps and different linear changes in the location parameter (Δa) and constant or additional changing scale parameter (b). The resulting changes in the expected value $\Delta\mu$ (see again (9.15)) are compared to mean least-squares estimator of the 100 time series $\overline{\Delta\mu}_{KQ}$, the standard deviation $\sigma_{\Delta\mu}$ of the estimator as well as the greatest and the smallest trend estimator ($\Delta\mu^+_{KQ}$ or $\Delta\mu^-_{KQ}$)

	$\Delta a = 15$ $b = 50$	$\Delta a = 15$ $b = 20$	$\Delta a = 8$ $b = 40$	$\Delta a = 0$ $b = 40$	$\Delta a = 8$ $b = 40 + \Delta 8$	$\Delta a = -15$ $b = 50$	$\Delta a = -8$ $b = 40 - \Delta 8$	$\Delta a = 0$ $b = 40 + \Delta 10$
$\Delta\mu$	15	15	8	0	12.6	−15	−12.6	5.77
$\overline{\Delta\mu}_{KQ}$	15.49	15.20	8.39	0.39	13.05	−14.51	−12.27	6.22
$\sigma_{\Delta\mu}$	6.97	2.79	5.57	5.57	6.16	6.70	5.02	6.31
$\Delta\mu^-_{KQ}$	0.78	9.31	−3.38	−11.38	0.08	−29.22	−22.83	−7.06
$\Delta\mu^+_{KQ}$	33.44	22.38	22.75	14.75	28.23	3.44	1.27	21.81

Table 9.3 Analogous to Table 9.2, but considering Gumbel-distributed time series with 100 time steps

	$\Delta a = 15$ $b = 50$	$\Delta a = 15$ $b = 20$	$\Delta a = 8$ $b = 40$	$\Delta a = 0$ $b = 40$	$\Delta a = 8$ $b = 40 + \Delta 8$	$\Delta a = -15$ $b = 50$	$\Delta a = -8$ $b = 40 - \Delta 8$	$\Delta a = 0$ $b = 40 + \Delta 10$
$\Delta\mu$	15	15	8	0	12.6	−15	−12.6	5.77
$\overline{\Delta\mu_{KQ}}$	12.85	14.14	6.28	−1.72	10.78	−17.15	−14.21	3.90
$\sigma_{\Delta\mu}$	23.82	9.53	19.06	19.06	20.91	23.82	17.31	21.39
$\Delta\mu_{KQ}^{-}$	−43.84	−8.54	−39.07	−47.07	−38.96	−73.84	−55.18	−49.93
$\Delta\mu_{KQ}^{+}$	59.52	32.81	43.62	35.62	51.44	29.52	22.13	46.37

time steps). A positive bias can be observed. Table 9.3 shows the analogous results for the smaller sample size (100 time steps). The absolute value of the bias has clearly increased but the trend estimator shows a negative bias now. The estimators' variance $\sigma_{\Delta\mu}$ has increased remarkably. It may be higher than the trend magnitude. Consequently, the greater sample size of precipitation totals taken into account within the statistical modelling approach (1 200 values instead of 100 in the present case) implies a smaller mean squared error of the maximum likelihood estimator given by the sum of the quadratic bias and the variance of the estimator.

9.5 Conclusions

A generalized consistent decomposition procedure of precipitation time series into a statistical and a deterministic part is introduced. The basis functions allowed to describe the deterministic components only contain trends, annual cycle, episodic component and extreme events in order to restrict to physically explainable functions. Under the additional assumption that climate change is not restricted to the mean value the signal detection technique is applied to two instead of one parameter of a PDF, which can be chosen without any further restriction.

In particular, we show that a time series decomposition technique based on a Gumbel distribution, with flexible location and scale parameters, succeeds to describe monthly precipitation total time series from German stations completely. The model provides a full analytical description of the time series. On this basis, probabilities of exceeding defined thresholds can be estimated reliably for every time step of the observation period. But the provided complete analytical description can be used to calculate the expected value for every time step, too. So, statistical modelling represents an alternative approach for estimating broadly used trends in observational precipitation time series. In this way non-Gaussian characteristics can be taken into account. Additionally, changes in different parameters can be considered and the mean squared error of the trend estimator is smaller using the statistical modelling.

Application of the method to a German station network of 132 time series covering 1901–2000 shows in winter and summer at several stations in the southern part of Germany an increase in the probability of exceeding the 95th percentile and a decrease in the probability of falling under the 5th percentile. In the western part, we observe the same phenomenon in the summer months, but these changes go along with smaller magnitudes. However, climate is getting more extreme in that region in winter: Probability for both exceeding the 95th percentile and for falling under the 5th percentile is increasing. In the eastern part of Germany, increases in the probability of occurrence of relatively low precipitation in winter as well as decreases in both probabilities (>95th percentile, <5th percentile) in summer and autumn prevail.

Exemplary, the trend map on the basis of the familiar least-squares method is compared to the trend map calculated on the basis of the Gumbel model. Both maps show positive trends in the western and southern part of Germany and negative trends in the eastern part. However, the robust method provides smaller amplitudes.

Acknowledgments We gratefully acknowledge that this work is supported by the Federal German Ministry of Education and Research (BMBF) under grant no. 01LD0032 within the context of the German Climate Research Programme (DEKLIM, `http://www.deklim.de`).

References

9.1. S. Coles, *An Introduction to statistical modeling of extreme values.* (Springer, Berlin, 2001)
9.2. J. Grieser, S.Trömel, C.D. Schönwiese, *Statistical time series decomposition into significant components and application to European temperature*. Theor. Appl. Climatol. **71**, 171–183 (2002)
9.3. Huber, P. J., *Robust Statistics.* (Wiley Series in Probability and Mathematical Statistics, New York, 1981)
9.4. M.R. Leadbetter, G. Lindgren, H. Rootzen, *Extremes and Related Properties of Random Sequences and Processes* (Springer, Berlin 1983)
9.5. H.B. Mann, *Nonparametric test against trend.* Econometrica, **13**, 245–259 (1945)
9.6. W.H. Press, S.A. Teukolsky, W.T. Vetterling, B.P. Flannery, *Numerical Recipes in Fortran 77* (Cambridge University Press, Cambridge 1992)
9.7. H. Rinne, *Taschenbuch der Statistik.* (Verlag Harri Deutsch, Thun/Frankfurt, 1997)
9.8. K. Schrader, T. Hettmansperger, *Robust Analysis of variance upon a likelihood ratio criterion.* Biometrika, **67**, 93–101 (1980)
9.9. H. v. Storch, F.W. Zwiers, *Statistical Analysis in Climate Research.* (Cambridge University Press, Cambridge 1999)
9.10. S. Trömel, *Statistische Modellierung monatlicher Niederschlagszeitreihen.* Report No. 2, Inst. Atm. Umwelt, Universität Frankfurt a.M., 2005
9.11. S. Trömel, C.D. Schönwiese, *A generalized method of time series decomposition into significant components including probability assessments of extreme events and application to observational German precipitation data.* Met. Z. 14, 417–427 (2005)
9.12. World data center for climate, Hamburg. http://cera-www.dkrz.de/ (2005)

Weir in the river Lahn (Giessen, Germany, 26.6.2005). Constructions and modifications of the river bed can lead to change points in river runoff

Chapter 10
A Review on the Pettitt Test

Diego Rybski and Jörg Neumann

Applying the Pettitt test we study long river runoff records from gauges in southern Germany and find significant change points. Theoretically, a change point represents a sudden change in the statistics of a record. Using detrended fluctuation analysis, we also find – in agreement with previous studies – pronounced long-term temporal autocorrelations in the considered records. The results of both approaches indicate a relation between the occurrence of change points and the strength of long-term correlations. In order to clarify a possible connection, we further analyse with both methods artificial long-term correlated records and find for weak long-term correlations already highly significant change points. The significance dramatically increases with the strength of the long-term correlations.

10.1 Introduction

In time series analysis the concept of change points, indicating a change in the statistical behaviour, is of basic interest. In geoscience the cause of such events can be anthropogenic interventions to the riverbed in the case of hydrologic records, modifications of instrumentation, or climate change [10.25]. Thus, it is a quest to identify whether such a point of change exists and if so, to locate it in time. At first, the usage of the term "change point" does not clarify what it really is, e.g. an abrupt change of the mean, the beginning of a linear trend, or the change of piecewise (linear) trends.

Recent studies reveal that many different records in nature exhibit not only short-term, but also long-term correlations, see, e.g. [10.15, 10.19] and references therein, which are characterised by a diverging correlation time. Since the Pettitt test is a common tool for the detection of change points [10.1, 10.7, 10.18, 10.27, 10.28], we want to give a review on the method and discuss the influence of long-term correlations.

D. Rybski (✉)
Institut für Theoretische Physik, Justus-Liebig-Universität, Giessen, 35 392 Giessen, Germany
e-mail: ca-dr@rybski.de

J.P. Kropp, H.-J. Schellnhuber (eds.), *In Extremis*,
DOI 10.1007/978-3-642-14863-7_10,

We examine the approach to the change point problem introduced by Pettitt in 1979 [10.22] by applying it to several hydrological records. Further we determine the long-term correlations of these series in the asymptotic regime and compare the strength with the significance of the change points. We find very significant change points and notice that the significance slightly increases with the strength of the long-term correlations. In order to clarify the influence of long-term correlations, we perform the Pettitt test on artificial long-term correlated sequences. We find that in many configurations, already for weak long-term correlations, a high significance of finding a change point is given by the Pettitt test. This result makes its applicability questionable.

In Sect. 10.2 we give a brief description of the analysed hydrological data. The Pettitt test itself and the detrended fluctuation analysis (DFA) are described in Sect. 10.3. The results for real and simulated data are given in Sect. 10.4. Finally, in Sect. 10.5 we give our conclusion.

10.2 Data

We select 23 river discharge records with monthly resolution of gauges located in southern Germany (Bavaria and Baden-Wuerttemberg). The records are prepared by subtracting the long-term seasonal mean, in order to obtain the anomalies. To normalise the fluctuations we further divide by the seasonal standard deviation (see Chap. 11). In Table 10.1 we give detailed information of the records and the locations of the gauges are shown in Fig. 10.1.

10.3 Methods

The Pettitt test [10.22] considers a sequence of random variables $X_1, X_2, \ldots, X_T$, which is said to have a change point at τ if X_t for $t = 1, \ldots, \tau$ have a common distribution function $F_1(x)$ and X_t for $t = \tau+1, \ldots, T$ have a common distribution function $F_2(x)$, and $F_1(x) \neq F_2(x)$. Pettitt does not make any assumptions about the functional forms of F_1 and F_2 except that they are continuous; see also [10.7]. The Pettitt test is based on the sign function, $\mathrm{sgn}(y) = 1$ if $y > 0$, 0 if $y = 0$, -1 if $y < 0$, where the sign of the difference of each pair of value of the sequence is given by

$$D_{ij} = \mathrm{sgn}(X_i - X_j)\,. \tag{10.1}$$

Next, parts of this matrix are summed up and define the quantity

$$U_{t,T} = \sum_{i=1}^{t} \sum_{j=t+1}^{T} D_{ij}\,. \tag{10.2}$$

Table 10.1 Details of considered gauges. Listed are the name of the gauge, its main river basin, its number, its basin area, the time range, and the length of the records. The gauges are mapped in Fig. 10.1

ID	Gauge	Main river basin	Gauge number	Basin area (km^2)	Begin	End	Months
1	Oberammergau	Ammer	16610709	114	1921	2001	972
2	Inkofen	Amper	16607001	3 043	1926	2002	924
3	Beuron	Donau	168	1 320	1926	2003	936
4	Hofkirchen	Donau	10088003	47 497	1901	2003	1 236
5	Ebnet	Dreisam	292	256	1941	2003	756
6	Bad Aibling	Glonn	18263005	144	1928	2000	876
7	Kempten	Iller	11402001	955	1901	2002	1 224
8	Wiblingen	Iller with channel	61405001	2 115	1921	2000	960
9	Oberaudorf	Inn	18000403	9 712	1901	2002	1 224
10	Mittenwald	Isar	16000708	404	1926	2003	936
11	Untergriesheim	Jagst	3470	1 826	1925	2003	948
12	Schwaibach	Kinzig	390	957	1914	2003	1 080
13	Gaildorf	Kocher	337	724	1885	1997	1 356
14	Garmisch	Loisach	16402009	392	1926	2002	924
15	Schweinfurt	Main	24022003	12 715	1901	2001	1 212
16	Heitzenhofen	Naab	14008006	5 426	1921	2002	984
17	Plochingen	Neckar	313	3 995	1919	2003	1 020
18	Regenstauf	Regen	15208500	2 660	1901	2001	1 212
19	Hof	Sächsische Saale	56001502	521	1921	2002	984
20	Stein at Altenmarkt	Traun	18483500	378	1926	2001	912
21	Pfronten-Ried	Vils	12183005	113	1911	2002	1 104
22	Biessenhofen	Wertach	12405005	450	1921	2002	984
23	Oberlauchringen	Wutach	357	617	1912	2003	1 104

The statistic $U_{t,T}$ is then considered for values of t with $1 \le t < T$. In [10.7] an iterative formula is suggested

$$U_{t,T} = U_{t-1,T} + \sum_{j=1}^{T} D_{tj} \ , \tag{10.3}$$

which uses $D_{ij} = -D_{ji}$ and $D_{ii} = 0$. Figure 10.2 is a sketch for the calculation of $U_{t,T}$. The square represents the matrix D_{ij} and the hatched area represents the entries whose sum is denoted $U_{t,T}$. The dotted line confines either the summation from top to bottom ($1 \le j < T$) after (10.2) or iteratively after (10.3), where the triangles cancel each other.

The maximum of the absolute sums $|U_{t,T}|$ gives the position of a possible change point $1 \le \tau < T$. The statistic is

$$K_T = \max_{1 \le t < T} |U_{t,T}| \, . \tag{10.4}$$

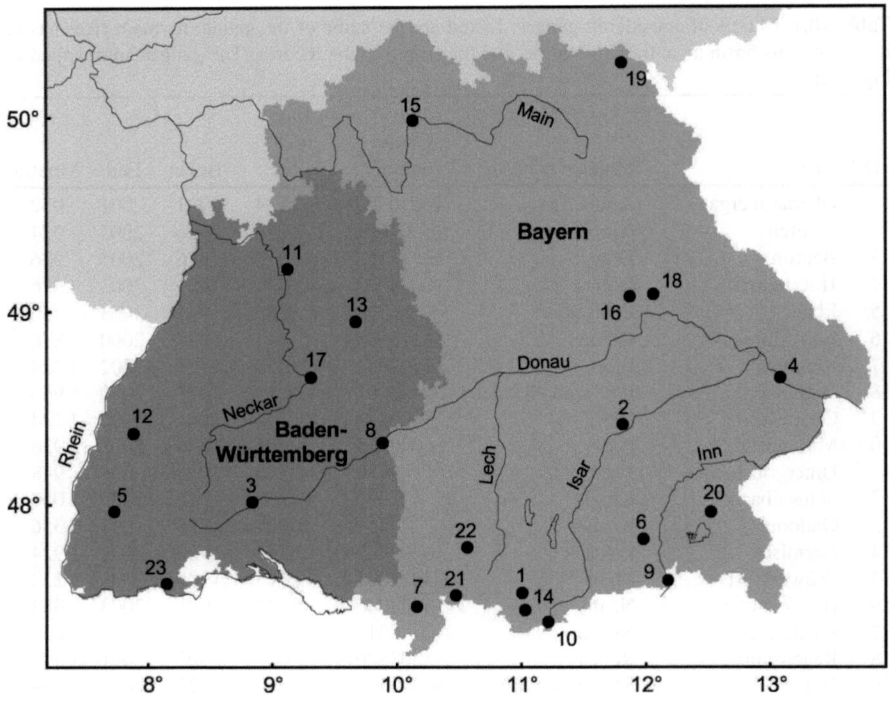

Fig. 10.1 Map of southern Germany (*Baden-Wuerttemberg and Bavaria, dark and light grey*). The *circles* represent gauges, which carry the IDs used in Table 10.1

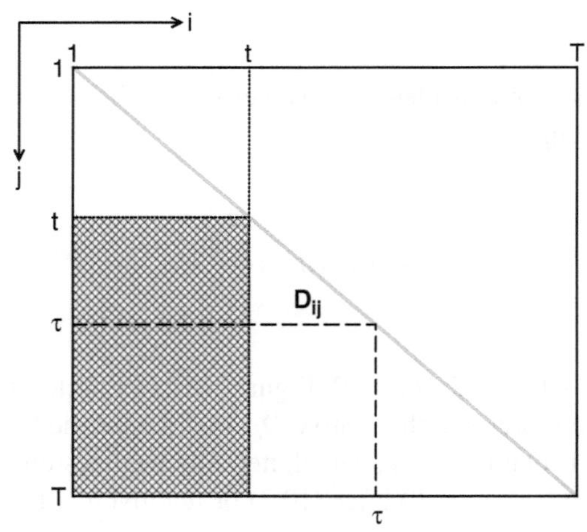

Fig. 10.2 Schematic illustration of the Pettitt test. The matrix D_{ij} (10.1) consists of the sign of the difference of any pair of values of the considered sequence X_i. The quantity $U_{t,T}$, *dotted line* (10.2), (10.3), corresponds to the sum over the entries of sub-matrices, see *hatched area*. The absolute value $|U_{t,T}|$ achieves its maximum at the change point, *dashed line* $t = \tau$

A "downward shift" in the level from the beginning of the series is indicated by a large $K_T^+ = \max_{1 \leq t < T} U_{t,T}$ and an "upward shift" by a large $K_T^- = -\min_{1 \leq t < T} U_{t,T}$. The significance probability associated with the value K_T is approximately given by

$$p \simeq 2\,\mathrm{e}^{\frac{-6K_T^2}{T^3+T^2}}\,, \tag{10.5}$$

where the approximation holds good, accurate to two decimal places, for $p \leq 0.5$ [10.22]. Note that large K_T (probable change point) lead to small p. However, Busuioc and von Storch [10.7] give $p = \exp(-3K_T^2/T^3 + T^2)$ as expression for the probability. We use the software package "HyStat" [10.18, 10.29]. Although Busuioc and von Storch [10.7] already emphasise the two assumptions of stationarity and the lack of serial correlation, in hydrology the test is used for the detection of change points.

We also analyse long-term correlations using the well-established detrended fluctuation analysis (DFA) [10.21] and concentrate on large time scales. In the presence of long-term correlations, the autocorrelation function decays with a power law (X normalised to zero mean and unit variance):

$$C(s) = \frac{1}{T-s}\sum_{i=1}^{T-s} X_i X_{i+s} \quad \sim \quad s^{-\gamma}\,, \quad 0 < \gamma \leq 1\,. \tag{10.6}$$

The direct determination of $C(s)$ and γ is hindered by non-stationarities such as trends. On large time scales $C(s)$ tends to fluctuate around zero level, thus it becomes impossible to determine the correlation exponent γ. The DFA method overcomes these problems, because it systematically removes trends on all time scales. For a detailed description of the method we refer to [10.5, 10.8, 10.11, 10.14, 10.19, 10.21], Chap. 11 in this book, and references therein. The fluctuation exponent α, provided by the DFA method, can have the following values:

$\alpha < 0.5$	anti-correlated
$\alpha \simeq 0.5$	uncorrelated
$0.5 < \alpha < 1.0$	long-term correlated
$1.0 \leq \alpha$	long-term correlated, non-stationary.

It is related to γ via [10.9, 10.14, 10.21]

$$\gamma = 2 - 2\alpha\,, \quad 0 < \gamma \leq 1. \tag{10.7}$$

The exponent α is similar to the Hurst exponent $H \gtrsim \alpha$, introduced by Hurst with his R/S-analysis [10.9, 10.13, 10.15]. An alternative method for the analysis of autocorrelations is among others the power spectral analysis, where a decaying power law with exponent $\beta = 2\alpha - 1$ can be observed [10.14, 10.23].

10.4 Results

10.4.1 River Runoff Records

The 23 monthly discharge records specified in Sect. 10.2 are studied for two main properties. First we apply the Pettitt test described in the previous section and obtain the position of a possible change point and its significance. The results are given in Table 10.2. In two cases (9%) of the here considered records, the significance for a change point is smaller than 0.8, in 18 (78%) it is bigger than 0.9, in 16 (70%) bigger than 0.95 and even 13 records (57%) have a change point with significance of 1.00. The position of the change points varies strongly, but we focus on the significance provided by the statistics.

As mentioned before, DFA systematically removes non-stationarities on all time scales. This enables the reliable characterisation of long-term correlations. The records discussed in this section show the reported long-term correlations, but they may also incorporate non-stationarities. DFA was tested for a variety of non-stationarities [10.8, 10.11, 10.14]. Our DFA results are also given in Table 10.2. We do not show examples of fluctuation functions, since these can be found in Chap. 11. Long-term correlations in hydrological records are a well-known and

Table 10.2 Results of the Pettitt-test and DFA for the monthly river discharge records presented in Table 10.1. We list the position τ of a change point and its significance $1-p$ after Eq. (10.5). Further the determined fluctuation exponent α and the with Eq. (10.7) resulting correlation exponent γ, Eq. (10.6), are given. The errors for α lay typically between 0.03 and 0.05

ID	Gauge	τ	$1-p$	α_{DFA2}	γ
1	Oberammergau	10/1946	0.61	0.77	0.45
2	Inkofen	01/1977	1.00	0.86	0.29
3	Beuron	10/1964	1.00	0.65	0.7
4	Hofkirchen	02/1965	0.99	0.83	0.34
5	Ebnet	02/1965	0.99	0.67	0.67
6	Bad Aibling	04/1947	1.00	0.73	0.53
7	Kempten	08/1964	0.42	0.63	0.75
8	Wiblingen	02/1965	1.00	0.73	0.53
9	Oberaudorf	09/1964	1.00	0.63	0.73
10	Mittenwald	10/1964	1.00	0.83	0.34
11	Untergriesheim	02/1978	1.00	0.84	0.32
12	Schwaibach	10/1977	0.98	0.66	0.68
13	Gaildorf	12/1921	0.88	0.72	0.55
14	Garmisch	08/1964	0.90	0.79	0.42
15	Schweinfurt	02/1978	1.00	0.87	0.26
16	Heitzenhofen	02/1978	1.00	0.9	0.21
17	Plochingen	12/1964	1.00	0.82	0.35
18	Regenstauf	02/1965	1.00	0.82	0.35
19	Hof	07/1946	0.93	0.82	0.35
20	Stein at Altenmarkt	03/1946	1.00	0.67	0.67
21	Pfronten-Ried	07/1946	0.89	0.61	0.78
22	Biessenhofen	11/1941	0.92	0.61	0.78
23	Oberlauchringen	06/1941	1.00	0.76	0.49

studied phenomenon [10.9, 10.13], Chap. 11 and references therein. Our DFA results agree with previous studies described in Chap. 11. Weakest long-term correlations were found for the record of Biessenhofen with $\alpha = 0.61$ and Heitzenhofen exhibits strongest long-term correlations with $\alpha = 0.90$. For the 23 records we obtain a mean value $\alpha = 0.75 \pm 0.09$.

We find that the records exhibit on the one hand significant change points (according to the Pettitt test) and on the other hand pronounced long-term correlations. This fact arises the question whether these two properties observed in the runoff records are related. In order to work out the pure influence of long-term correlations on the Pettitt test, in the absence of non-stationarities, we perform simulations with artificial data in the next section.

We would like to remark that we have also performed the analogue study based on the test of Bernier [10.2, 10.29] and found very similar results showing a strong dependence of the significance on the strength of long-term correlations.

10.4.2 Simulations

The Pettitt test was developed under the assumptions of stationarity and a lack of serial correlation [10.7, 10.22], although it is a common tool for the detection of change points in natural records, e.g. in hydrology. In [10.7] already the influence of short-term correlations on the Pettitt test is reported, using an AR(1) process. We extend this work by the application to long-term correlated sequences. This was done by creating long-term correlated data using Fourier filtering method [10.20] with given fluctuation exponent $\alpha = \frac{1}{2}(\beta + 1)$ and length $T = 10\,000$. By construction, these sequences are stationary and do not exhibit change points. For each value of α we use 100 configurations (only 10 for $\alpha > 0.7$) and determine the probability for the change point found by the method. Histograms of the occurring probabilities are given in Fig. 10.3 and show how the outcome is affected by the strength of long-term correlations. For uncorrelated records ($\alpha = 0.5$), the significance is more or less equally distributed. Already for weak long-term correlations ($\alpha = 0.55$), higher probability occurs more frequently. At $\alpha = 0.65$ almost always $p = 1.00$ is found.

In Fig. 10.4 we plot the mean probability against the specified fluctuation exponent α. The probability increases with the strength of long-term correlations and achieves quickly the value 1. Further the results of the analysed river runoff records are plotted, which follow more or less the simulated results. We therefore conclude that the Pettitt test is strongly affected by autocorrelations, especially by long-term correlations.

10.4.3 Reasoning

In Fig. 10.5 we show three numerical examples as an illustration of the Pettitt test. If the sequence X_i is uncorrelated, then D_{ij} is homogeneous and $|U_{t,T}|$ does not achieve large values. A change point in the form of a sudden offset causes

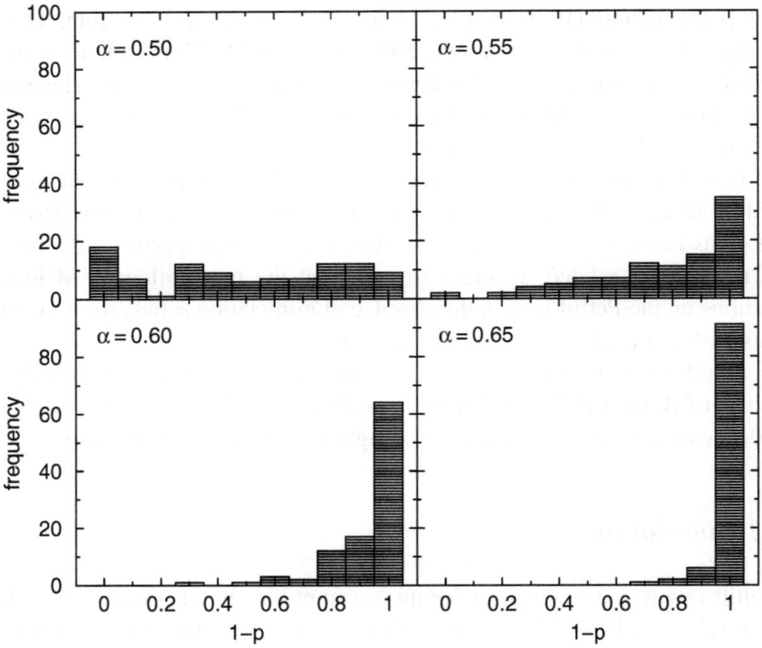

Fig. 10.3 Frequencies of probabilities for change points found in artificial stationary long-term correlated sequences. For each value of $\alpha = 0.5, 0.55, 0.6, 0.65$ 100 configurations were analysed

concentrations of ± 1 since the difference in a certain direction becomes more often positive or negative. In the case of correlated noise, natural clustering occurs and $|U_{t,T}|$ reaches large values, even in the stationary case and in the absence of a change point.

Let us assume X_k being the maximum value of the complete record. Then the sum in (10.3) is equal to $T - 1$. Accordingly, for the second largest value the sum is equal to $T - 3$, and so on. Then $U_{t,T}$ is the cumulative sum over the sequence of these integer values in the corresponding order. The cumulative sum, also called profile, represents a random walk [10.3, 10.4]. It is known that the mean square displacement of such a random walk strongly depends on the serial correlations of the initial record. In addition, this property is used in fluctuation analysis [10.6, 10.19]. Thus, it is clear why the significance for finding change points with the Pettitt test is drastically increased for long-term correlated records. The fact that the results for real records do not follow exactly those of the simulated records in Fig. 10.4, we attribute to additional properties such as multifractality; see [10.15, 10.16] and Chap. 11.

In addition a series can be separated at a once found change point and the method can be applied to each of both pieces. As long as change points are found this procedure can be repeated stepwise. Segmentation procedures are described, e.g. in [10.12] or [10.10]. In 1974 Klemeš [10.17] proposed a model for long-term correlations, where a series consists of independent constant offsets with a certain distribution of lengths and overlaid white noise. So we see a strong relation between the occurrence of look alike change points and long-term correlations.

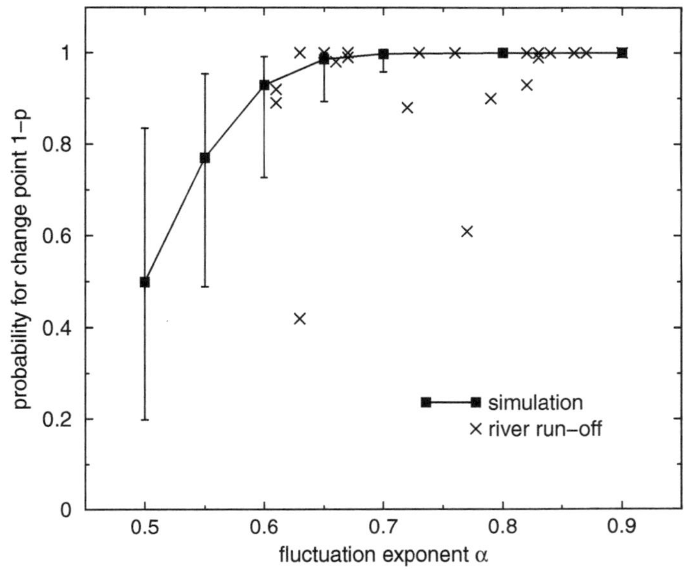

Fig. 10.4 Dependence of the probability of change point after Pettitt-test on the strength of long-term correlations. The square-symbol represents values of artificial sequences of 100 configurations (10 for $\alpha > 0.7$) of length $T = 10\,000$ and specified fluctuation exponent α, compare Fig. 10.3. The results for real data, discussed in the previous Sect. 10.4, are shown with the symbol $\times$. The two outlying gauges have the ID 1 and 7, see Table 10.2

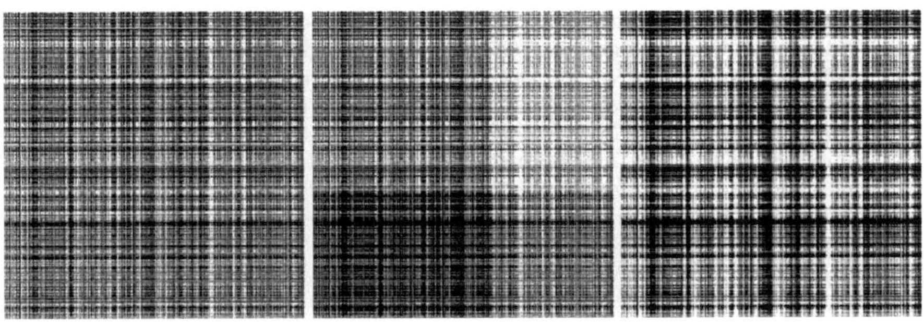

Fig. 10.5 The matrix D_{ij} (10.1) for three examples of Gaussian noise with length $T = 512$ and normalised variance ($\sigma^2_{X_i}$=1). *Black* and *white* correspond to $D_{ij} = \pm 1$, zero is *grey* (*diagonal*). *Left*: uncorrelated series X_i lead to homogeneous D_{ij}. *Central*: uncorrelated series with change point at $\tau = 300$ in form of an offset of strength 1. *Darker or brighter areas* indicate that differences in one direction occur more frequently. The statistic $U_{t,T}$ (10.2), (10.3) evaluates this by summation. *Right*: correlated noise with $\alpha = 0.85$ causes a natural patchiness

10.5 Conclusion

In this work we have studied the influence of long-term correlations on the Pettitt test. For this purpose we have analysed 23 runoff records from gauges in southern Germany with (a) the Pettitt test and (b) DFA. The analysis provides for (a) in most of the cases significant change points and for (b) typical long-term correlations.

The significance of the change points seems to increase with the strength of the long-term correlations. This outcome is supported by the same analysis for artificial long-term correlated records, which proves a clear dependence between long-term correlations and a change point after [10.22].

We find that the application of the Pettitt test leads to enormous uncertainties and is not suitable for records which exhibit long-term correlations. Our results therefore confirm [10.7] who discussed similar effects for short-term correlated data. We would like to remark that the fact, that the detected change points are found at similar years for different records, is not an indicator for the correctness of those results, but rather due to the cross-correlations between the records [10.24, 10.26].

Basically, the mechanism behind the Pettitt test is plausible as depicted in Fig. 10.5. In order to overcome the influence of autocorrelations, we suggest to modify the expression for the significance p, (10.5), by taking into account long- and short-term correlations.

Acknowledgments We are grateful to Armin Bunde, Hans Weber, Michael Becker and Peter Braun for having initiated this work. Further we would like to thank the BMBF for financial support and the Water Management Authority (especially KLIWA) for providing the observational data. We additionally thank Jan F. Eichner for many useful discussions.

References

10.1. A.J. Back, Aplicação de análise estatística para identificação de tendências climáticas. Pesquisa Agropecuária Brasileira, **36**(5), 717–726 (2001)
10.2. J. Bernier, *Engineering risk in Natural Resources Management,* chapter Statistical detection of changes in geophysical series, pages 159–176. NATO Science Series E. Springer, 1994. Proceedings of the NATO Advanced Study Institute on ‘Engineering Risk and Reliability in a Changing Physical Environment, New Developments in Resources Management with Applications to Non-Steady Conditions’, Deauville, France, May 24–June 4, 1993
10.3. A. Bunde, S. Havlin, (eds.), *Fractals and Disordered Systems* (Springer, New York, 1991)
10.4. A. Bunde, S. Havlin (eds.), *Fractals in Science* (Springer, New York, 1994)
10.5. A. Bunde, S. Havlin, J.W. Kantelhardt, T. Penzel, J.H. Peter, K. Voigt. Correlated and uncorrelated regions in heart-rate fluctuations during sleep. Phys. Rev. Lett. **85**(17), 3736–3739 (2000)
10.6. A. Bunde, J. Kropp, H.-J. Schellnhuber (eds.), *The Science of Disasters* (Springer, Berlin, 2002)
10.7. A. Busuioc, H. von Storch, Changes in the winter precipitation in Romania and its relation to the large-scale circulation. Tellus, **48**A(4), 538–552 (1996)
10.8. Z. Chen, P.C. Ivanov, K. Hu, H.E. Stanley, Effect of nonstationarities on detrended fluctuation analysis. Phys. Rev. *E*, **65**(4), 041107 (2002)
10.9. J. Feder, *Fractals.* Physics of Solids and Liquids (Plenum Press, New York, NY, 1988)
10.10. K. Fukuda, H.E. Stanley, L.A.N. Amaral, Heuristic segmentation of a nonstationary time series. Phys. Rev. E, **69**(2), 021108 (2004)
10.11. K. Hu, P.C. Ivanov, Z. Chen, P. Carpena, H.E. Stanley, Effect of trends on detrended fluctuation analysis. Phys. Rev. E, **64**(1), 011114 (2001)
10.12. P. Hubert, The segmentation procedure as a tool for discrete modeling of hydrometeorological regimes. *Stochastic Environmental Research and Risk Assessment*, **14**(4-5), 297–304 (2000)

10.13. H.E. Hurst, Long-term storage capacity of reservoirs. Trans. Am. Soc. Civil Eng. **116**(2447), 770–808 (1951)
10.14. J.W. Kantelhardt, E. Koscielny-Bunde, H.H.A. Rego, S. Havlin, A. Bunde, Detecting long-range correlations with detrended fluctuation analysis. Physica A, **295**(3-4), 441–454 (2001)
10.15. J.W. Kantelhardt, E. Koscielny-Bunde, D. Rybski, P. Braun, A. Bunde, S. Havlin. Long-term persistence and multifractality of precipitation and river runoff records. J. Geophys. Res. – Atm. **111**(D1), D01106 (2006)
10.16. J.W. Kantelhardt, D. Rybski, S.A. Zschiegner, P. Braun, E. Koscielny-Bunde, V. Livina, S. Havlin, A. Bunde, Multifractality of river runoff and precipitation: Comparison of fluctuation analysis and wavelet methods. Physica A, **330**(1–2), 240–245 (2003)
10.17. V. Klemeš, The hurst phenomenon: A puzzle? Water Resourc. Res., **10**(4), 675–688 (1974)
10.18. Arbeitskreis KLIWA. Langzeitverhalten der Hochwasserabflüsse in Baden-Württemberg und Bayern. report Heft 2, Landesanstalt für Umweltschutz Baden-Württemberg, Bayerisches Landesamt für Wasserwirtschaft, Deutscher Wetterdienst, Karlsruhe, München, Offenbach, 2002. KLIWA-Projekt A 2.1.3 "Analyse zum Langzeitverhalten der Hochwasserabflüsse"
10.19. E. Koscielny-Bunde, A. Bunde, S. Havlin, H.E. Roman, Y. Goldreich, H.-J. Schellnhuber, Indication of a universal persistence law governing atmospheric variability. Phys. Rev. Lett., **81**(3), 729–732 (1998)
10.20. H.A. Makse, S. Havlin, M. Schwartz, H.E. Stanley, Method for generating long-range correlations for large systems. Phys. Rev E, **53**(5), 5445–5449 (1996)
10.21. C.-K. Peng, S.V. Buldyrev, S. Havlin, M. Simons, H.E. Stanley, A.L. Goldberger, Mosaic organization of DNA nucleotides. Phys. Rev. E, **49**(2), 1685–1689 (1994)
10.22. A.N. Pettitt, A non-parametric approach to the change-point problem. Appl. Stat., **28**(2), 126–135 (1979)
10.23. G. Rangarajan, M. Ding, Integrated approach to the assessment of long range correlation in time series data. Phys. Rev. E, **61**(5), 4991–5001 (2000)
10.24. D. Rybski. *Untersuchungen von Korrelationen, Trends und synchronem Verhalten in Klimazeitreihen*. Dissertation zur Erlangung des Doktorgrades der Naturwissenschaften im Fachbereich Physik, Justus-Liebig-Universität Giessen, 2006
10.25. D. Rybski, A. Bunde, S. Havlin, and H. von Storch. Long-term persistence in climate and the detection problem. Geophys. Res. Lett. **33**(6), L06718 (2006)
10.26. D. Rybski, S. Havlin, A. Bunde, Phase synchronization in temperature and precipitation records. Physica *A*, **320**, 601–610 (2003)
10.27. A. Tarhule, M.K. Woo, Changes in rainfall characteristics in northern Nigeria. Int. j. Climatol., **18**(11), 1261–1271 (1998)
10.28. R. Tomozeiu, A. Busuioc, V. Marletto, F. Zinoni, C. Cacciamani, Detection of changes in the summer precipitation time series of the region Emilia-Romagna, Italy. Theor. Appl. Climatol., **67**(3-4), 193–200 (2000)
10.29. W. Willems, Th. Dick, Hystat – Benutzerhandbuch. manual v1.02, Institut für Angewandte Wasserwirtschaft und Geoinformatik, Ottobrunn, May 2002

Part III
Long-Term Phenomena and Nonlinear Properties

Creek in Ein Advat (Ramon Nature Reserve near Mize Ramon, Israel, 16.2.2010, Courtesy of L. Costa). Rivers being dry in summer or frozen in winter exhibit much weaker persistence than other rivers

Chapter 11
Detrended Fluctuation Studies of Long-Term Persistence and Multifractality of Precipitation and River Runoff Records

Diego Rybski, Armin Bunde, Shlomo Havlin, Jan W. Kantelhardt, and Eva Koscielny-Bunde

We studied and compared the autocorrelation behaviour and the temporal multifractal properties of long daily river discharge and precipitation records from 42 hydrological stations and 99 meteorological stations around the globe. To determine the scaling behaviour in the presence of trends, we applied detrended fluctuation analysis (DFA) and multifractal DFA. We found that the daily runoffs are characterised by a power-law decay of the autocorrelation function above some crossover time that usually is several weeks. The corresponding correlation exponent varies from river to river in a wide range. Below the crossovers pronounced short-term correlations occur. In contrast, most of the precipitation series show scaling behaviour corresponding to a rapid decay of the autocorrelation function. For the multifractal characterisation of the data we determined the generalised Hurst exponents and applied three operational models. We found that the universal multifractal model fits well the scaling behaviour of the positive moments in nearly all runoff and precipitation records, while positive as well as negative moments are consistent with two-parameter fits from an extended version of the multiplicative cascade model for all runoff records and most of the precipitation records. Some weakly multifractal precipitation records, however, are better characterised by a simple bifractal model.

11.1 Introduction

The analysis of the persistence in river flows and precipitation has been initiated by H. E. Hurst who found about half a century ago that runoff records from various rivers exhibit long-range statistical dependencies [11.27]. Later, similar long-term

D. Rybski (✉)
Institut für Theoretische Physik Justus-Liebig-, Universität Giessen, 35392 Giessen, Germany
e-mail: ca-dr@rybski.de

J.P. Kropp, H.-J. Schellnhuber (eds.), *In Extremis*,
DOI 10.1007/978-3-642-14863-7_11,

correlated fluctuation behaviour was reported in many other geophysical records including temperature and precipitation data [11.28, 11.34, 11.44, 11.47, 11.49], see also [11.16]. For a critical discussion of the 'Hurst phenomenon' and alternative explanations we refer to [11.5, 11.6, 11.17, 11.33, 11.39, 11.50, 11.59, 11.63] and references therein.

In stationary time series the scaling of the fluctuations with time is reflected by the scaling of the power spectrum $E(f)$ with frequency f, $E(f) \sim f^{-\beta}$. The exponent β is related to the decay of the corresponding autocorrelation function $C(s)$. For β between 0 and 1, $C(s)$ decays by a power law, $C(s) \sim s^{-\gamma}$, with $\gamma = 1 - \beta$. In this case, the mean correlation time diverges, and the system is regarded as long-term correlated. For $\beta = 0$, the runoff or precipitation data are uncorrelated ('white noise'). The exponents β and γ can also be determined from a fluctuation analysis, where the departures from the mean daily runoffs or precipitation are considered as increments of a random walk process. If the record is uncorrelated, the fluctuation function $F_2(s)$, which is equivalent to the root-mean-square displacement of the random walk, increases as the square root of the time scale s, $F_2(s) \sim \sqrt{s}$. For long-term correlated data, the random walk becomes anomalous, and $F_2(s) \sim s^{h(2)}$. The fluctuation exponent $h(2)$ is nearly identical to the Hurst exponent (as determined by Hurst's original rescaled range analysis) and related to the exponents β and γ via $\beta = 1 - \gamma = 2h(2) - 1$. Many early studies exclusively focused on either the absolute values or the variances of the full distribution of the fluctuations, which can be regarded as the first moment $F_1(s)$ and the second moment $F_2(s)$ of the fluctuations, respectively, in given time segments of length s.

The conventional methods, however, may fail when trends are present in the system. Trends are systematic deviations from the average runoff or precipitation that are caused by external processes, e.g. the construction of a water regulation device, the seasonal cycle, or a changing climate (e.g. global warming). Monotonous trends may lead to an overestimation of the fluctuation exponent and thus to an underestimation of γ. It is even possible that uncorrelated data, under the influence of a trend, look like long-term correlated ones when using the above analysis methods [11.5]. In addition, long-term correlated data cannot simply be detrended by common regression or the technique of moving averages, since these methods destroy the correlations on long time scales (above the window size used). Furthermore, it is difficult to distinguish trends from long-term correlations, because stationary long-term correlated time series exhibit persistent behaviour and a tendency to stay close to the momentary value. This causes positive or negative deviations from the average value for long periods that might look like a trend. For a discussion of these problems, see [11.8, 11.29, 11.36, 11.37, 11.62]. Further we would like to remark that the pronounced mountain–valley structure of long-term correlated records leads to a clustering of extreme events; see Chap. 1.

In the last years, several methods such as wavelet techniques (WT) and detrended fluctuation analysis (DFA) have been developed that are able to determine long-term correlations in the presence of trends. For details and applications of the methods to a large number of meteorological, climatological and biological records we refer to

[11.3, 11.7, 11.9, 11.29, 11.56, 11.69]; for early DFA applications in hydrology, see [11.31, 11.49, 11.52]. These methods, described in Sect. 11.3, consider fluctuations in the cumulated runoff or precipitation records (often called the 'profile' or 'landscape' of the record). They differ in the way the fluctuations are determined and in the type of polynomial trend that is eliminated in each time window of size s.

Using the detrending methods, we have studied 42 long daily runoff records and 99 long daily precipitation records from all over the world [11.30, 11.31, 11.35]. We found that $F_2(s)$ scales as $s^{h(2)}$ for large time lags s. For the precipitation data, $h(2)$ is close to 0.5, indicating rapidly decaying autocorrelations. We like to note, however, that this result is not synonymous with white noise behaviour, since the precipitation data exhibit weak short-term correlations and pronounced multifractal behaviour (see below). Our study has also been motivated by the fact that the results for precipitation data have been discussed controversially in literature. Based on power spectra analysis [11.18] observed white noise behaviour at intermediate scales and long-term correlations on time scales above 3 years, References [11.44, 11.49, 11.57, 11.58] reported fractal scaling and long-term correlations on time scales below three years. For more recent works on the crossovers in the scaling behaviour of rainfall we refer to [11.14, 11.48].

For the runoff data, in contrast, the fluctuations show a pronounced crossover at intermediate scales (typically several weeks) [11.30, 11.31, 11.35]. Below the crossover, for small time windows s, $h(2)$ is close to 1.5, characterising a highly correlated regime similar to Brownian noise in agreement with the findings of [11.49]. Well above the crossover, at large times, the scaling exponent $h(2)$ varies from river to river between 0.55 and 0.95 in a non-universal manner always remaining in the stationary regime ($h(2) < 1$). Our findings are not consistent with the hypothesis that the scaling is universal with an exponent close to 0.75 [11.16, 11.28] for both small and large time scales.

In the last decade it has been realized that a multifractal description is required for both precipitation and runoff records [11.14, 11.45, 11.55, 11.71, 11.72], and all moments $F_q(s)$ need to be studied for a full characterisation of the fluctuations in these records. For each precipitation or runoff record, this multifractal description can be regarded as a 'fingerprint', which, among other things, can serve as an efficient non-trivial test bed for the performance of state-of-the-art precipitation–runoff models. We have applied the recently developed multifractal detrended fluctuation analysis [11.32] that can systematically distinguish between long-term correlations and trends. We determined the multifractal spectrum and compared the long-term scaling behaviour of both daily runoff and daily precipitation records on the same time scales [11.30, 11.31, 11.35].

Our approach differs from the multifractal approach introduced into hydrology by Schertzer et al. [11.10, 11.11, 11.42, 11.55, 11.64, 11.65, 11.71, 11.72] that was based on the concept of scale invariance in turbulence [11.19], and trace moment techniques, see also [11.60]. We performed the multifractal analysis by studying how all moments of the fluctuations $F_q(s) \sim s^{h(q)}$ scale with time s in the asymptotic regime [11.30–11.32, 11.35]. Our approach includes also negative q values and is not based on any specific model assumptions. To describe the numerical outcome of the analysis for $h(q)$ of both runoff and precipitation records, we employed three fitting formulas derived from operational models: (i) the universal multifractal model in [11.65], which is constrained to $q > 0$; (ii) an extended multiplicative cascade model [11.31]; and (iii) a bifractal model [11.30]. We found that the universal multifractal model (three fitting parameters) can be used to describe well the scaling behaviour for the positive moments q of runoff and precipitation records. Both positive and negative moments, however, follow rather closely the formula from the extended multiplicative cascade model (two fitting parameters) for all runoff records and 55% of the analysed precipitation records. Some of the precipitation records exhibit rather weak multifractality and 27% of them can be better characterised by the bifractal description.

This chapter is organized as follows: In Sect. 11.2 we describe the runoff and precipitation records we studied. Sections 11.3 and 11.4 are devoted to our methods for autocorrelation analysis and multifractal time series analysis, respectively; we relate to other multifractal formalisms in Sect. 11.4.2. Sections 11.5 and 11.6 summarize our results for runoff data and precipitation data and include the discussion. Finally, the main results are summarized in Sect. 11.7 and additional model considerations are discussed in the Appendix.

11.2 Data

We have analysed long daily runoff records $\{W_i\}$ from 42 hydrological stations and long daily precipitation records $\{P_i\}$ from 99 meteorological stations [11.30, 11.35]. The stations are representative for different rivers and different climate zones. The 99 precipitation records we analysed were measured at sites in Europe (40), Asia (34), North-America (15), Australia (5), and South-America (5). The latitude of the locations varies between 52.6S (Campbell Island, New Zealand) and 71.3N (Barrow Post, USA). The sites are concentrated on the northern hemisphere, thus the average latitude is 41N. Further, the sites are located at elevations from sea level up to 3 650 m (Lhasa, China) with an average of about 400 m. Concerning the climate [11.41], 33 sites are located in maritime regions, 56 in continental climate, and 10 in high-continental climate. In other categories, the stations are sited in tropical (12), sub-tropical (24), warm (44), cold (13), and polar (6) climate. In terms of humidity, the climate is really arid at only one considered site (Kizil Arvat, Turkmenistan), but semi-arid at 20 sites, semi-humid at 59 sites, and humid at 19 sites. The duration of the records ranges from 34 to 189 years (average 86 years), yielding approx. 12 000–69 000 data points each.

Of the 42 daily runoff records (see Table 11.1 for a full list), 18 are from the southern part of Germany and 24 are from North and South America, Africa, Australia, Asia, and Europe. The duration of the records ranges from 39 to 111 years (average 77 years) for the records from the southern part of Germany and from 51 to 171 years (average 92 years) for the international records, yielding

Table 11.1 Table of investigated international river basins [11.21] and investigated south German river basins. We list the river and station name, the duration of the investigated daily record, the size of the basin area, and the results of our analysis, $H \equiv h(2)$, as well as the multifractal quantities a, b, and $\Delta\alpha$

River name	Gauge name	Period of observation (years)	Basin area (km^2)	$h(2)$	a	b	$\Delta\alpha$
Zaire	Kinshasa, Zaire	81	3 475 000	0.95	0.52	0.52	0.00
Mississipi	St. Louis, USA	59	1 805 000	0.91	0.44	0.61	0.48
Orinoco	Puente Angostura, Venezuela	65	836 000	0.73	0.50	0.69	0.46
Columbia River	The Dallas, USA	114	613 830	0.59	0.54	0.76	0.50
Danube	Orsova, Romania	151	576 232	0.85	0.50	0.60	0.26
Dvina	UST-Pinega, Russia	89	348 000	0.56	0.53	0.79	0.58
Fraser River	Hope, USA	84	217 000	0.69	0.53	0.70	0.38
Rhine	Rees, Germany	143	159 680	0.76	0.52	0.65	0.32
Niger	Koulikoro, Mali	79	120 000	0.60	0.51	0.78	0.62
Susquehanna	Marietta, USA	61	67 310	0.60	0.53	0.79	0.57
Susquehanna	Harrisburg, USA	96	62 419	0.58	0.55	0.77	0.48
Elbe	Dresden, Germany	151	53 096	0.80	0.48	0.65	0.42
Labe	Decin, Czechia	102	51 104	0.80	0.45	0.68	0.61
Maas	Borgharen, Netherland	80	21 300	0.76	0.49	0.68	0.48
Weser	Vlotho, Germany	171	17 618	0.76	0.50	0.68	0.43
Tana	Polmak, Norway	51	14 005	0.56	0.50	0.81	0.69
Thames	Kingston, England	113	9 948	0.80	0.47	0.67	0.51
Grand River	Gallatin, USA	72	5 830	0.72	0.42	0.76	0.87
Mary River	Miva, Australia	76	4 830	0.60	0.52	0.78	0.57
Severn	Bewdley, England	71	4 330	0.63	0.54	0.73	0.43
Gaula	Haga Bru, Norway	90	3 080	0.55	0.57	0.77	0.43
Barron River	Myola, Australia	79	1 940	0.60	0.50	0.79	0.65
Mitta Mitta River	Hinnomunije, Australia	67	1 530	0.75	0.47	0.68	0.53
Johnston River	Upstream Central Mill, Australia	74	390	0.58	0.52	0.78	0.58
Southern Germany							
Danube	Achleiten	97	76 653	0.82	0.49	0.63	0.35
Danube	Kehlheim	97	22 950	0.85	0.48	0.63	0.39
Danube	Donauwörth	74	15 037	0.81	0.49	0.63	0.37
Neckar	Plochingen	79	3 995	0.80	0.49	0.65	0.39
Kocher	Stein	111	1 929	0.75	0.53	0.64	0.26
Jagst	Untergriesheim	73	1 826	0.76	0.45	0.69	0.61
Isar	Bad Tölz	39	1 554	0.68	0.53	0.71	0.41
Vils	Grafenmühle	58	1 436	0.61	0.50	0.78	0.62
Danube	Beuron	70	1 309	0.65	0.53	0.72	0.45
Amper	Fürstenfeldbruck	77	1 235	0.81	0.47	0.65	0.47
Wutach	Oberlauchringen	85	1 129	0.75	0.52	0.67	0.37
Neckar	Horb	65	1 118	0.68	0.44	0.75	0.78
Tauber	Bad Mergentheim	66	1 018	0.80	0.44	0.70	0.68
Kinzig	Schwaibach	82	921	0.67	0.52	0.72	0.47
Loisach	Kochel	87	684	0.82	0.48	0.65	0.44
Murg	Rotenfels	77	469	0.70	0.53	0.70	0.41
Wertach	Biessenhofen	77	450	0.66	0.56	0.70	0.31
Würm	Leutstetten	77	413	0.90	0.39	0.66	0.77

approx. 14 000–62 000 data points each. The basin area sizes vary between 390 and 3 475 000 km^2; the averages are 7 400 and 350 000 km^2 for the rivers in southern Germany and the international rivers, respectively.

11.3 Correlation Analysis

11.3.1 General

We consider a record of daily water runoff values W_i or precipitation P_i measured at a certain gauge. The index i counts the days in the record, $i = 1,2,\ldots,N$. To eliminate the periodic seasonal trends, we concentrate on the departures

$$\phi_i^{(\mathrm{W})} = W_i - \overline{W_i} \quad \text{and} \quad \phi_i^{(\mathrm{P})} = P_i - \overline{P_i} \tag{11.1}$$

from the mean daily runoff $\overline{W_i}$ or precipitation $\overline{P_i}$. $\overline{W_i}$ or $\overline{P_i}$ are calculated for each calendar date i (e.g. 1st April) by averaging over all years in the record. In addition, one may also eliminate seasonal trends in the variance and analyse

$$\tilde{\phi}_i^{(\mathrm{W})} = (W_i - \overline{W_i})/(\overline{W_i^2} - \overline{W_i}^2)^{1/2} \quad \text{and} \quad \tilde{\phi}_i^{(\mathrm{P})} = (P_i - \overline{P_i})/(\overline{P_i^2} - \overline{P_i}^2)^{1/2} \tag{11.2}$$

instead of ϕ_i. In order to avoid repetitions we will skip the indices (P) and (W) in all equations that apply to both $\phi_i^{(\mathrm{P})}$ and $\phi_i^{(\mathrm{W})}$ from now on. We also checked whether the results are modified if the seasonal trend $\overline{W_i}$ is smoothed by a continuous approximation based on polynomial fits.

The autocorrelation function $C(s)$ describes, how the persistence decays in time. If the ϕ_i are uncorrelated, $C(s)$ is zero for all $s > 0$. If correlations exist only up to a certain number of days $s_\times$, the correlation function will vanish above $s_\times$. For long-term correlations, $C(s)$ decays by a power law

$$C(s) = \frac{\langle \phi_i \phi_{i+s} \rangle}{\langle \phi^2 \rangle} \sim s^{-\gamma}, \qquad 0 < \gamma < 1, \tag{11.3}$$

where the average $\langle \cdots \rangle$ is over all pairs with the same time lag s.[1] For large values of s, a direct calculation of $C(s)$ is hindered by the level of noise present in the finite records and by non-stationarities in the data. There are several alternative methods for calculating the correlation function in the presence of long-term correlations, which we describe in the following sections.

[1] In this chapter we are only interested in the asymptotical scaling behaviour and do not consider parametric models for the river runoffs like the FARIMA model (see, e.g. [11.20, 11.51, 11.52]), which by definition cannot account for multifractality. Accordingly we do not need to employ approximate maximum likelihood estimators [11.26, 11.76].

11.3.2 Standard Fluctuation Analysis (FA)

In the standard fluctuation analysis (FA), we consider the 'profile'

$$Y(i) \equiv \sum_{k=1}^{i} \phi_k \ , \qquad i = 1, 2, \ldots, N \tag{11.4}$$

and study how the fluctuations of the profile, in a given time window of size s, increase with s. We can consider the profile $Y(i)$ as the position of a random walker on a linear chain after i steps. The random walker starts at the origin and performs, in the kth step, a jump of length ϕ_k to the right, if ϕ_k is positive, and to the left, if ϕ_k is negative.

To find how the square fluctuations of the profile scale with s, we first divide each record of N elements into $N_s = \text{int}(N/s)$ non-overlapping segments of size s starting from the beginning and N_s non-overlapping segments of size s starting from the end of the considered record. Then we determine the fluctuations in each segment ν.

In the FA, we obtain the fluctuations just from the values of the profile at both endpoints of each segment ν,

$$F^2(\nu, s) = [Y(\nu s) - Y((\nu - 1)s)]^2 \ , \tag{11.5}$$

and average $F^2(\nu, s)$ over all subsequences to obtain the mean fluctuation $F_2(s)$,

$$F_2(s) \equiv \left[\frac{1}{2N_s} \sum_{\nu=1}^{2N_s} F^2(\nu, s) \right]^{1/2} \ . \tag{11.6}$$

By definition, $F_2(s)$ can be viewed as the root-mean-square displacement of the random walker on the chain, after s steps. For uncorrelated ϕ_i values, we obtain Fick's diffusion law $F_2(s) \sim s^{1/2}$. For the relevant case of long-term correlations, where $C(s)$ follows the power-law behaviour of (11.3), $F_2(s)$ increases by a power law (see, e.g. [11.9]),

$$F_2(s) \sim s^{h(2)} \ , \tag{11.7}$$

where the fluctuation exponent $h(2)$ is related to the correlation exponent γ and the power spectrum exponent β by

$$h(2) = 1 - \gamma/2 = (1 + \beta)/2 \ , \tag{11.8}$$

see, e.g. [11.29]. For power-law correlations decaying faster than $1/s$, we have $h(2) = 1/2$ like for uncorrelated data.

11.3.3 The Detrended Fluctuation Analysis (DFA)

As described in the Introduction, the results of FA are unreliable if trends occur in the data. To overcome this problem, the detrended fluctuation analysis (DFA) was developed [11.56]. There are different orders of DFA that are distinguished by the way the trends in the data are eliminated.

In lowest order (DFA1) we determine, for each segment ν, the best *linear* fit of the profile and identify the fluctuations by the variance $F^2(\nu, s)$ of the profile from this straight line. This way, we eliminate the influence of possible linear trends on scales larger than the segment. Note that linear trends in the profile correspond to stair-like trends in the original record. DFA1 has been proposed originally by Peng et al. [11.56] when analysing correlations in DNA. It can be generalised straightforwardly to eliminate higher order trends [11.7, 11.29].

In second-order DFA (DFA2) one calculates the variances $F^2(\nu, s)$ of the profile from best *quadratic* fits of the profile (Fig. 11.1), this way eliminating the influence of possible linear and parabolic trends on scales larger than the segment considered. In general, in nth-order DFA, we calculate the variances of the profile from the best nth-order polynomial fit, this way eliminating the influence of possible $(n-1)$ th-order trends on scales larger than the segment size, since the profile represents the integrated record.

Explicitly, we calculate the best polynomial fit $p_\nu(i)$ of the profile in each of the $2N_s$ segments ν and determine the variance

$$F^2(\nu, s) \equiv \frac{1}{s}\sum_{i=1}^{s}\left[Y((\nu-1)s+i) - p_\nu(i)\right]^2 \ . \tag{11.9}$$

Then we employ (11.6) to determine the mean fluctuation $F_2(s)$.

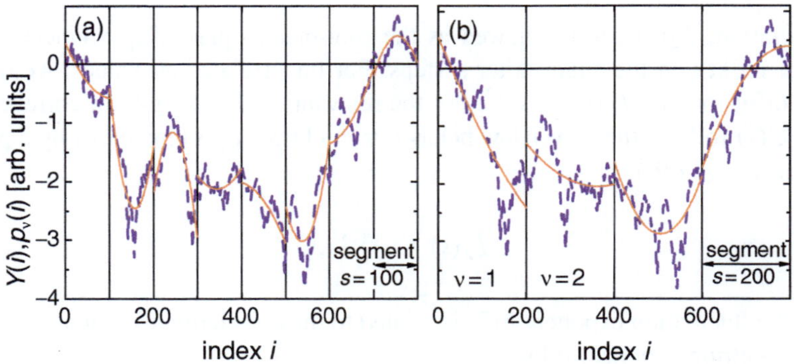

Fig. 11.1 Illustration of the detrending procedure in the second order detrended fluctuation analysis (DFA2). The profile $Y(i)$ (*dashed lines, blue*) calculated by summation of the time series is split into non-overlapping segments of equal duration s (time scale). This step is illustrated for $s = 100$ days in (**a**) and for $s = 200$ days in (**b**). Least square quadratic fits (*continuous lines, red*) to the profile are calculated in each segment. The squares of the differences between the profile and the fits are used to calculate the fluctuation function $F_2(s)$ of the DFA procedure. After [11.15]

Since FA and the various stages of the DFA have different detrending capabilities, a comparison of the fluctuation functions obtained by FA and DFAn can yield insight into both long-term correlations and types of trends [11.29, 11.61]. This cannot be achieved by the conventional methods, like the spectral analysis. We like to note that DFA1 is somewhat similar to rescaled range analysis introduced by Hurst (for a review see, e.g. [11.16]). For monofractal data, $h(2)$ is identical to the Hurst exponent.

11.3.4 Wavelet Transform (WT)

The wavelet methods we employ here are based on the determination of the mean values $\overline{Y}_\nu(s)$ of the profile in each segment ν (of length s) and the calculation of the fluctuations between neighbouring segments. The different order techniques we have used in analysing data fluctuations differ in the way the fluctuations between the average profiles are treated and possible non-stationarities are eliminated. The first-, second-, and third-order wavelet methods are described below:

(i) In the first-order wavelet method (WT1), one simply determines the fluctuations from the first-order difference $F^2(\nu, s) = [\overline{Y}_\nu(s) - \overline{Y}_{\nu+1}(s)]^2$ to be inserted in (11.6). WT1 corresponds to FA where constant trends in the profile are eliminated, while linear trends are not eliminated.

(ii) In the second-order wavelet method (WT2), one determines the fluctuations from the second-order difference $F^2(\nu, s) = [\overline{Y}_\nu(s) - 2\overline{Y}_{\nu+1}(s) + \overline{Y}_{\nu+2}(s)]^2$. So, if the profile consists of a trend term linear in s and a fluctuating term, the trend term is eliminated. Regarding trend elimination, WT2 corresponds to DFA1.

(iii) In the third-order wavelet method (WT3), one determines the fluctuations from the third-order difference $F^2(\nu, s) = [\overline{Y}_\nu(s) - 3\overline{Y}_{\nu+1}(s) + 3\overline{Y}_{\nu+2}(s) - \overline{Y}_{\nu+3}(s)]^2$. By definition, WT3 eliminates linear and parabolic trend terms in the profile. In general, in WTn we determine the fluctuations from the nth-order difference, this way eliminating trends described by $(n-1)$-order polynomials in the profile.

Methods (i–iii) are called wavelet methods, since they can be interpreted as transforming the profile by discrete wavelets representing first-, second-, and third-order cumulative derivatives of the profile. The first-order wavelets are known in the literature as Haar wavelets. One can also use different shapes of the wavelets (e.g. Gaussian wavelets with width s), which have been used by Arneodo et al. [11.3] to study, for example, long-range correlations in DNA. Since the various stages of the wavelet methods WT1, WT2, WT3, etc. have different detrending capabilities, a comparison of their fluctuation functions can yield insight into both long-term correlations and types of trends.

At the end of this section, before describing the results of the FA, DFA, and WT analysis, we note that for very large s values, $s > N/4$ for DFA and $s > N/10$ for FA and WT, the fluctuation functions become inaccurate due to statistical errors. The difference in the statistics is due to the fact that the number of independent segments of length s is larger in DFA than in WT, and the fluctuations in FA are larger than in DFA. Hence, in the analysis we will concentrate on s values lower than $s_{\max} = N/4$ for DFA and $s_{\max} = N/10$ for FA and WT. We manually chose an appropriate (shorter) fitting range of typically two orders of magnitude and assign a scaling exponent $h(2)$ to the curve when, in this fitting rage, the curve can be reasonably well approximated by a straight line.

11.4 Multifractal Analysis

11.4.1 Multifractal DFA (MF-DFA)

For multifractal time series, a single scaling exponent like $h(2)$ or γ does not completely characterise the record, since many subsets of the series have different scaling behaviour, e.g. large fluctuations are less correlated than small fluctuations. In order to study these multifractal scaling properties, the DFA procedure can be generalised to higher moments [11.32]. This generalisation (multifractal DFA, MF-DFA) is equivalent to the wavelet transform modulus maxima (WTMM) method [11.3, 11.53], but the MF-DFA is much easier to implement on the computer; see [11.31] for a comparison of the two methods (for wavelet methods see also [11.38]). In this procedure, the variance $F^2(\nu, s)$ in (11.6) is replaced by its q/2nd power and the square root is replaced by the $1/q$th power, where $q \neq 0$ is a real parameter,

$$F_q(s) \equiv \left[\frac{1}{2N_s} \sum_{\nu=1}^{2N_s} \left[F^2(\nu, s) \right]^{q/2} \right]^{1/q} . \tag{11.10}$$

Analogous to (11.7) one defines then the generalised fluctuation exponent $h(q)$ by

$$F_q(s) \sim s^{h(q)} , \tag{11.11}$$

We note that $h(1)$ corresponds to the classical Hurst exponent H determined by rescaled range analysis (since first moments are considered in both cases) and that multifractal DFA is identical to standard DFA if $q = 2$ (hence the notation $h(2)$ in (11.7)).

For monofractal time series, $h(q)$ is independent of q, since the scaling behaviour of the variances $F^2(\nu, s)$ is identical for all segments ν. If, on the other hand, small and large fluctuations scale differently, there will be a significant dependence of $h(q)$ on q: If we consider positive values of q, the segments ν with large variance $F^2(\nu, s)$

(i.e. large deviations from the corresponding fit) will dominate the average $F_q(s)$. Thus, for positive values of q, $h(q)$ describes the scaling behaviour of the segments with large fluctuations. Usually large fluctuations are characterised by a smaller scaling exponent $h(q)$. On the contrary, for negative values of q, the segments ν with small variance $F^2(\nu, s)$ will dominate the average $F_q(s)$. Hence, for negative values of q, $h(q)$ describes the scaling behaviour of the segments with small fluctuations, which are usually characterised by a larger scaling exponent.

11.4.2 Comparison with Related Multifractal Formalisms

The multifractal fluctuation exponents $h(q)$ defined in (11.11) are directly related to the classical Renyi exponents $\tau(q)$ (see, e.g. [11.16, 11.60]) via [11.32]

$$h(q) = [\tau(q) + 1]/q \; . \tag{11.12}$$

In the geophysics literature also other multifractal quantities have been used that $h(q)$ can easily be related to the following:

(1) The 'generalised variogram' $C_q(s)$ (see, e.g. (3.82–3.84) in [11.60] and references therein) is defined as

$$C_q(s) \equiv \langle |Y(i+s) - Y(i)|^q \rangle \sim s^{K(q)} \; , \tag{11.13}$$

where the average is taken over all values of i. Comparing (11.10), (11.11), and (11.13) one can easily verify that $h(q)$ and $K(q)$ are related by

$$K(q) = qh(q) \; . \tag{11.14}$$

(2) In several geophysics and turbulence papers (see, e.g. [11.10, 11.11, 11.42, 11.55, 11.64, 11.65, 11.71, 11.72]), the structure function

$$S_q(s) \equiv \langle |\phi_{i+s} - \phi_i|^q \rangle \sim s^{\zeta(q)} = s^{qH(q)} \tag{11.15}$$

has been analysed directly without employing the profile $Y(i)$ as done in (11.13). It is easy to see that $H(q)$ is related to $h(q)$ by

$$H(q) = \zeta(q)/q = h(q) - 1 \; . \tag{11.16}$$

Accordingly, the multifractal exponents $H(q)$ defined by Davis et al. [11.10] and the exponents $h(q)$ defined here in (11.11) differ only by 1. This difference is due to the fact that here we analyse the cumulative sum of ϕ_i, while Davis et al. as well as the Lovejoy–Schertzer group analyse the ϕ_i directly.

For modelling the multifractal behaviour, one can employ, for example, a particular multifractal process where $H(q)$ (or $h(q) = H(q) + 1$) is known and adjust the parameters (see, e.g. Appendix). However, in most multifractal processes $H(1) = 0$ is fixed. Hence, one has to use a fractional integration and shift the whole function $H(q)$ to adjust the value for $q = 1$. When doing this, Lovejoy and Schertzer obtained an interesting formula for $\zeta(q) = qH(q)$ for positive q-values (Schertzer, D., private communication),

$$\zeta(q) = qH' - \frac{C_1}{\alpha' - 1}(q^{\alpha'} - q), \qquad q \geq 0. \tag{11.17}$$

The great advantage of the MF-DFA method used here is that it includes also negative q values, such that the basic Renyi exponents $\tau(q)$ can be calculated for both negative and positive q values. This allows a further characterisation of the multifractal series by the singularity spectrum $f(\alpha)$, which is related to $\tau(q)$ via a Legendre transform (see, e.g. [11.16, 11.60]),

$$\alpha = \frac{\mathrm{d}\tau(q)}{\mathrm{d}q} \quad \text{and} \quad f(\alpha) = q\alpha - \tau(q). \tag{11.18}$$

Here, α is the singularity strength or Hölder exponent, while $f(\alpha)$ denotes the dimension of the subset of the series that is characterised by α. Using (11.12), we can directly relate α and $f(\alpha)$ to $h(q)$,

$$\alpha = h(q) + q\frac{\mathrm{d}h(q)}{\mathrm{d}q} \quad \text{and} \quad f(\alpha) = q[\alpha - h(q)] + 1. \tag{11.19}$$

The strength of the multifractality of a time series can be characterised by the difference between the maximum and minimum singularity strength α, $\Delta\alpha = \alpha_{\max} - \alpha_{\min}$, which fulfil $f(\alpha) \to 0$ for $\alpha \to \alpha_{\max}$ and $\alpha \to \alpha_{\min}$.

11.5 Results of the Correlation Behaviour

After the description of the methods we turn to our results for the autocorrelation behaviour in this section and our results of the multifractal analysis in the next section. We begin with the results regarding runoff data.

11.5.1 River Runoff

In our study we analysed 42 runoff records, 18 of them are from the southern part of Germany, and the rest is from North and South America, Africa, Australia, Asia, and Europe (see Table 11.1). We begin the analysis with the runoff record for the river Weser in the northern part of Germany, which has the longest record (171 years) in this study. Figure 11.2(a) shows the fluctuation functions $F_2(s)$ obtained from FA and DFA1–DFA3. In the log–log plot, the curves are approximately straight lines

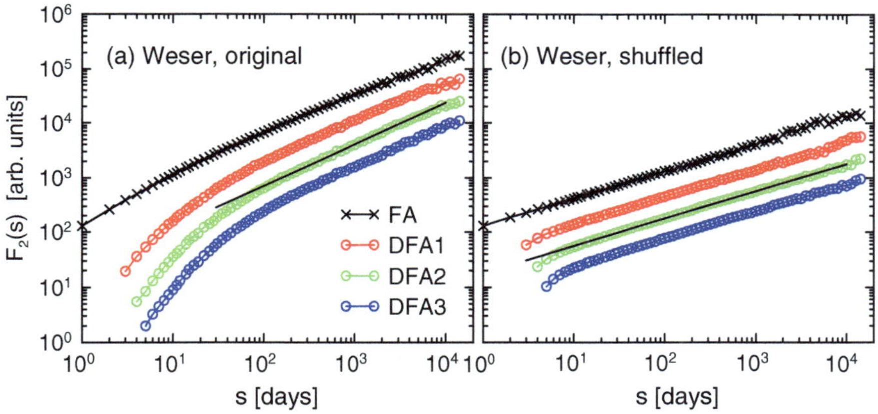

Fig. 11.2 (**a**) The fluctuation functions $F_2(s)$ versus time scale s obtained from FA and DFA1–DFA3 in double logarithmic plots for daily runoff departures $\phi_i = W_i - \overline{W_i}$ from the mean daily runoff $\overline{W_i}$ for the river Weser, measured from 1823 till 1993 by the hydrological station Vlotho in Germany. (**b**) The analogue curves to (**a**) when the ϕ_i are randomly shuffled. The *straight lines* have slopes $h(2) = 0.76$ and 0.5 in (**a**) and (**b**), respectively

for s above 30 days, with a slope $h(2) \approx 0.75$. The error bar is of the order of 0.03 [11.32]. This result for the Weser suggests that there exists long-term persistence expressed by the power-law decay of the correlation function, with an exponent $\gamma \approx 0.5$ (see (11.8)).

To show that the slope $h(2) \approx 0.75$ is due to long-term correlations and not due to a broad probability distribution (Joseph versus Noah phenomenon, see [11.46]), we have eliminated the correlations by randomly shuffling the ϕ_i. This shuffling has no effect on the probability distribution function of ϕ_i. Figure 11.2b shows $F_2(s)$ for the shuffled data. We obtain $h(2) = 1/2$, showing that the exponent $h(2) \approx 0.75$ is due to long-term correlations.

To show that the slope $h(2) \approx 0.75$ is not an artefact of the seasonal dependence of the variance and skew, we also considered records where ϕ_i was divided by the variance of each calendary day and applied further detrending techniques that take into account the skew [11.43]. In addition, we smoothed the seasonal trend $\overline{W_i}$ using polynomial fits. In all cases, we found just insignificant changes in the scaling behaviour for large times. This can be understood easily, since these kinds of seasonal trends cannot effect the fluctuation behaviour on time scales well above 1 year. It is likely, however, that the seasonal dependencies of the variance and possibly also of the skew contribute to the behaviour at small times, where the slope $h(2)$ is much larger than 0.75 in most cases.

Figure 11.3 and Table 11.2 summarize our results for the DFA2 scaling exponents $h(2)$ on large time scales for all rivers we studied; the mean value is $h(2) = 0.72$. Based on some of the data we estimate that the error bars are of the order of 0.03 [11.32]. One can see clearly that the fluctuation exponents vary strongly from river to river reflecting the fact that there exist different mechanisms for floods where each may induce different scaling. For example, rain-induced floods and

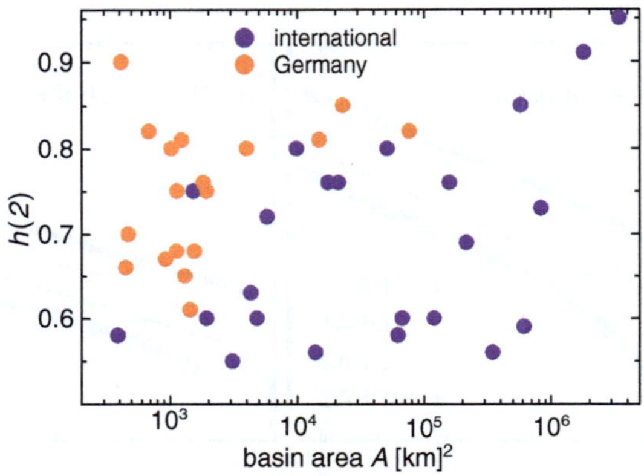

Fig. 11.3 Long-term fluctuation exponents $h(2)$ for all international runoff records (*violet symbols*, [11.21]) and all runoff records from southern Germany (*orange symbols*) that were analysed, as a function of the basin area A. Each symbol represents the result for one hydrological station. After [11.35]

Table 11.2 Details of the records analysed in Figs. 11.8 and 11.9 and average values for all records we studied. The recording period and the fitting parameters obtained within the extended multiplicative cascade model, (11.21), and the Lovejoy–Schertzer model, (11.20), are reported. Further, our two main parameters, the fluctuation exponents $h(2) = 1 - \gamma/2 = (\beta + 1)/2$ and the resulting multifractality strength $\Delta\alpha$, are given. In the case of Jena, (11.24) has been used instead of (11.21) and the corresponding parameters are $\alpha_1 = 0.57$, $\alpha_2 = 0.45$, and $q_\times = 3.3$. The values for all analysed runoff records are detailed in Table 11.1

	Period	$h(2)$	$\Delta\alpha$	a	b	H'	C_1	α'
Precipitation								
Spokane	1881–1994	0.51	0.30	0.63	0.77	−0.48	0.008	1.9
Edinburgh	1896–1997	0.52	0.34	0.61	0.77	−0.47	0.012	1.8
Jena	1827–2000	0.56	0.11	–	–	−0.43	0.006	2.3
Average (99 records)	86 years	0.53	0.29	–	–	−0.45	0.012	2.0
Std. deviation	33 years	0.04	0.14	–	–	0.06	0.010	0.4
Runoff								
Weser	1823–1993	0.76	0.43	0.50	0.68	−0.24	0.023	1.7
Fraser	1913–1996	0.69	0.38	0.53	0.70	−0.29	0.017	1.7
Susquehanna	1891–1986	0.58	0.48	0.55	0.77	−0.40	0.018	1.7
Average (42 records)	86 years	0.72	0.49	–	–	−0.25	0.039	1.4
Std. deviation	27 years	0.11	0.16	–	–	0.10	0.028	0.5

snow-induced floods may introduce different spatial scaling behaviour [11.23], which might also result in different temporal scaling behaviour (see below). The $h(2)$ values spread from 0.55 to 0.95. Since the correlation exponent γ is related to $h(2)$ by $\gamma = 2 - 2h(2)$, the exponent γ spreads from almost 0 to almost 1, covering the whole range from very weak to very strong long-term correlations. Accordingly, there is no universal scaling behaviour. This is in contrast to temperature records, where universal long-term persistence at land stations, i.e. rather identical values of $h(2) = 0.66 \pm 0.06$ were observed [11.15, 11.34, 11.68, 11.75].

Figure 11.3 shows that the exponents do not depend systematically on the basin area A. This is in line with the conclusions of [11.24] for the flood peaks, where a systematic dependence on A could also not be found. In addition, here is also no pronounced regional dependence: the rivers within a local area (such as southern Germany) tend to have nearly the same range of exponents as the international rivers.

Figure 11.4 shows the fluctuation functions $F_2(s)$ of three more rivers, from Africa, Australia, and Europe. The panels on the left-hand side show the FA and DFA1–DFA3 curves, while the panels on the right-hand side show the results from the analogous wavelet analysis WT1–WT4. Evidently, the results are equivalent for both methods. On large time scales, the fluctuation functions (from DFA1–DFA3 and WT2–WT4) show power-law behaviour, with exponents $h(2) \simeq 0.95$ for the Zaire, $h(2) \simeq 0.60$ for the Mary river, and $h(2) \simeq 0.55$ for the Gaula river. The Mary river in Australia is rather dry in the summer. The Gaula river in Norway is frozen in the winter, and a constant runoff value has been assumed in the data during

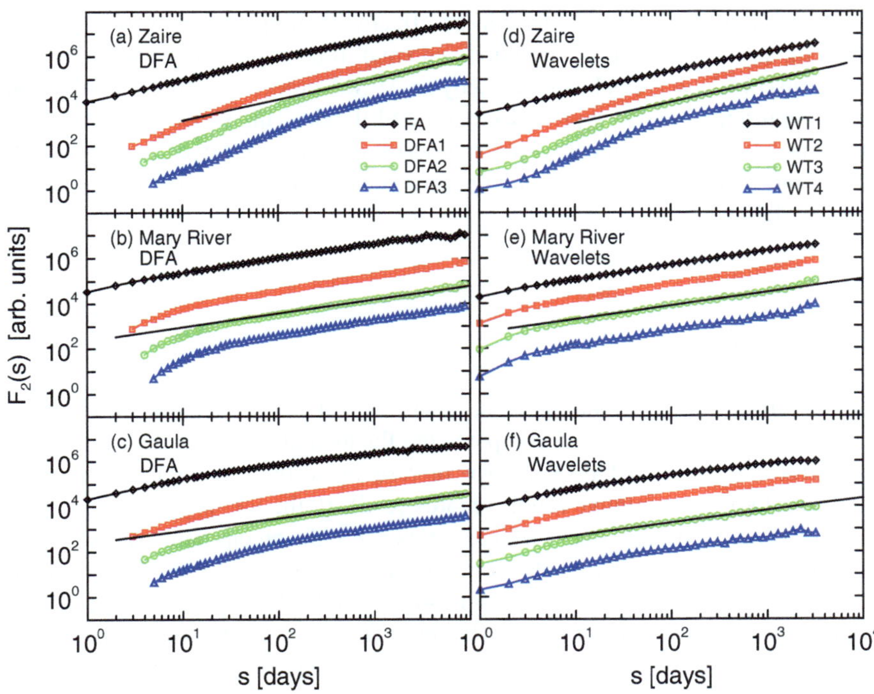

Fig. 11.4 The fluctuation functions $F_2(s)$ versus time scale s obtained from FA and DFA1–DFA3 in double logarithmic plots for three additional representative hydrological stations: (**a**) the Zaire in Kinshasa, Zaire, (**b**) the Mary River in Miva, Australia, and (**c**) the Gaula River in Haga Bru, Norway. (**d–f**) The fluctuation functions $F_2(s)$ obtained for the same rivers as in (**a–c**), from first- to fourth-order wavelet analysis (WT1–WT4). The *straight lines* are best linear fits to the DFA2 and the WT3 results on large time scales. The *straight lines* have slopes $h(2) = 0.95$, 0.60, and 0.55 in (**a, d**), (**b, e**), and (**c, f**), respectively

freezing time [11.21]. For the Mary river, the long-term exponent $h(2) \simeq 0.60$ is well below the average value, while for the Gaula river, the long-term correlations are not pronounced ($h(2) = 0.55$) and even hard to distinguish from the uncorrelated case $h(2) = 0.5$. We obtained similar results for the other two 'frozen' rivers (Tana from Norway and Dvina from Russia) that we analysed. For interpreting this distinguished behaviour of the frozen rivers we like to note that on permafrost ground the lateral inflow (and hence the indirect contribution of the water storage in the catchment basin) contributes to the runoffs in a different way than on normal ground, see also [11.23]. Our results (based on three rivers only) seem to suggest that the contribution of snow melting leads to less correlated runoffs than the contribution of rainfall, but more comprehensive studies will be needed to confirm this result.

Most curves in Fig. 11.4 show a crossover at time scales of several weeks. A similar crossover has been reported by Tessier et al. [11.71] for small French rivers without artificial dams or reservoirs. We note that the crossover time scale is similar to the period of planetary waves, which are oscillations of very predominantly tropospheric origin with typical periods of about 2–30 days (see, e.g. [11.40]). Below the crossover we find an effective scaling exponent $h(2) \approx 1.5$, indicating strong short-term correlations on small time scales. Approximately, the short-term correlations can be modelled by an ARMA process, where the correlation time is represented by the typical decay time of floods. This yields $h(2) = 1.5$ on short time scales in agreement with our observation. Note, however, that long-term correlations (and multifractality on large time scales) require an additional and different modelling approach.

11.5.2 Precipitation

Now we turn to the precipitation data, where we studied the records of 99 meteorological stations. Figure 11.5 compares the fluctuation functions $F_2(s)$ obtained from DFA1, DFA2, and DFA3 for two representative daily precipitation records (a–b) with two representative runoff records (c–d). In the log–log plot, the curves are approximately straight lines on large scales s. For the precipitation records there is – contrary to the runoff records – only a very weak crossover in $F_2(s)$, and the $h(2)$ values are rather close to 0.5, indicating rapidly decaying autocorrelations. Specifically, we find $h(2) \simeq 0.55$ for Hamburg and $h(2) \simeq 0.50$ for Vienna, corresponding to a correlation exponent $\gamma \approx 0.9$ for Hamburg and $\gamma \geq 1$ for Vienna. For the precipitation data, the higher slopes at very small scales are partly a methodical artefact (see [11.29]).

Figure 11.6 shows the distributions of the $h(2)$ values obtained with DFA2 on large time scales for all 99 precipitation records using (11.1) for seasonal detrending. Based on some of the data we estimate that the error bars are of the order of 0.03 [11.32]. One can clearly see that most precipitation records exhibit no long-term correlations ($h(2) \approx 0.5$) or only very weak long-term correlations ($h(2) \approx 0.55$);

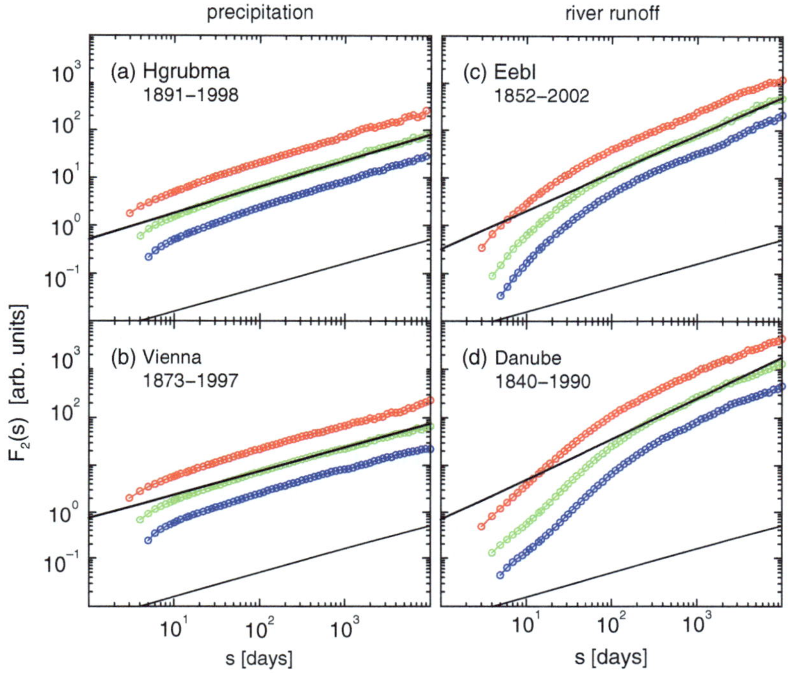

Fig. 11.5 The fluctuation functions $F_2(s)$ versus time scale s obtained from DFA1 to DFA3 (*from the top to the bottom*) in double logarithmic plots for representative precipitation and runoff records (*shifted vertically for clarity*). Precipitation records: (**a**) Hamburg, Germany ($h(2) = 0.55$) and (**b**) Vienna, Austria ($h(2) = 0.50$); runoff records: (**c**) Elbe river in Dresden, Germany ($h(2) = 0.80$) (**d**) Danube river in Orsova, Romania ($h(2) = 0.85$). The *straight lines* through the data have the reported slopes, and *lines with slope* $h(2) = 0.5$ are shown below the data for comparison with the uncorrelated case

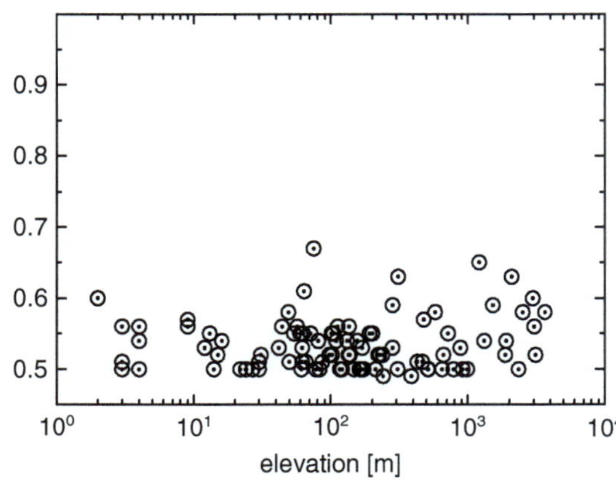

Fig. 11.6 Long-term fluctuation exponents $h(2)$ for all international precipitation records that we analysed, as a function of the elevation. Each symbol represents the result for one meteorological station

the mean value is $h(2) = 0.53 \pm 0.04$. The figure also shows that the exponents $h(2)$ do not depend systematically on the elevation of the meteorological station. Further we find no systematic dependence of the $h(2)$ values on climate zone or geographical region. In order to confirm the absence of long-term autocorrelations in the precipitation series, we have also calculated the autocorrelation function $C(s)$ (see (11.3)). We found that after a very small number of days all correlations vanish (not shown). Note, however, that this result is obtained for the second moment, and thus it refers only to two-point correlations; higher order correlations are not excluded by our result.

Figure 11.7 gives further examples of DFA studies of precipitation records. The data from the summit Zugspitze is an example with relatively strong long-term correlations with $\alpha \simeq 0.6$. The large slope of FA indicates a trend in the records. A similar effect is found for the record of Irkutsk.

Our results for $h(2)$ for precipitation data are not in agreement with [11.49], where, using DFA1 and seasonal detrending based on procedures similar to (11.1) for nine precipitation records with 15 min resolution, $h(2) \approx 1.0$ ($\alpha = h(2)$ in the notation used there) was found on time scales below 10 days and $h(2) = 0.6 \ldots 0.8$ on time scales from 10 days to 16 months. We cannot exactly pinpoint the reason for

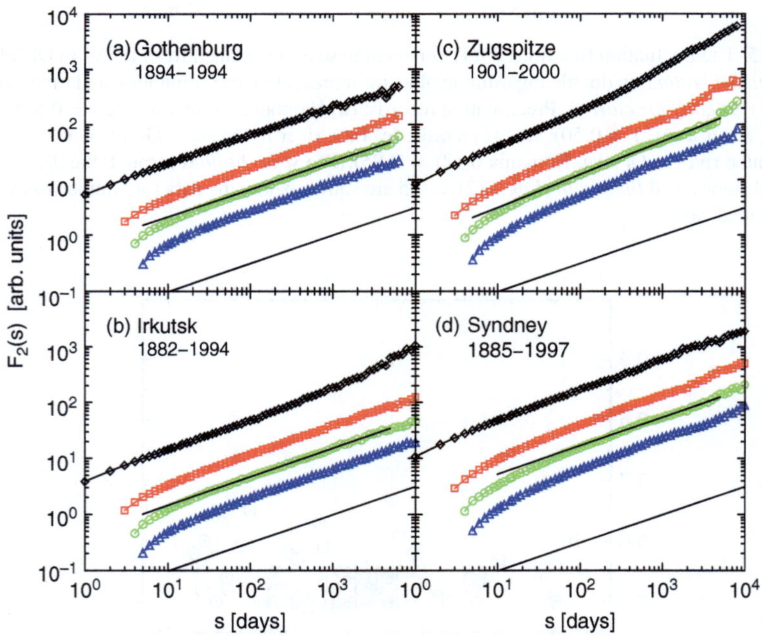

Fig. 11.7 The fluctuation functions $F_2(s)$ versus time scale s obtained from FA and DFA1–DFA3 (*from top*) in double logarithmic plots for four additional representative precipitation stations: (**a**) Gothenburg (USA), (**b**) Irkutsk (RUS), (**c**) Zugspitze (GER), and (**d**) Sydney (AUS). The *straight lines* are best linear fits to the DFA2 results on large time scales and have the slopes 0.5, 0.51, 0.6, and 0.51 (a–d). The *line* at the bottom of each panel has slope 0.5

the disagreement, in particular since the binned spectral analysis by Matsoukas et al. [11.49] agrees with our results. However, the DFA1 discrepancy might be caused by the shorter time series considered by Matsoukas et al. [11.49]. Our $h(2)$ values are also not in agreement with [11.57, 11.58], where records of length 6 months were considered and an exponent $h(1) \approx 0.76$ ($H = h(1) \gtrsim h(2)$) was found using rescaled range analysis. We believe that the reason for this disagreement is that the seasonal trend cannot be eliminated in these short records and acts like a long-term correlation. Thus it seems that the larger exponent obtained by Peters et al. is due to the seasonal trend but not to long-term correlations. For intermediate time scales up to 3 years, our finding is in agreement with [11.18], who, however, report long-term correlations on even larger time scales in disagreement with our conclusion. This observation of correlations on large time scales might also have been caused by non-stationarities in the data. Finally, we like to note that our results do not change significantly, when also the seasonal trend in the variance of the data was eliminated by using (11.2) instead of (11.1). Although $h(2)$ changes by about ± 0.03 for individual records, the average $h(2)$ remains constant.

Since the autocorrelations of rainfall data decay rapidly in time, their temporal correlations can neither account for the long-term correlations nor for the broad distribution of the correlation exponents of the runoff data. These results seem to indicate that the persistence of the runoffs is rather caused by storage effects than by long-term memory of the precipitation; the integration of rainfall in time–space might produce long-term memory in river flows. If this interpretation is correct, i.e. runoff persistence is caused by catchment storage, we would expect it to be less emphasized in arid regions. Indeed, for the Mary river in Australia, the only river with arid catchment in our study, we find a quite low fluctuation exponent, $h(2) \approx 0.6$, corresponding to weak persistence (see Fig. 11.4). Further studies are needed to confirm our interpretation. Note that the integration hypothesis cannot explain the differences in the multifractality of runoff and rainfall that we discuss now.

11.6 Results of the Multifractal Behaviour

Next we compare the multifractal behaviour of the 42 daily river runoff records and of the 99 daily precipitation records. For all records, we found that MF-DFA2, MF-DFA3, and MF-DFA4 (corresponding to second-, third-, and fourth-order polynomials in the detrending procedure, respectively, see Sects. 11.3.3 and 11.4.1) yield similar results for the fluctuation function $F_q(s)$. Therefore we present only the results for MF-DFA2. Again we begin with the seasonal detrending according to (11.1).

Figure 11.8 shows two representative examples for the MF-DFA2 fluctuation functions $F_q(s)$, for (a) the precipitation record at Spokane, USA, and (b) the runoff record of the Weser river. The standard fluctuation function $F_2(s)$ is plotted with crosses. The crossover in $F_2(s)$ for the runoff data that were discussed in the previous section can also be seen in the other moments, and additionally a similar crossover can be seen in the negative moments for the precipitation data.

The position of the crossover increases monotonically with decreasing q and the crossover becomes more pronounced. While most of the recent literature focussed on short-term multifractality and the crossovers occurring on time scales below 1 year [11.12–11.14, 11.25, 11.45, 11.54, 11.55, 11.71], here we are mainly interested in the asymptotic behaviour of $F_q(s)$ at large times s. One can see clearly that above the crossover, the $F_q(s)$ functions are straight lines in the double logarithmic plot, and the slopes increase slightly when going from high positive moments towards high negative moments (from the bottom to the top). For the precipitation at Spokane (Fig. 11.8(a)), for example, the slope changes from 0.42 for $q = 8$ to 0.57 for $q = -8$. The monotonous increase of the slopes, $h(q)$, is the signature of multifractality. We obtain similar increases for the runoff records, although all $h(q)$ are larger on the average; for the Weser river the results are $h(8) = 0.63$ and $h(-8) = 0.84$ (see Fig. 11.8b).

When the data are shuffled (see Fig. 11.8c,d), all functions $F_q(s)$ show asymptotic scaling of approximately $F_q(s) \sim s^{1/2}$. This indicates that the multifractality

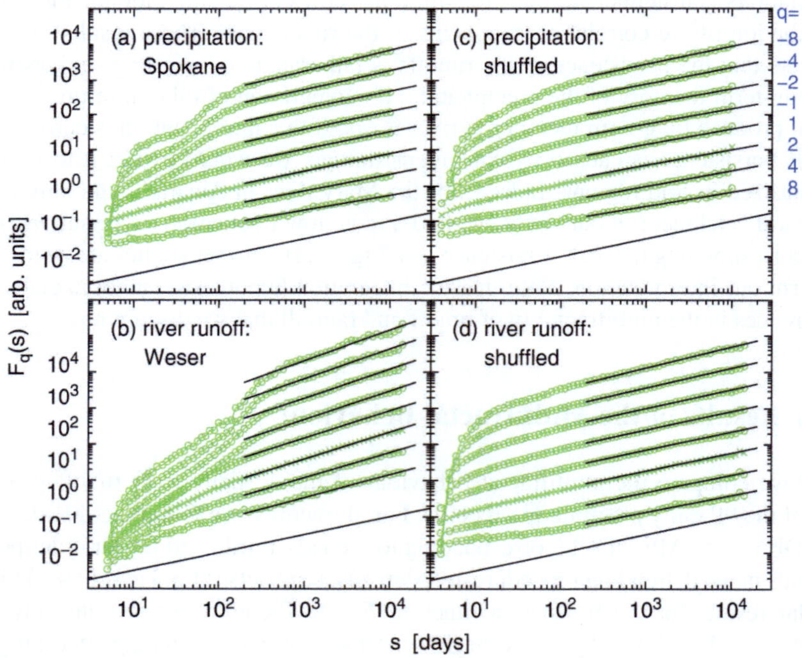

Fig. 11.8 The multifractal fluctuation functions $F_q(s)$ versus time scale s obtained from multifractal DFA2 for two representative records: (**a**) precipitation at Spokane (USA) and (**b**) river Weser in Vlotho (GER). The *curves* correspond to different values of q, $q = -8, -4, -2, -1, 1, 2, 4, 8$ (*from the top to the bottom*) and are shifted vertically for clarity. Panels (**c, d**) show the corresponding results for shuffled records. The curve for $q = 2$ (*standard* DFA) is plotted with crosses ($\times$). The *straight lines* through the data are the linear fits we use to calculate $h(q)$ (as reported in Fig. 11.9), and additional lines with slope $h(2) = 0.5$ are shown below the data for comparison with the uncorrelated case

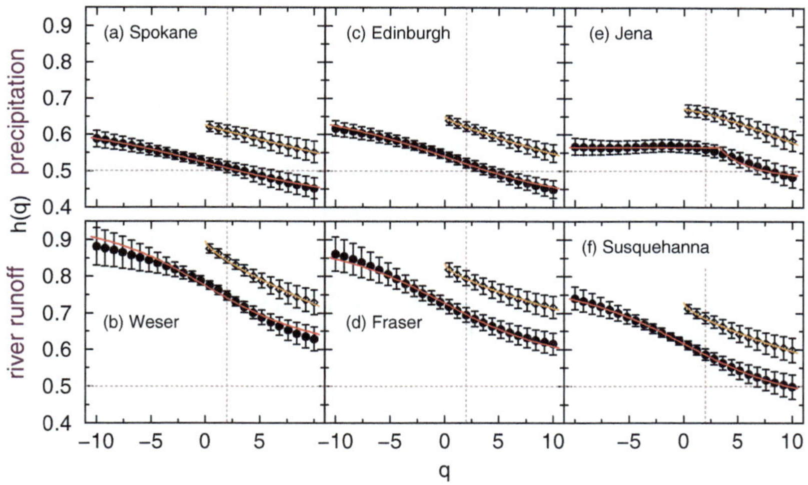

Fig. 11.9 The generalised Hurst exponents $h(q)$ for three representative daily precipitation records: (**a**) Spokane (USA), (**c**) Edinburgh (UK), (**e**) Jena (GER), and three representative daily runoff records: (**b**) Weser in Vlotho (GER), (**d**) Fraser in Hope (USA), and (**f**) Susquehanna in Marietta (USA). The $h(q)$ values (*full symbols*) have been determined by straight line fits of log–log plots of $F_q(s)$ on large time scales s. The *solid lines* through the full symbols are fits to the data using the operational extended multiplicative cascade model, (11.21), except for (**c**), where the bifractal model, (11.24), has been used instead. In this case the corresponding parameters are $\alpha_1 = 0.57$, $\alpha_2 = 0.45$, and $q_\times = 3.3$. In the *upper right corner* of each plot $h(q)$ is shown (*open symbols, shifted* upwards by 0.1) together with fits using the Lovejoy–Schertzer model, (11.20). The *straight dotted lines* indicate uncorrelated behaviour $h = 0.5$ and $q = 2$, respectively

vanishes under shuffling. Accordingly the observed multifractality originates in the long-term correlations of the record and is not caused by singularities in the distribution of the daily records (see also [11.46]). A reshuffling-resistant multifractality would indicate a so-called 'statistical' type of non-linearity as discussed in [11.67]. We obtain similar patterns for all runoff and precipitation records.

From the asymptotic slopes of the curves in Fig. 11.8 (see linear fits in the plots), we obtained the generalised Hurst exponents $h(q)$, which are plotted versus the moment q in Fig. 11.9 for three representative precipitation records and three representative runoff records; see Table 11.2 for details. Our results for $h(q) = [\tau(q) + 1]/q$ may be compared with several functional forms used in the literature to describe multifractality.

11.6.1 Fits by the Universal Multifractal Model

First we consider the formula (see (11.16) and (11.17))

$$h(q) = H' + 1 - \frac{C_1}{\alpha' - 1}(q^{\alpha' - 1} - 1)\,, \qquad q \geq 0, \tag{11.20}$$

with the three parameters H', C_1, and α', that has been successfully used by Lovejoy, Schertzer, and coworkers [11.42, 11.55, 11.64, 11.65, 11.71, 11.72] (see also [11.14]) to describe the multifractal behaviour of rainfall and runoff records at smaller time scales. Of course, the functional form (11.20) does apply only to positive moments q. The lines in the upper right parts of Fig. 11.9 indicate the corresponding fits, and the values of the three parameters are listed in Table 11.2 together with their means for all precipitation and runoff records.

Figure 11.10 summarizes our results for H', C_1, and α', for both precipitation and runoff records. For the precipitation data, we obtained $H' = -0.45 \pm 0.06$, $C_1 = 0.01 \pm 0.01$, and $\alpha' = 2.0 \pm 0.4$, while the runoff data yielded $H' = -0.25 \pm 0.10$, $C_1 = 0.04 \pm 0.03$, and $\alpha' = 1.4 \pm 0.5$.

Our values for H' are in marginal agreement with earlier studies by Tessier et al. [11.71] who obtained $H' = -0.35 \pm 0.2$ for precipitation records and $H' = -0.05 \pm 0.2$ for runoff records. The same conclusion holds for [11.55] who reported $H' = -0.03 \pm 0.14$ for runoff records. The agreement is surprising, since in these earlier studies a seasonal detrending (our Eqs. (11.1) or (11.2)) has not been performed. The multifractality is characterised by the parameters C_1 and α'. Here, Tessier et al. [11.71] report $\alpha' = 1.6 \pm 0.2$ and $C_1 = 0.10 \pm 0.05$ for precipitation and $\alpha' = 1.45 \pm 0.2$ and $C_1 = 0.2 \pm 0.1$ for river flows, while Pandey et al. [11.55] obtained $\alpha' = 1.7 \pm 0.11$ and $C_1 = 0.12 \pm 0.03$ for river flows. In a more recent study, Tchiguirinskaia et al. [11.70] analysed several runoff records in Russia, finding $\alpha' \approx 1.7$ and $C_1 \approx 0.03$. Our results for the exponent α' are, within the error bars, in agreement with these earlier studies. The result for C_1 is in agreement with [11.70], but disagrees with [11.55, 11.71], who did not explicitly focus on the asymptotic regime and therefore obtained larger exponents C_1 (as in [11.70] for the short-term regime).

Tessier et al. [11.71] concluded that the H' values for precipitation and for runoff differ by $\Delta H' = \Delta h \approx 0.3$, while the α' and C_1 values are compatible. We find a similar difference in the Hurst exponents, $\Delta h \approx 0.2$. Our α' and C_1 values for precipitation and runoff seem to be marginally compatible, since the corresponding

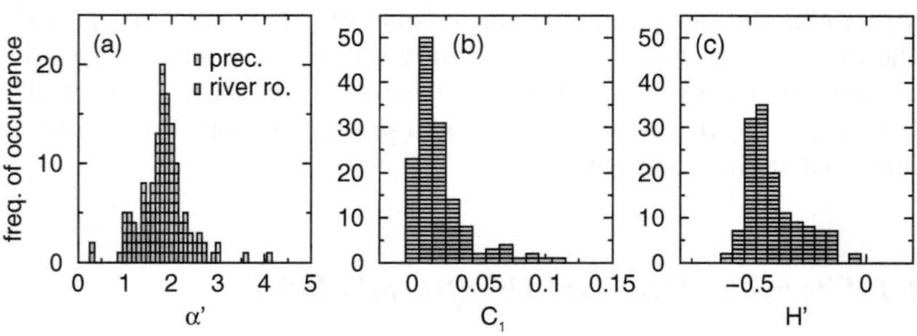

Fig. 11.10 Histograms of the parameters α', C_1, and H' for all 99 daily precipitation records (*grey*) and all 42 daily runoff records (*blue*). The values have been determined by applying non-linear fits of (11.20) to the generalised Hurst exponents $h(q)$. Each *box* represents the result for one meteorological or hydrological station

histograms in Fig. 11.10 are shifted to the left and to the right, respectively, for the runoff data. The average of C_1 is more than three times larger for runoff. Thus, we conclude that the runoff fluctuations cannot be generated by a simple fractional (time) integration of rainfall series. We believe that storage effects and the highly intermittent spatial behaviour of rainfall as well as the highly non-linear interaction between rainfall and runoff have to be taken into account here.

11.6.2 Fits by the Extended Multiplicative Cascade Model

Next we consider the functional form

$$\tau(q) = -\frac{\ln(a^q + b^q)}{\ln 2} \quad \text{or} \quad h(q) = \frac{1}{q} - \frac{\ln(a^q + b^q)}{q \ln 2}, \tag{11.21}$$

which can be derived from a modification of the multiplicative cascade model [11.31]; see Appendix. The advantage of this formula is that it extends also to negative q-values, and thus can be used for obtaining the multifractal spectrum $f(\alpha)$. By (11.21), the infinite number of exponents $h(q)$ can be described by only two independent parameters, a and b; the width of the corresponding singularity spectrum (see (11.19)) is given by $\Delta\alpha = |\ln a - \ln b| / \ln 2$. Our fitting results for the examples shown in Fig. 11.9 are listed in Table 11.2, except for the Jena record, where (11.21) does not provide a reasonable fit. The two parameters a and b can then be regarded as multifractal fingerprint for a considered runoff or precipitation record. This is particularly important for evaluating precipitation–runoff models, for example, artificial rainfall data (input for the models) can be generated with these two parameters and then the runoff (output of the models) can be checked based on them.

It is important to emphasize that the parameters a and b have been obtained from the *asymptotic* part of the generalised fluctuation function and are therefore not affected by seasonal dependencies, which cannot be fully eliminated from the data. We would like to note that different multifractal models for high-resolution precipitation time series, which include the crossover on short time scales, have been suggested [11.12, 11.13, 11.66, 11.73, 11.74]. For a model based on self-organized criticality we refer to [11.2].

However, for the precipitation records, there are several cases where (11.21) cannot be used to fit $h(q)$ for all q values. In some of these cases (like the Jena precipitation record), a simple bifractal model fits much better to the $h(q)$ data.

11.6.3 Fits by the Bifractal Model

For bifractal records the Renyi exponents $\tau(q)$ are characterised by two distinct slopes α_1 and α_2,

$$\tau(q) = \begin{cases} q\alpha_1 - 1 & q \le q_\times \\ q\alpha_2 + q_\times(\alpha_1 - \alpha_2) - 1 & q > q_\times \end{cases} \tag{11.22}$$

or

$$\tau(q) = \begin{cases} q\alpha_1 + q_\times(\alpha_2 - \alpha_1) - 1 & q \le q_\times \\ q\alpha_2 - 1 & q > q_\times \end{cases}. \tag{11.23}$$

If this behaviour is translated into the $h(q)$ picture using (11.12), we obtain that $h(q)$ exhibits a plateau from $q = -\infty$ up to a certain $q_\times$ and decays hyperbolically for $q > q_\times$,

$$h(q) = \begin{cases} \alpha_1 & q \le q_\times \\ q_\times(\alpha_1 - \alpha_2)\frac{1}{q} + \alpha_2 & q > q_\times \end{cases}, \tag{11.24}$$

or vice versa,

$$h(q) = \begin{cases} q_\times(\alpha_2 - \alpha_1)\frac{1}{q} + \alpha_1 & q \le q_\times \\ \alpha_2 & q > q_\times \end{cases}. \tag{11.25}$$

Both versions of this bifractal model require three parameters. The multifractal spectrum is degenerated to two single points, thus its width can be defined as $\Delta\alpha = \alpha_1 - \alpha_2$. One example of a bifractal fit is shown in Fig. 11.9(e).

11.6.4 Results

We have fitted the $h(q)$ spectra in the range $-10 \le q \le 10$ for all 99 precipitation records and all 42 runoff records by (11.21) and (11.24) (or (11.25)) for some precipitation records. For all runoff records, the extended multiplicative cascade model fits best. For the precipitation records, (11.21) fits best in just 54 cases. Either (11.24) or (11.25) could be used to fit the results of 27 precipitation records. However, 18 of the precipitation records could fit neither by the extended multiplicative cascade model nor by the bifractal approach.

We have determined the multifractality strength $\Delta\alpha$ for the 81 precipitation records where either the extended multiplicative cascade model or the bifractal description could be used and for all 42 runoff records. The corresponding histograms are shown in Fig. 11.11. We find that there is no systematic dependence of the $\Delta\alpha$ values on the geographic region or on the climate zone. Again, the rivers in southern Germany also show a broad distribution. However, $\Delta\alpha$ seems to show a slightly decreasing tendency with increasing basin area (not shown). The average multifractality strength is significantly smaller for the precipitation records ($\Delta\alpha = 0.29 \pm 0.14$) than for the runoff records ($\Delta\alpha = 0.49 \pm 0.16$). Note that for 18 precipitation records $\Delta\alpha$ could not be determined, since neither the extended multiplicative cascade model nor the simple bifractal approach fit.

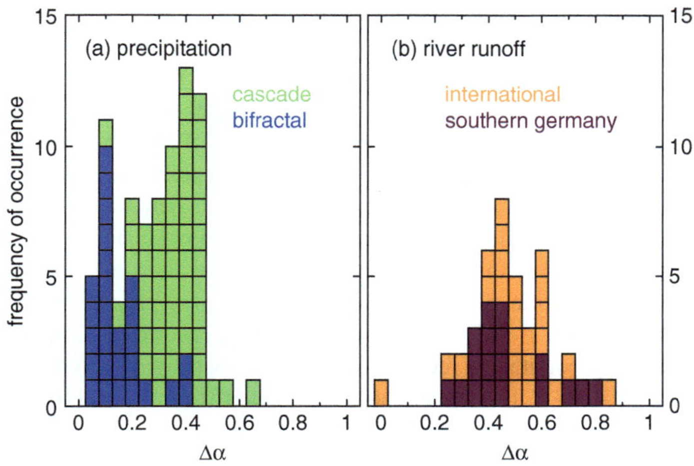

Fig. 11.11 Histograms of the multifractality strength $\Delta\alpha$ for (**a**) 81 of 99 daily precipitation records and (**b**) all 42 daily runoff records, 18 from southern Germany (*brown*) and 24 from international hydrological stations (*orange*). The values have been determined by applying non-linear fits to the generalised Hurst exponents $h(q)$. Each *box* represents the result for one meteorological or hydrological station. For all *boxes*, (11.21) was used, except for the *blue boxes* in (**a**) where the bifractal models, (11.24) or (11.25), have been used since they fitted better

Finally, we like to note that $h(q)$ for the precipitation records remains practically unchanged, when the seasonal detrending according to (11.2) instead of (11.1) is employed. This does not hold for rivers with strong seasonal trends like those that nearly dry up in summer or freeze in winter. In this case, $h(q)$ is shifted towards higher values by a constant, which ranges between 0.1 and 0.25, but such that the strength $\Delta\alpha$ of the multifractality remains unchanged. We are not convinced, however, that (11.2) is the better way of removing seasonal periodicities in case of a multifractal analysis. In the multifractal description, $h(q)$ for negative q characterises the scaling behaviour of the small fluctuations, while $h(q)$ for positive q characterises the scaling behaviour of the large fluctuations. Thus, the multifractal analysis can distinguish the scaling of small and large fluctuations. Now, if we divide by the seasonal trend in the standard deviation, following (11.2), *some* small fluctuations are rendered large and *some* large fluctuations are rendered small. Hence, we believe that some of the information provided by a multifractal analysis is obscured this way. The problem of seasonality effects is addressed in Chap. 13.

11.7 Conclusion

In summary, we have analysed long precipitation records and long river discharge records using detrended fluctuation analysis (DFA) and its multifractal extension (MF-DFA). We find that the precipitation records are mainly characterised by asymptotic scaling with fluctuation exponents $h(2) \approx 1/2$, indicating a fast decay of the autocorrelation function, while the runoff records are long-term correlated above a crossover time scale of several weeks with $h(2)$ varying in a broad range. This result is rather surprising, since the fluctuations of rainfall are often claimed to play an important role in the long-term correlated (persistent) fluctuations of the runoff. Our study seems to indicate that the persistence of the runoff is not so much related with persistence in precipitation, but is rather caused by storage processes occurring in the soil and the highly intermittent spatial behaviour of the rainfall. Furthermore, extensive studies are needed to prove this conclusion.

In addition we studied the multifractal properties of the time series using the multifractal generalisation of the DFA method that was crosschecked with the WTMM technique. We found that the multifractal spectra of all 42 runoff records and 54 of the 99 precipitation records can be described by a 'universal' function $\tau(q) = -\ln(a^q + b^q)/\ln 2$, which can be obtained from a generalisation of the multiplicative cascade model and has solely two parameters a and b or, equivalently, the fluctuation exponent $h(2) = \frac{1}{2} - \ln(a^2 + b^2)/\ln 4$ and the width $\Delta\alpha = \ln \frac{a}{b}/\ln 2$ of the singularity spectrum. Since our function for $\tau(q)$ applies also for negative q values, we could derive the singularity spectra $f(\alpha)$ from the fits. In contrast to the runoff records, 45 of the precipitation records seem to require a different description, and a simple bifractal fit can be used in 27 cases. For positive moments, the three-parameter Lovejoy–Schertzer approach always yields a good fit to both runoff and precipitation data.

We found that there are no significant differences between the distributions of the fluctuation exponent $h(2)$ and the width $\Delta\alpha$ of the singularity spectrum for rivers in southern Germany and for international rivers. We also found that there is no significant dependence of these parameters on the size of the basin area. Thus, the multifractality occurring in all runoff records supports the idea of a 'universal' multifractal behaviour of river runoffs suggested by Lovejoy and Schertzer in a different context. However, there is a slight decrease of the multifractal width $\Delta\alpha$ with increasing basin area. We suggest that the values of $h(2)$ and $\Delta\alpha$ can be regarded as 'fingerprints' for each station or river, which can serve as an efficient non-trivial test bed for the state-of-the-art precipitation–runoff models.

Apart from the practical use of (11.21) with the parameters a and b, we presently are lacking a physical model for this behaviour. It will be interesting to see if physically based models, e.g. the random tree-structure model presented in [11.22], can be related to the multiplicative cascade model presented here. If so, this would give a physical explanation for how the multiplicative cascade model is able to simulate river flows.

We have also investigated the origin of the multifractal scaling behaviour by comparison with the corresponding shuffled data. We found that the multifractality is removed by shuffling that destroys the time correlations in the series while the distribution of the values is not altered. After shuffling, we obtain $h(q) \approx 1/2$ for all values of q, indicating monofractal behaviour. Hence, our results suggest that the multifractality is not due to the existence of a broad, asymmetric (singular) probability density distribution [11.1], but due to a specific dynamical arrangement of the values in the time series, i.e. a self-similar 'clustering' of time patterns of values on different time scales. We believe that our results will be useful also to improve the understanding of extreme values (singularities) in the presence of multifractal long-term correlations and trends.

Acknowledgments We would like to thank the BMBF, the DFG, the DAAD, and the Minerva Foundation for financial support. We are particularly grateful to Daniel Schertzer and Alberto Montanari for critical reading of the manuscript and very useful discussion of the multifractal formalism used by Schertzer, Lovejoy, and coworkers. We are grateful to Dr. Peter Braun for many discussions. We also would like to thank H. Österle (PIK Potsdam, Germany), the Water Management Authorities of Bavaria and Baden-Württemberg (Germany), and the Global Runoff Center (GRDC) in Koblenz (Germany) for providing the observational data.

Appendix

Extended Multiplicative Cascade Model

In the following, we like to motivate the two-parameter formula (11.21) and show how it can be obtained from the well-known multifractal cascade model [11.4, 11.16, 11.32]. In the model, a record ϕ_k of length $N = 2^{n_{\max}}$ is constructed recursively as follows: In generation $n = 0$, the record elements are constant, i.e. $\phi_k = 1$ for all $k = 1, \ldots, N$. In the first step of the cascade (generation $n = 1$), the first half of the series is multiplied by a factor a and the second half of the series is multiplied by a factor b. This yields $\phi_k = a$ for $k = 1, \ldots, N/2$ and $\phi_k = b$ for $k = N/2 + 1, \ldots, N$. The parameters a and b are between 0 and 1, $0 < a < b < 1$. Note that we do not restrict the model to $b = 1 - a$ as is often done in the literature [11.16]. In the second step (generation $n = 2$), we apply the process of step 1 to the two subseries, yielding $\phi_k = a^2$ for $k = 1, \ldots, N/4$, $\phi_k = ab$ for $k = N/4 + 1, \ldots, N/2$, $\phi_k = ba = ab$ for $k = N/2 + 1, \ldots, 3N/4$, and $\phi_k = b^2$ for $k = 3N/4 + 1, \ldots, N$. In general, in step $n + 1$, each subseries of step n is divided into two subseries of equal length, and the first half of the ϕ_k is multiplied by a while the second half is multiplied by b. For example, in generation $n = 3$ the values in the eight subseries are a^3, a^2b, a^2b, ab^2, a^2b, ab^2, ab^2, b^3. After $n_{\max}$ steps, the final generation has been reached, where all subseries have length 1 and no more splitting is possible. We note that the final record can be written as $\phi_k = a^{n_{\max} - n(k-1)} b^{n(k-1)}$, where $n(k)$ is the number of digits 1 in the binary representation of the index k, e.g. $n(13) = 3$, since 13 corresponds to binary 1101.

For this multiplicative cascade model, the formula for $\tau(q)$ has been derived earlier [11.4, 11.16, 11.32]. The result is $\tau(q) = [-\ln(a^q + b^q) + q\ln(a + b)]/\ln 2$ or

$$h(q) = \frac{1}{q} - \frac{\ln(a^q + b^q)}{q\ln 2} + \frac{\ln(a + b)}{\ln 2}. \tag{11.26}$$

Since for $a + b = 1$ the last term vanishes; in this special case it is often missing in literature. It is easy to see that $h(1) = 1$ for all values of a and b. Thus, in this form the model is limited to cases where $h(1)$, which is the exponent Hurst defined originally in the R/S method, is equal to 1. In order to generalise this multifractal cascade process such that any value of $h(1)$ is possible, we have subtracted the offset $\Delta h = \ln(a+b)/\ln(2)$ from $h(q)$, compare with (11.21). The constant offset Δh corresponds to additional long-term correlations incorporated in the multiplicative cascade model. For generating records without this offset, we rescale the power spectrum. First, we fast Fourier transform (FFT) the simple multiplicative cascade data into the frequency domain. Then, we multiply all Fourier coefficients by $f^{-\Delta h}$, where f is the frequency. This way, the slope β of the power spectra $E(f) \sim f^{-\beta}$ (the squares of the Fourier coefficients) is decreased from $\beta = 2h(2)-1 = [2\ln(a+b) - \ln(a^2+b^2)]/\ln 2$ into $\beta' = 2[h(2) - \Delta h] - 1 = -\ln(a^2+b^2)/\ln 2$, which is consistent with (11.21). Finally, backward FFT is employed to transform the signal back into the time domain. A similar Fourier filtering technique has been used by Tessier et al. [11.71] when generating surrogate runoff data.

Comparison with Model Data

In order to see how well the extended multiplicative cascade model fits to the real data (for a given river), we generate the model data as follows [11.35]: (i) we determine a and b for the given river (by best fit of (11.21)); (ii) we generate the simple multiplicative cascade model with the obtained a and b values; and (iii) we implement the proper second moment long-term correlations as described above.

Figure 11.12a shows the MF-DFA of the model data with parameters a and b determined for the river Weser. By comparing with Fig. 11.8b we see that the extended model gives the correct scaling of the multifractal fluctuation functions $F_q(s)$ on time scales above the crossover. Below the crossover, however, the model

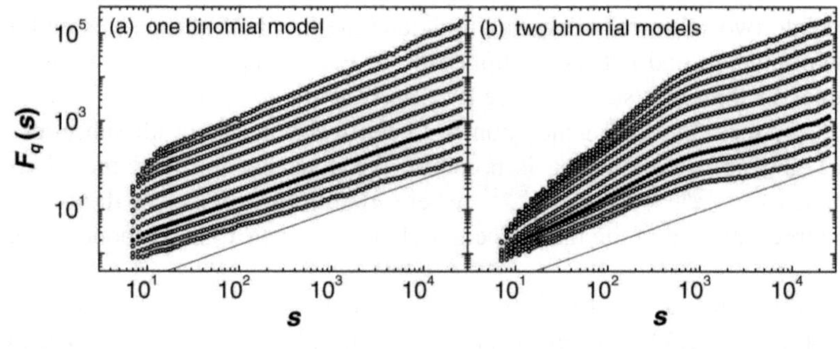

Fig. 11.12 The fluctuation functions $F_q(s)$ obtained from the multifractal DFA for surrogate series generated by the extended multiplicative cascade model with parameters $a = 0.50$ and $b = 0.68$ that correspond to the values we obtained for the river Weser. The fluctuation function $F_q(s)$ for (**a**) the original ϕ_i series and (**b**) modified values of the parameters a and b ($a = 0.26$, $b = 0.59$) have been used on scales $s \leq 256$ to simulate the apparent stronger multifractality on smaller scales observed for most rivers. For the figure, results from 10 surrogate series of length 140 years were averaged

does not yield the observed $F_q(s)$ in the original data. In order to obtain the proper behaviour below the crossover, a different type of multifractality below the crossover, represented by different values of a and b, can be introduced, as was done for Fig. 11.12b.

References

11.1. P.L. Anderson, M.M. Meerschaert, Modeling river flows with heavy tails. Water Resourc. Res. **34**(9), 2271–2280 (1998)

11.2. R.F.S. Andrade, H.-J. Schellnhuber, M. Claussen, Analysis of rainfall records: Possible relation to self-organized criticality. Physica A, **254**(3–4), 557–568 (1998)

11.3. A. Arneodo, B. Audit, N. Decoster, J.F. Muzy, and C. Vaillant, *The Science of Disasters*, Chapter 2. Wavelet Based Multifractal Formalism: Applications to DNA Sequences, Satellite Images of the Cloud Structure, and Stock Market Data, (Springer, Berlin, 2002) pp. 27–102

11.4. A.L. Barabási, T. Vicsek, Multifractality of self-affine fractals. Phys. Rev. A, **44**(4), 2730–2733 (1991)

11.5. R.N. Bhattacharya, V.K. Gupta, E. Waymire, The hurst effect under trends. J. Appl. Prob. **20**(3), 649–662 (1983)

11.6. R.N. Bhattacharya, E.C. Waymire, *Stochastic Processes With Applications*, pages 53–62. Wiley Series in Probability and Mathematical Statistics: Applied Probability and Statistics. (Wiley, New York, NY, 1990)

11.7. A. Bunde, S. Havlin, J.W. Kantelhardt, T. Penzel, J.H. Peter, K. Voigt, Correlated and uncorrelated regions in heart-rate fluctuations during sleep. Phys. Rev. Lett. **85**(17), 3736–3739 (2000)

11.8. A. Bunde, J.W. Kantelhardt, Langzeitkorrelationen in der natur: von klima erbgut und herzrhythmus. Physikalische Blätter, **57**(5), 49–54 (2001)

11.9. A. Bunde, J. Kropp, H.-J. Schellnhuber, (eds.), *The Science of Disasters* (Springer-Verlag, Berlin, 2002)

11.10. A. Davis, A. Marshak, W. Wiscombe, R. Cahalan, Multifractal characterizations of nonstationarity and intermittency in geophysical fields: Observed, retrieved, or simulated. J. Geophys. Res. **99**(D4), 8055–8072 (1994)

11.11. A. Davis, A. Marshak, W. Wiscombe, R. Cahalan, *Current Topics in Nonstationary Analysis: Proceedings of the 2nd Workshop on Nonstationary Random Processes and Their Applications*, (World Scientific Pub. Co. Inc., Singapore, 1996) pp. 97–158

11.12. R. Deidda. Rainfall downscaling in space-time multifractal framework. Water Res. Res., **36**(7), 1779–1794 (2000)

11.13. R. Deidda, R. Benzi, and F. Siccardi. Multifractal modeling of anomalous scaling laws in rainfall. Water Resourc. Res., **35**(6), 1853–1867 (1999)

11.14. E.M. Douglas, A.P. Barros, Probable maximum precipitation estimation using multifractals: Application in the eastern united states. J. Hydrometeorol. **4**(6), 1012–1024 (2003)

11.15. J.F. Eichner, E. Koscielny-Bunde, A. Bunde, S. Havlin, H.-J. Schellnhuber. Power-law persistence and trends in the atmosphere: A detailed study of long temperature records. Phys. Rev. E, **68**(4), 046133 (2003)

11.16. J. Feder, *Fractals*. Physics of Solids and Liquids (Plenum Press, New York, NY, 1988)

11.17. W. Feller, The asymptotic distribution of the range of sums of independent random variables. Ann. Math. Stati., **22**, 427–432 (1951)

11.18. K. Fraedrich, C. Larnder, Scaling regimes of composite rainfall times series. *Tellus*, **45**A(4), 289–298 (1993)

11.19. U. Frisch, G. Parisi, *Turbulency and Predictability in Geophysical Fluid Dynamics*, (North-Holland, Amsterdam, 1985) pp. 84–92

11.20. G. Galeati, A comparison of parametric and non-parametric methods for runoff forecasting. Hydrol Sci J / J. Sci. Hydrolo. **35**(1), 79–94 (1990)
11.21. GRDC, Runoff data provided by Global Runoff Data Center (GRDC) in Koblenz (Germany), (2002)
11.22. V.K. Gupta, S.L. Castro, T.M. Over, On scaling exponents of spatial peak flows from rainfall and river network geometry. J. Hydrol., **187**(1–2), 81–104 (1996)
11.23. V.K. Gupta, D.R. Dawdy, *Scale Issues in Hydrological Modelling*, chapter Physical Interpretations of Regional Variations in the Scaling Exponents of Flood Quantiles, pages 106–119. (Wiley, Chichester, 1995)
11.24. V.K. Gupta, O.J. Mesa, D.R. Dawdy, Multiscaling theory of flood peaks: Regional quantile analysis. *Water Resourc. Res.* **30**(12), 3405–3421 (1994)
11.25. D. Harris, M. Menabde, A. Seed, G. Austin, Multifractal characterization of rain fields with strong orographic influence. J. Geophys. Res., **101**(D21), 26405–26414 (1996)
11.26. J.R.M. Hosking, Modeling persistence in hydrological time series using fractional differencing. Water Resourc. Res., **20**(12), 1898–1908 (1984)
11.27. H.E. Hurst. Long-term storage capacity of reservoirs. Trans. Am. Soc. Civil Engi. **116**(2447), 770–808 (1951)
11.28. H.E. Hurst, R.P. Black, Y.M. Simaika, *Long-term Storage, An Experimental Study* (Constable & Co. Ltd., London, 1965) ASIN B0007J0Z4U.
11.29. J.W. Kantelhardt, E. Koscielny-Bunde, H.H.A. Rego, S. Havlin, A. Bunde, Detecting long-range correlations with detrended fluctuation analysis. Physica A, **295**(3–4), 441–454 (2001)
11.30. J.W. Kantelhardt, E. Koscielny-Bunde, D. Rybski, P. Braun, A. Bunde, S. Havlin. Long-term persistence and multifractality of precipitation and river runoff records. J. Geophys. Res. – Atm. **111**(D1), D01106, 2006.
11.31. J.W. Kantelhardt, D. Rybski, S.A. Zschiegner, P. Braun, E. Koscielny-Bunde, V. Livina, S. Havlin, and A. Bunde. Multifractality of river runoff and precipitation: comparison of fluctuation analysis and wavelet methods. Physica A, **330**(1–2), 240–245 (2003)
11.32. J.W. Kantelhardt, S.A. Zschiegner, E. Koscielny-Bunde, S. Havlin, A. Bunde, H.E. Stanley, Multifractal detrended fluctuation analysis of nonstationary time series. Physica A, **316** (1–4), 87–114 (2002)
11.33. V. Klemeš, The hurst phenomenon: A puzzle? Water Resourc. Res., **10**(4), 675–688 (1974)
11.34. E. Koscielny-Bunde, A. Bunde, S. Havlin, H.E. Roman, Y. Goldreich, H.-J. Schellnhuber, Indication of a universal persistence law governing atmospheric variability. Phys. Rev. Lett. **81**(3), 729–732 (1998)
11.35. E. Koscielny-Bunde, J.W. Kantelhardt, P. Braun, A. Bunde, and S. Havlin, Long-term persistence and multifractality of river runoff records. J. Hydrol., **322**(1–4), 120–137 (2006)
11.36. D. Koutsoyiannis, Climate change, the hurst phenomenon, and hydrological statistics. Hydrol. Sci. J. – J. Sci. Hydrol., **48**(1), 3–24 (2003)
11.37. D. Koutsoyiannis, Nonstationarity versus scaling in hydrology. J. Hydrol., **324**(1–4): 239–254 (2006)
11.38. P. Kumar, E. Foufoula-Georgiou, Wavelet analysis for geophysical applications. Rev. Geophys. **34**(4), 385–412 (1997)
11.39. H. Künsch, Discrimination between monotonic trends and long-range dependence. J. Appl. Proba. **23**(4), 1025–1030 (1986)
11.40. J. Laštovička, P. Križan, P. Šauli, D. Novotná, Persistence of planetary wave type oscillations in *fo*F2 over europe. Ann. Geophys., **21**(7), 1543–1552 (2003)
11.41. W. Lauer, P. Frankenberg, In Diercke Weltatlas Westermann Schulbuchverlag, Braunschweig 4th edition, p. 220–221 (1996)
11.42. D. Lavallee, S. Lovejoy, D. Schertzer, P. Ladoy, *Fractals in Geography*, chapter 8. Nonlinear Variability of Landscape Topography: Multifractal Analysis and Simulation, (Prentice Hall, 1993) pp. 158–192

11.43. V. Livina, Y. Ashkenazy, Z. Kizner, V. Strygin, A. Bunde, S. Havlin, A stochastic model of river discharge fluctuations. Physica A, 330(1–2):283–290 (2003)
11.44. S Lovejoy, B.B. Mandelbrot, Fractal properties of rain, and a fractal model. Tellus, **37**A, 209–232 (1985)
11.45. S. Lovejoy, D. Schertzer, *New Uncertainty Concepts in Hydrology and Water Resources*, chapter III.2 Multifractals and rain, International Hydrology Series. (Cambridge University Press, Cambridge, 1995) pp. 61–103
11.46. B.B. Mandelbrot, J.R. Wallis, Noah, Joseph, and operational hydrology. Water Resourc. Res., **4**(5), 909–918 (1968)
11.47. B.B. Mandelbrot, J.R. Wallis, Some long-run properties of geophysical records. Water Resourc. Res., **5**(2), 321–340 (1969)
11.48. M. Marani, On the correlation structure of continuous and discrete point rainfall. Water Resourc. Res., **39**(5), 1128 (2003)
11.49. C. Matsoukas, S. Islam, Rodriguez-Iturbe I, Detrended fluctuation analysis of rainfall and streamflow time series. J. Geophys. Res., **105**(D23), 29165–29172 (2000)
11.50. O.J. Mesa, G. Poveda, The hurst effect: The scale of fluctuation approach. Water Resourc. Res., **29**(12), 3995–4002 (1993)
11.51. A. Montanari, R. Rosso, M.S. Taqqu, Fractionally differenced arima models applied to hydrologic time series: Identification, estimation, and simulation. Water Resourc. Res., **33**(5), 1035–1044 (1997)
11.52. A. Montanari, R. Rosso, M.S. Taqqu, A seasonal fractional arima model applied to the nile river monthly flows at Aswan. Water Resourc. Res., **36**(5), 1249–1259 (2000)
11.53. J.F. Muzy, E. Bacry, A. Arneodo, Wavelets and multifractal formalism for singular signals: Application to turbulence data. Phys. Rev. Lett. **67**(25), 3515–3518 (1991)
11.54. J. Olsson, V.P. Singh, K. Jinno, Effect of spatial averaging on temporal statistical and scaling properties of rainfall. J. Geophys. Res., **104**(D16), 19117–19126 (1999)
11.55. G. Pandey, S. Lovejoy, D. Schertzer, Multifractal analysis of daily river flows including extremes for basins of five to two million square kilometres, one day to 75 years. J. Hydrol., **208**(1–2), 62–81 (1998)
11.56. C.-K. Peng, S.V. Buldyrev, S. Havlin, M. Simons, H.E. Stanley, A.L. Goldberger. Mosaic organization of DNA nucleotides. Phys. Rev. E, **49**(2), 1685–1689 (1994)
11.57. O. Peters, K. Christensen, Rain: Relaxations in the sky. Phys. Rev. E, **66**(3), 036120, (2002)
11.58. O. Peters, C. Hertlein, K. Christensen, A complexity view of rainfall. Phys. Rev. Lett. **88**(1), 018701 (2002)
11.59. K.W. Potter, Evidence for nonstationarity as a physical explanation of the hurst phenomenon. Water Resourc. Res., **12**(5), 1047–1052 (1976)
11.60. I. Rodríguez-Iturbe, A. Rinaldo, *Fractal River Basins: Chance and Self-Organization.* (Cambridge University Press, Cambridge, 1997)
11.61. D. Rybski, A. Bunde, Methods of trend detection in long-term correlated records. *preprint*, 2007.
11.62. D. Rybski, A. Bunde, S. Havlin, H. von Storch, Long-term persistence in climate and the detection problem. Geophys. Res. Lett., **33**(6), L06718 (2006)
11.63. J.D. Salas, D.C. Boes, V. Yevjevich, G.G.S. Pegram, Hurst phenomenon as a pre-asymptotic behavior. J. Hydrol., **44**(1/2), 1–5 (1979)
11.64. D. Schertzer, S. Lovejoy. Physical modelling and analysis of rain and clouds by anisotropic scaling multiplicative processes. J. Geophys. Res., **92**(D8), 9693–9714 (1987)
11.65. D Schertzer, S. Lovejoy (eds), *Nonlinear Variability in Geophysics: Scaling and Fractals.* (Kluwer, Dordrecht, 1991)
11.66. F. Schmitt, S. Vannitsem, A. Barbosa, Modeling of rainfall time series using two-state renewal processes and multifractals. J. Geophys. Res., **103**(D18), 23181–23193 (1998)
11.67. M. Sivapalan, C. Jothityangkoon, M. Menabde, Linearity and nonlinearity of basin response as a function of scale: Discussion of alternative definitions. Water Resourc. Res., 38(2), 1012 (2002)

11.68. R. Talkner and R.O. Weber. Power spectrum and detrended fluctuation analysis: Application to daily temperatures. Phys. Rev E, **62**(1) 150–160 (2000)
11.69. M.S. Taqqu, V. Teverovsky, and W. Willinger. Estimators for long-range dependence: An empirical study. Fractals **3**(4), 785–798 (1995)
11.70. I. Tchiguirinskaia, P. Hubert, D. Schetzer. *Preventing & Fighting Hydrological Disasters*. (Orizonturi Universitare Publishing House, Timisoara, Rumänien, 2002)
11.71. Y. Tessier, S. Lovejoy, P. Hubert, D. Schertzer, and S. Pecknold. Multifractal analysis and modeling of rainfall and river flows and scaling, causal transfer functions. J. Geophys. Res., **101**(D21), 26427–26440 (1996)
11.72. Y. Tessier, S. Lovejoy, D Schertzer. Universal multifractals: Theory and observations for rain and clouds. J. Appl. Meteorol., **32**(2), 223–250 (1993)
11.73. D. Veneziano, R.L. Bras, J.D. Niemann, Nonlinearity and self-similarity of rainfall in time and a stochastic model. J. Geophys. Res., **101**(D21), 26371–26392 (1996)
11.74. D. Veneziano, V. Iacobellis. Multiscaling pulse representation of temporal rainfall. Water Resourc. Res., **38**(8), 1138 (2002)
11.75. R.O. Weber, P. Talkner, Spectra and correlations of climate data from days to decades. J. Geophys. Res. Atm., **106**(D17), 20131–20144 (2001)
11.76. P. Whittle, Estimation and information in stationary time series. Arkiv For Matematik, **2**(23), 423–434 (1953)

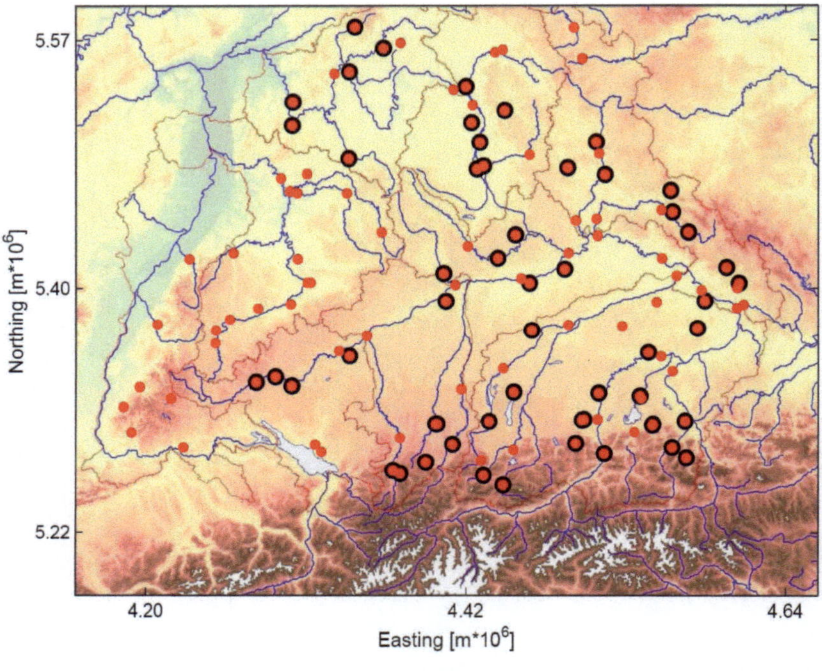

Overview of the geographical location of the 118 river runoff gauges used in this study. The map contains only the main rivers so that the gauges corresponding to tributaries are shown without rivers. A subset of hydrologically independent gauges are marked by *black circles* around the *red spots*

Chapter 12
Long-Term Structures in Southern German Runoff Data

Miguel D. Mahecha, Holger Lange, and Gunnar Lischeid

Hydrological discharge time series are known to depict low-frequency oscillations, long-range statistical dependencies, and pronounced nonlinearities. A better understanding of this runoff behaviour on regional scales is crucial for a variety of water management purposes and flood risk assessments. We aimed at extracting long-term components which influence simultaneously a set of southern German runoff records.

The methodological approach was to perform a "double dimensionality reduction." First, we prefiltered the multidimensional time series by a nonlinear dimensionality reduction method (isometric feature mapping; Isomap) and its special linear case (classical multidimensional scaling; CMDS). This analysis led to the extraction of univariate but highly representative time series. Second, we analysed the generated time series by "singular system analysis" (SSA), a "dimensionality reduction method in the time domain" to identify the subsignals contained in the original series.

With the combined approach Isomap-SSA (and CMDS-SSA) it was possible to identify significant long-term components, influencing the overall discharge behaviour. It turned out that increasing nonlinearity in the Isomap projection stabilized the SSA performance. The overall findings indicated that the southern German runoff records were influenced by periods of $\sim$4, $\sim$5, $\sim$7, $\sim$9 and $\sim$10 years.

We conclude that the application of new, nonlinear methods of dimensionality reduction contributes to a better extraction of low-frequency modes from high-dimensional time series. The recovered long-term modes should be considered in future hydrological assessments, since they contribute significantly to the overall runoff behaviour.

M.D. Mahecha (✉)
Biogeochemical Model-Data Integration Group, Max-Planck Institut für Biogeochemie, 07701 Jena, Germany
e-mail: miguel.mahecha@bgc-jena.mpg.de

J.P. Kropp, H.-J. Schellnhuber (eds.), *In Extremis*,
DOI 10.1007/978-3-642-14863-7_12,

12.1 Introduction

A well-known phenomenon in river discharge records is the presence of low-frequency oscillations (e.g. [12.26]) as well as long-range statistical dependencies in the time domain (e.g. [12.18]). A better understanding of these runoff characteristics at long time scales impacts a variety of management tasks: it could improve the quality of flood risk assessments, lay foundations for investigating runoff behaviour under climate change, or help clarifying the connection between atmospheric cycles, precipitation patterns and freshwater availability. In this context, a bunch of techniques has been developed in succession to the pioneering work by Hurst [12.16]. For instance, detrended fluctuation analysis (DFA) has been used as an effective tool for the characterization of memory effects in runoff time series [12.18] and the separation of trends from long-range correlations. Quantities like the rescaled range Hurst exponent [12.23] or the DFA exponents describe the statistical properties (in particular long-range correlations) of (single) time series. They are, however, not suitable to extract periodic components. For the latter task, powerful methods like singular system analysis (SSA) [12.4] can be applied. SSA is not limited to the extraction of sinusoidal oscillations [12.15], the technique also identifies trends [12.13] and performs a signal-to-noise distinction [12.12].

Our goal in this study was to identify long-term components that are present simultaneously in a large set of discharge series of southern Germany. It was attempted to extract components with periods >3 years directly from the multidimensional discharge series, rather than from single gauges. This is meaningful since individual river flows exhibit synchronizations among these components on regional scales [12.20], and a common origin of these patterns is plausible. In this study, 118 gauges from southern Germany were investigated, where each time series contained gapless daily mean discharge values for 51 years (1951–2001). A previous study [12.20] focussed on a subset of these gauges.

For a set of L gauges and for each given moment in time, individual measurements are combined into a vector with dimension L. Hence, the multivariate series is represented as a trajectory embedded in a high-dimensional *ambient space* $\mathbb{R}^L$. However, given that rivers form networks, discharges from tributaries and a main river (as well as, obviously, discharges from different gauges of the same river) may safely be assumed as dependent, and common driving forces such as climate patterns render them correlated as well. Thus, the discharge behaviour may be represented by fewer than L variables. This (generally unknown) number of effectively independent variables is called the *intrinsic dimensionality* of the data [12.17]. In other words, the data sample an M-dimensional manifold, which is a subspace of $\mathbb{R}^L$, where $M < L$; a very compact description of the data is achieved if $M << L$ [12.5]. *Dimensionality reduction* aims at estimating this intrinsic dimension and characterizes topological properties of the data manifold, ideally without loss of relevant information, i.e. preserving overall time series characteristics. Peculiarities of individual gauges, however, may be lost in this effective representation.

This chapter presents a combination of dimensionality reduction in the space of gauges and SSA, which is an example for dimensionality reduction in time. In a first step, a linear and a nonlinear method were applied to identify representative low-dimensional modes. The second step consisted in analysing these modes by (univariate) SSA. Proceeding in this manner, we gain quantitative information on the *common* long-term properties of the whole set of river discharges.

This chapter is organized as follows. In Sect. 12.2, the linear dimensionality reduction technique "classical multidimensional scaling" (CMDS) and a recently developed nonlinear CMDS variant, the "isometric feature mapping" (Isomap), are explained. Then, the basics of SSA, which may be considered as a principal component analysis (PCA) in the time domain, are shortly described. The subsequent section shows the results of this "double dimensionality reduction", focussing on the extraction of common long-term components. The discussion is devoted to the advantages and limitations of the combined method as well as to possible implications for hydrological practitioners.

12.2 Methods

12.2.1 Data Sets and Preparation

We denote our set of time series as $x_{tl} : t = 1, \ldots, T; l = 1, \ldots, L$, where T is the length of the series and L is the number of gauges. We will work with methods requiring considerable computational resources. To meet this constraint, all data were aggregated to 7 days. To obtain scale homogeneity within a data vector, it is customary to perform first a z-transformation (every single time series obtains zero mean and unit standard deviation). Downstream gauges sample part of the same runoff behaviour as upstream gauges in the same basin and cannot be regarded as independent signals. Thus, gauges were selected which were "hydrologically independent" to the degree possible, i.e. from the same river, only the gauge with the largest basin was chosen, leaving $L = 53$ gauges in the total set of 118. All subsequent analyses were performed on both data sets, to see whether the elimination of redundant discharge signals gives a different picture of the runoff behaviour.

12.2.2 Dimensionality Reduction in Space

12.2.2.1 Classical Multidimensional Scaling (CMDS)

Any real-valued L-dimensional time series $\boldsymbol{x}_t$, where t denotes time and the components of the vector span the set of measurements (gauges), can be conceptualized as a trajectory in the space $\mathbb{R}^L$. However, in the presence of correlations between the individual series and autocorrelations within a single series, ensembles of trajectories tend to form a lower dimensional manifold $\Omega \subset \mathbb{R}$. To uncover properties of Ω, in particular its dimension and topological attributes such as connectedness and inter-point distances, is the aim of dimensionality reduction based on projection methods.

Classical (linear) multidimensional scaling (CMDS), known also as principal coordinate analysis [12.6], is a non-iterative method which tries to preserve the linear inter-point distances in a low-dimensional Euclidean space by appropriate projection. It starts with the matrix $\mathbf{D}^{(\mathbf{X})}$ of Euclidean distances $d_{ij} = \|\boldsymbol{x}_i - \boldsymbol{x}_j\|$ in $\mathbb{R}^L$. Then, a double-centred (rows as well as columns sum up to zero) matrix of inner products, $\tau(\mathbf{D})$, is calculated. For z-transformed time series, its elements are the scalar products of columns of $\mathbf{D}^{(\mathbf{X})}$, appropriately centred, and τ is positive definite. Diagonalization of τ by singular value decomposition then leads to a set of eigenvectors (principal coordinates). Choosing M of these with the largest eigenvalues, which explain a certain fraction of the total variance, leads to a subspace Y. The calculation of τ is repeated in Y, and the procedure minimizes the cost function

$$\eta = \|\tau\left(\mathbf{D}^{(X)}\right) - \tau\left(\mathbf{D}^{(Y)}\right)\|. \tag{12.1}$$

In this equation, $\|\mathbf{Q}\|$ denotes the L^2 matrix norm defined as $\sqrt{\sum_{ij} q_{ij}^2}$. For a more detailed description, see [12.3, 12.6].

CMDS is successful if $M << L$. Then, CMDS has achieved a low-dimensional representation of the data space with similar topology.

12.2.2.2 Nonlinear CMDS: Isometric Feature Mapping (Isomap)

CMDS tries to preserve the *linear* inter-point distances in the low-dimensional representation space, spanned by the extracted principal coordinate vectors $Y_1, \ldots, Y_M$ with $M << L$. The natural extension of this idea is to find an embedding method which preserves the distance structure as measured on *nonlinear* spaces. This approach was introduced to a wide scientific community by Tenenbaum et al. [12.32] and makes use of the shortest inter-point distances on curved surfaces or geodesic distances. The metric of these distances is defined as

$$g_{ij} = \sum_{l=1}^{L} \frac{\partial X_l}{\partial \Theta_i} \frac{\partial X_l}{\partial \Theta_j}. \tag{12.2}$$

In (12.2), X_l are local Euclidean coordinates, and Θ_i refers to a global curved coordinate system for the manifold Θ. In the case of a Euclidean space, the metric is trivially the identity matrix. The geodesic metric in (12.2) is not calculable, since the assumed nonlinear manifold Θ is unknown for measured data. Isomap circumvents this calculation by a locally linear approach, approximating the global non-Euclidean properties of Θ. For this, a threshold expressed either by a number k of nearest neighbours (k-NN) or a radius ϵ in the $\mathbb{R}^L$ is defined. The connection of each pair of points within the k-NN or ϵ-radius leads to a connectivity graph G. Applying e.g. Dijkstra's algorithm [12.9], it is then possible to find the shortest inter-point distances $d_{ij}^{(G)}$ on the graph. These distances then approximate g_{ij}.

The distance matrix $\mathbf{D}^{(G)}$ obtained in this manner is used as input to MDS instead of $\mathbf{D}^{(X)}$. The cost function (12.1) is modified to [12.32]

$$\eta = \|\tau\left(\mathbf{D}^{(G)}\right) - \tau\left(\mathbf{D}^{(Y)}\right)\|. \tag{12.3}$$

For a well-chosen parameter (k or ϵ), the derived principal coordinates are good approximations for the vectors $\{\theta_1, \ldots, \theta_M\} \approx \{Y_1, \ldots, Y_M\}$ which span the manifold [12.2]. The choice of the threshold parameter is difficult and must be based on heuristics, where the analyst must find a compromise between a low value maximizing the nonlinearity and a sufficiently large value to warrant the overall graph connection [12.28]. In most cases, the manifold is not homogeneously sampled and clusters of data points occur. Then, a small threshold parameter connects points within the clusters, but not between clusters. On the other hand, too large threshold values could lead to shortcuts on the approximated manifold [12.29]. CMDS is the special case of Isomap where the threshold is maximal, i.e. $k \geq T - 1$ or $\epsilon \geq \max(d_{ij})$, respectively [12.31].

12.2.3 Dimensionality Reduction in Time

The presence of common temporal structures, such as periodic components, in a multivariate data set leads to the idea of a compact description of its dynamics using (far) less than all measurements in time. This is the central task of singular system analysis (SSA), a technique derived from dynamical system analysis [12.4]. Working with embedding techniques, SSA seeks to decompose a time series into "simple" (e.g. oscillatory) modes, quantifying the amount of variance explained by each mode. Similar to PCA, SSA is successful when few modes suffice to represent a substantial part of the total series.

Time series analysis requires tests on the significance of extracted signals (as opposed to noise), investigation of signal separability and the analysis of different temporal scales [12.13]. We present procedures for each of these tasks in this section.

12.2.3.1 Description of Singular System Analysis

SSA can be interpreted as extension of the PCA framework to the time domain. The idea is that a time series is a (linear) combination of multiple signals of different period lengths. The goal is to identify the individual signals and to relate them to corresponding temporal scales via a subsequent Fourier analysis. A crucial parameter of SSA is the embedding dimension P, which has to be chosen with care. It is a compromise between two conflicting goals: The first is to maximize the information content of the analysis which requires a maximum window length P. The second is to maximize the statistical confidence, achieved through a high number of overlapping windows, $\widetilde{N} = T - P + 1$. Empirical findings in literature suggest an optimal choice of T/P ranging from $T/P = 11$ [12.12] to $T/P = 2.5$ [12.20]. From the centred values of the time series, a *lag covariance matrix* **C** is set up which is estimated as [12.12]

$$c_{ij} = \frac{1}{T - \mid i - j \mid} \sum_{t=1}^{T-|i-j|} x_t x_{t+|i-j|}. \tag{12.4}$$

The entries of the matrix depend on the lag $\mid i - j \mid$, and its dimensions are $P \times P$. **C** is then diagonalized by SVD [12.10]

$$\mathbf{C} = \mathbf{E}\boldsymbol{\Lambda}\mathbf{E}^{\mathrm{T}}. \tag{12.5}$$

From the resulting eigenvectors (empirical orthogonal functions) e^{κ}, the principal components U^{κ} are obtained through projection

$$U_t^{\kappa} = \sum_{j=1}^{P} x_{t+j-1} e_j^{\kappa} \tag{12.6}$$

and represent the extracted signals. Each PC is itself a time series of length $\widetilde{N}$. The PCs are often easily interpretable visually or through their periodograms, which usually contain just one or two peaks each. It is also possible to reconstruct the part of the original series which is represented by a collection of U^{κ}'s:

$$\mathrm{RC}_t^{\mathscr{K}} = \frac{1}{\mathscr{M}_t} \sum_{\kappa \in \mathscr{K}} \sum_{j=\mathscr{L}_t}^{\mathscr{U}_t} U_{i-j+1}^{\kappa} e_j^{\kappa}. \tag{12.7}$$

In this equation, $\mathscr{K}$ is any index set describing the combination of PCs chosen. $\mathscr{M}_t$ is a normalization factor and $\mathscr{L}_t$ and $\mathscr{U}_t$ are boundaries which depend on the position within the time series [12.12].

The amount of variance contained in the reconstructed component RC is given by the sum of the corresponding eigenvalues of the PCs contained. In this study, the reconstruction step was carried out selecting for each RC a single component only.

An extension of SSA to the decomposition of multivariate time series is straightforward [12.12], where the PCs are then space–time matrices. This multichannel or M-SSA is, however, computationally highly demanding when the number of channels is not small and was not feasible in our study.

12.2.3.2 Test of Significance

To investigate whether the PCs or the reconstructed components of an SSA are significant or not, one needs an appropriate null hypothesis. This is done by testing the power spectrum of the RCs versus a noise assumption. It has been demonstrated in previous studies (see, e.g. [12.1, 12.30]) that red noise, or the theoretical spectrum of an AR[1] process, is a well-suitable noise model for geoscientific records. Hydro-meteorological records often exhibit high spectral power at low frequencies and low power at high frequencies, which resemble the $1/f^2$ behaviour of red noise. The null hypothesis of the test is that the original time series is compatible with a red noise process [12.20]. In our application of SSA, the reconstructed components were considered significant when they were outside the 95% confidence range of a corresponding AR[1] process.

12.2.3.3 Weak Signal Separability

The eigenvectors of the SSA decomposition form an orthogonal system. This is not necessarily the case for the reconstructed components. The significance test described also does not provide information on the separability of the extracted signals. Thus, a qualitative criterion of signal separability has to be introduced. Golyandina et al. [12.13] proposed the term "weak separability" for non-orthogonal systems. Different signals should only be regarded as part of essentially different underlying processes when they are not interrelated. The criterion to test this is the so-called w-orthogonality (12.8), where two series (reconstructed components) RC_i and RC_j are said to be w-orthogonal if

$$(\mathrm{RC}_i, \mathrm{RC}_j)_w \overset{\text{def}}{=} \sum_{t=1}^{T} w_t \mathrm{RC}_{i,t} \mathrm{RC}_{j,t} \tag{12.8}$$

is compatible with zero.

The weights w_t are shaped like a trapezoid to assure that the effect of the SSA embedding dimension P is taken into account. Defining $P^* = \min(P, \widetilde{N})$ and $\widetilde{N}^* = \max(P, \widetilde{N})$, the weights are given by

$$w_t = \begin{cases} t+1 & \text{for } 0 \leq t \leq P^* - 1, \\ P^* & \text{for } P^* \leq t < \widetilde{N}^*, \\ T-t & \text{for } \widetilde{N}^* \leq t \leq T-1. \end{cases} \tag{12.9}$$

The weighting of the terms in the correlation sum is especially important for large values of P [12.13].

The deviation from the w-orthogonality is quantified by

$$\varrho_{ij} = \frac{|(\mathrm{RC}_i, \mathrm{RC}_j)_w|}{\|\mathrm{RC}_i\|_w \|\mathrm{RC}_j\|_w} \tag{12.10}$$

and has a range between 0 and 1.

No rigorous significance thresholds have been developed for ϱ_{ij} so far.

12.3 Results

12.3.1 Dimensionality Reduction

The dimensionality reduction of the hydrological time series was highly effective in terms of explained variance. Both the CMDS and the Isomap approach required just two dimensions (out of 118 and 53, resp.) to represent ~95% of the data variance. For Isomap, the k-nearest neighbour method was chosen rather than ϵ, because the radius threshold produced unconnected subgraphs from the time series data, indicating strong clustering. The search for an optimal value of k was difficult in this case since the results were robust for a large range of k values.

The linear CMDS was superior to the nonlinear Isomap concerning the explanatory power of only the first principal coordinate, recovering 84% of the variance (see Fig. 12.1). The first Isomap dimension did only recover 71% of the variance. One may conclude that the subsequent SSA of the leading mode is best performed on base of the first CMDS coordinate. However, one could also argue that both methods recovered approximately equal amounts of variance in the first two modes and inclusion of nonlinearity just led to a different partitioning of the variance on the required two modes. Figure 12.1 compares the performance of an Isomap variant (k=17) with CMDS.

Gámez et al. [12.11] indicated that visual interpretation of the Isomap coordinates might elucidate interesting patterns of a spatially reduced time series such as the identification of well-known climate patterns in sea surface temperature (SST) data in their case. However, the temporal visualization of the runoff Isomap modes did

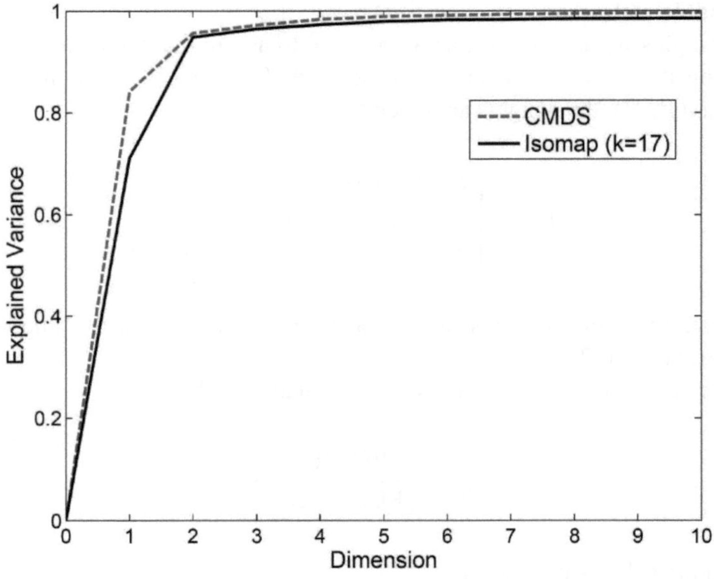

Fig. 12.1 Cumulative explained variances of the first 10 CMDS (*dashed line*) and Isomap (*solid line*) dimensions for the hydrologically independent series

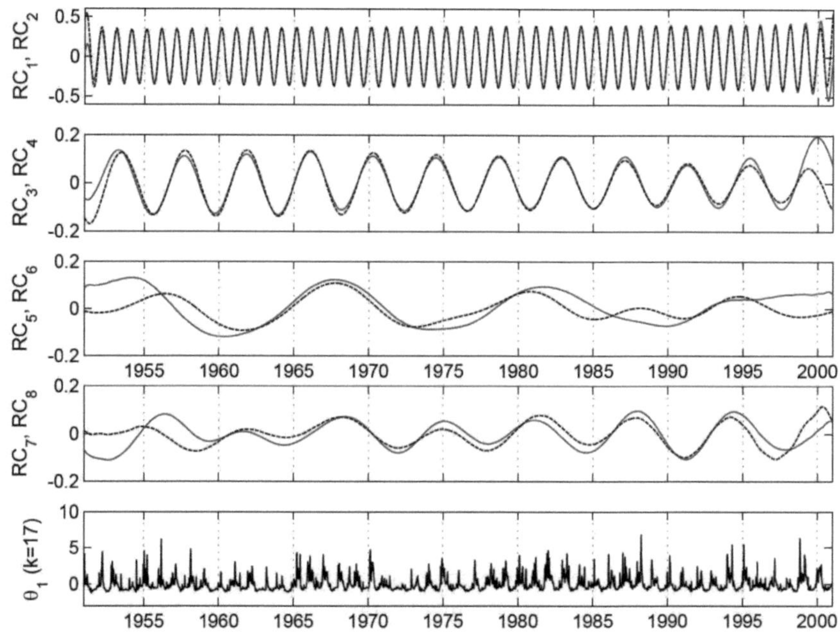

Fig. 12.2 The eight leading reconstructed components from the 118 gauges, based on the first Isomap ($k = 17$) coordinate (*lower panel*). RC1 and RC2 contain the annual cycle not discussed further, RC3–8 are long-term components

not give immediate substantial insights (Fig. 12.2), which is another motivation for a subsequent SSA.

Another important aspect of the Isomap and CMDS projection is whether the dominant modes preserve long-range statistical properties of the time series. As an estimator of the strength of a possible long-term memory, the Hurst coefficient $\mathcal{H}$ [12.16] has been determined. We chose the standard rescaled range or $\frac{R}{S}$ analysis [12.23] for its estimation. There are more sophisticated alternatives (cf. [12.25] for a hydrological application) but it has been shown that $\frac{R}{S}$ is a suitable algorithm for long time series and in the absence of trends [12.18]. Our intention is only a comparison of H values from the original time series and from the first mode of the reduction procedure, and not its absolute values. The average Hurst coefficient for all individual time series was estimated to be $\mathcal{H} = 0.66$ with a standard deviation of 0.06. The first CMDS mode showed a value of $\mathcal{H} = 0.7$, while Isomap found precisely $\mathcal{H} = 0.66$ which means that both methods preserve the long-term characteristics of the series already in the first principal coordinate.

12.3.2 Dimensionality Reduction in the Time Domain

Since the spatial reduction of the multidimensional runoff series from southern Germany was successful in terms of explained variance, only the leading dimension was used as input to a univariate SSA. An embedding dimension P corresponding

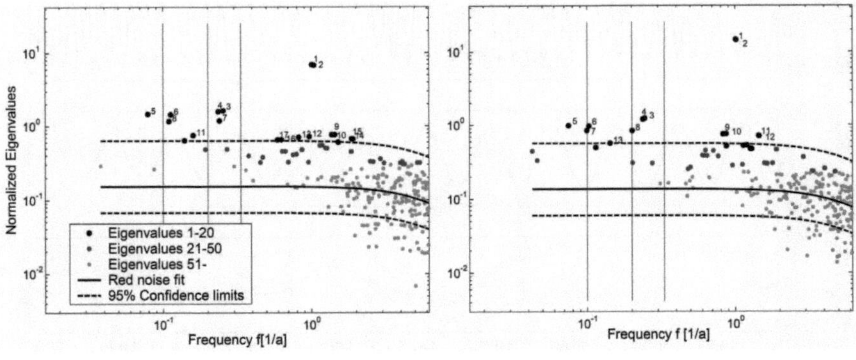

Fig. 12.3 Testing the reconstructed components versus the red noise assumptions. The left plot shows the results for the RCs derived from an SSA analysis of CMDS mode 1. The right plot shows the red noise test based on the RCs derived from SSA on Isomap ($k = 17$)

to $\frac{N}{P} = 2$–2.5 provided stable results in terms of signal separability and sensitivity to the parameters used to generate the input series (i.e. choice of k in Isomap).

First, we combined a CMDS reduction with SSA. This approach was also suggested by Ghil et al. [12.12]. By applying SSA to the first PC, seven long-term components could be identified which were not compatible with the red noise assumption. The periods of these components correspond to (in descending order of eigenvalues) 4, 4.3, 12.7, 8.9, 4.2, 9.1 and 6.3 years. However, the signal separability as estimated by the weak separability correlation (12.10) was unsatisfying for these components. Combining Isomap with SSA did not have that problem. In this case, we found significant long-term components with the periods 4.1, 4.2, 13.4, 9.8, 10, 5 and 7 years (see Fig. 12.3). A typical feature of SSA is the extraction of two components of almost identical period, which is an effect of the linear decomposition of the trajectory matrix: to represent oscillating modes SSA requires at least two linear modes with opposite parity [12.15].

The significant long-term components did not show any harmonics (integer multiple overtones). They would be clearly correlated, but in fact the w-correlations indicate that the SSA led to well-separable signals (see Fig. 12.4), at least better than the CMDS-prefiltered data.

To test for the effect of hydrological dependence, we repeated the signal extraction for a subset of the 53 "independent" gauges. The results are not distinguishable from those for the full data set. We found equal periods and a comparable signal separability. The only interesting feature was that even more long-term components were extracted. However, their amplitude was close to the 95% significance interval and thus debatable. The redundancy contained in the full data set thus did not spoil the performance of the reduction techniques.

Another general observation was that it was not possible to identify significant trends. It is possible that periods of half the time series length or more are in fact trends, since there is no chance to distinguish between these oscillations and

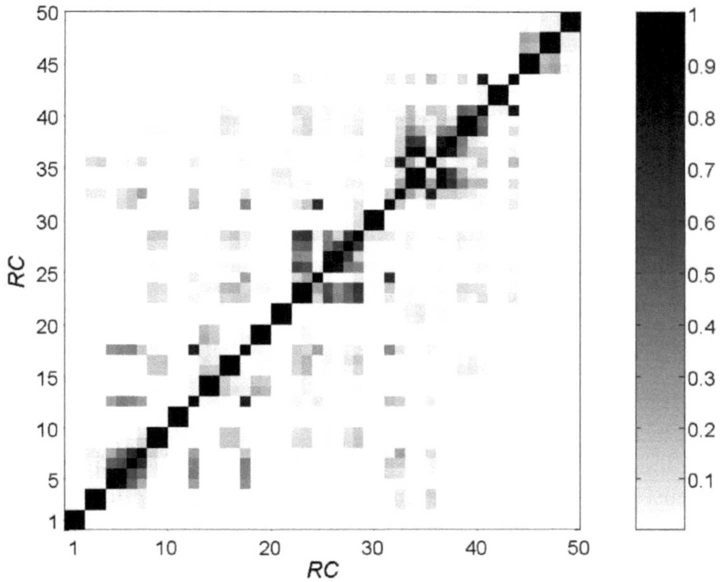

Fig. 12.4 The w-correlation matrix of the first 50 reconstructed components from the SSA analysis of the first Isomap ($k = 17$) mode

trends [12.13]. These periods seem to be present (Fig. 12.3); they are, however, non-significant.

12.3.3 Regional Patterns and Hydrological Dependencies

Are the long-term components found for the whole data set present everywhere? In addition to the common periods found for the whole region, we investigated spatial patterns of these components. We divided the investigated geographical area in non-overlapping stripes from north (A) to south (G) of 50 km width each. The gauges within each strip were investigated again by the combined approach of Isomap and SSA. In this case, the data were aggregated to monthly means, but a sensitivity analysis indicated that the aggregation level is not critical for the identification of long-term components.

The results led to a complicated picture (see Table 12.1). Although the main patterns found by the analysis of the whole data set are confirmed, the significance of individual periods varies in space. Some additional periods appear to affect geographical subareas only. It is known [12.24] that precipitation patterns in Germany show differences in dominating periods along a southwest to northeast gradient, attributable to opposite correlations with the North Atlantic oscillation. Further investigations could clarify whether there are similar phenomena for the runoff data as well.

Table 12.1 Significant long-term periods found by the combined approach of Isomap and SSA in different geographical north–south stripes of southern Germany (A, north to G, south) for monthly means. The periods are sorted by their respective eigenvalues

A	B	C	D	E	F	G
11.2	11.3	5.0	4.9	15.7	3.9	4.7
5.5	7.9	7.0	4.6	13.0	4.7	4.2
3.8	4.0	4.3	6.4	5.1	14.6	8.6
6.9	7.2	4.2	6.0	7.0	4.0	10.4
5.4	4.0	6.8	7.2	4.9	7.3	5.2
3.8	3.5	5.6	4.3	16.6	8.0	
3.5	4.9	19.0	24.4	3.1	4.6	

12.4 Discussion

The results indicate that both CMDS and Isomap allow for a significant reduction of dimensionality for discharge time series. However, the principal coordinates obtained lacked obvious patterns with an easy interpretation apart from the annual cycle. This indicates that dimensionality reduction alone is not very useful for the identification of long-term runoff behaviour. In addition, success in dimensionality reduction necessarily implies that plenty of structure is moved into just the first component, rendering a subsequent decomposition technique almost a necessity if interest is in interpretation of individual modes. Fortunately, the reduction maintained the persistence of the input series quantitatively as measured by the Hurst parameter. Thus, the subsequent analysis with SSA was both mandatory and justified.

One crucial result was that CMDS explained more variance in the first mode than Isomap when the local geometry was emphasized by selecting a small threshold parameter. This effect was reduced by increasing k and did not extend to the second dimension. This finding confirms the suggestion of Ghil et al. [12.12], who assumed that PCA is a suitable way to reduce the dimensionality of multidimensional series. A variety of reasons could attribute to these findings. One possibility for recovering less of the data variance by Isomap could be attributed to the constraints of the methods. For example, de Silva and Tenenbaum [12.7] claim that Isomap is superior only in special nonlinear manifolds. It may also be that the underlying manifold is simply linear and thus the CMDS method outperforms the Isomap variants for small threshold parameters. The latter is forced to find geodesics between far away points which might be connected by a linear neighbourhood relation. This could lead to an overestimation of the inter-point distances in cases where the manifold is unevenly sampled. Unfortunately, it is neither possible to decide whether the underlying manifold is nonlinear, nor whether it is sufficiently sampled, nor whether the topological constraints of Isomap limit its application in the present case.

The combined approach of Isomap and SSA found a variety of long-term modes. The respective periods were more or less also obtained by combining CMDS and SSA. However, the separability of the extracted signals was much better when a small Isomap threshold parameter was applied. This could be a hint that nonlinear decomposition is more suitable to single out independent components (and thus processes). Time series prefiltering with Isomap is already a first signal-selective process.

12.4.1 Methodological Outlook

This study has shown that the new methods of nonlinear dimensionality reduction are suitable to improve traditional methods. In particular, Isomap should replace conventional PCA to generate low-dimensional modes for a subsequent SSA analysis. In the context of this study, we also tested the potential of other nonlinear dimensionality reduction techniques [12.21]. For example, we prefiltered the time series via nonlinear principal component analysis [12.14, 12.19] and locally linear embedding [12.27], but both were unsuitable for finding long-term components by the subsequent application of SSA. However, the field of nonlinear dimensionality reduction provides much more interesting methods, which could not be explored sufficiently.

Another interesting feature arises from the interpretation that SSA is simply a variant of PCA [12.12]. If the traditional PCA can be replaced by other methods of nonlinear dimensionality reduction, the SSA decomposition step could also be replaced by Isomap or other methods. The subsequent PCs can equally be obtained from the EOFs as from the Isomap coordinates. In fact some experiments on this question have been carried out, indicating that for this task it is very difficult to find an adequate threshold parameter. But once a good parameter was found (where ϵ proved to be much better than k), the signal separability was much better than the one identified by standard SSA. This shows the potential of the presented methodological framework which has to be studied in more detail.

Although there is room for improvement, we already have provided a powerful tool for handling multidimensional time series. High-dimensional data sets will be increasingly used by hydrological practitioners. Regionalized runoff data as obtained by coupling runoff time series, remote sensing, land use models and other available data [12.8] can be used for obtaining better spatially resoluted flood risk estimates. The output time series are of much higher dimensionality than available through gauged sites. Thus, powerful methods to extract the main patterns of the data are required and the combination of Isomap and SSA presents an example.

12.5 Conclusion

We investigated the long-term behaviour of river runoff from 118 gauges in southern Germany. As a multichannel SSA for the full data set was computationally excluded and a high degree of redundancies could be expected, a combined approach of nonlinear dimensionality reduction and SSA was developed. This combination proved to be a computationally effective, methodological flexible tool for the analysis of high-dimensional time series.

The reduction efficiency was very high, pointing towards synchronous dynamics on a range of time scales. Independent of spatial or temporal resolution as well as unaffected by hydrological dependencies, it was possible to find significant long-term components unambiguously with characteristic periods of $\sim$4, $\sim$6, and $\sim$11 years. The extracted modes are consistently reappearing in the series; however, we also observed regional differences indicating that the long-term behaviour of river discharges is varying in space. The reduction techniques preserved the persistence of the original series.

The approach allows future extensions and applications, and thus provides a new perspective for large-scale assessments, e.g. in the context of regionalized discharge data.

Acknowledgments The authors are grateful to the Bavarian Environment Agency and the Global Runoff Data Centre for providing the runoff data investigated. This work has been supported by the German Federal Ministry of Education and Research (BMBF) within the research project "Scaling Analysis of Hydrometeorological Time Series" (grant no. 03330271).

References

12.1. M.R. Allen, L.A. Smith, *Monte Carlo SSA: Detecting irregular oscillations in the presence of colored noise* J. Clim. **9**, 3373–3404 (1996)
12.2. M. Bernstein, V. de Silva, J.C. Langford, Y.B. Tenenbaum, *Graph Approximations to Geodesics on Embedded Manifolds*, Technical report, Stanford University, Stanford (2000); [available from `http://Isomap.stanford.edu/BdSLT.pdf`] (Accessed Jan. 2006)
12.3. I. Borg, P. Groenen, *Modern Multidimensional Scaling* (Springer, New York, NY, 1997), p. 471
12.4. D.S. Broomhead, G.P. King, *Extracting qualitative dynamics from experimental data.* Physica D, **20**, 217–236 (1986)
12.5. F. Camastra, *Data dimensionality estimation methods: A survey.* Pattern Recognit., **36**, 2945–2954, (2002)
12.6. T.F. Cox, M.A.A. Cox, *Multidimensional Scaling*, (Chapman & Hall/CRC, Boca Raton, FL, 2001), p. 308
12.7. V. de Silva, Y.B. Tenenbaum, Global versus local methods in nonlinear dimensionality reduction, in S. Becker, S. Thrun, K. Obermayer ed. by *Advances in Neural Information Processing Systems 15*, (MIT Press, Cambridge, 2002), pp. 705–712
12.8. B. Diekrüger, M.J. Kirkby, U. Schröder (eds.), *Regionalization in Hydrology* vol 254 (IAHS Publications, Wallingford CT, (1999)
12.9. E.W. Dijkstra, *A note on two problems in connexion with graphs.* Numerische Mathematik, **1**, 269–271 (1959)
12.10. J.B.Elsner, A.A. Tsonis, *Singular Spectrum Analysis. A New Tool in Time Series Analysis*, (Plenum Press, New York, NY, 1996) p. 164
12.11. A.J. Gámez, C.S. Thou, A. Timmermann, J.Kurths, *Nonlinear dimensionality reduction in climate data.* Nonlinear Processes Geophys., **11**, 393–398 (2004)
12.12. M. Ghil, M.R. Allen, M.D. Dettinger, K. Ide, D. Kondrashov, M.E. Mann, A.W. Robertson, A. Saunders, Y. Tian, F. Varadi, P. Yiou, *Advanced spectral methods for climatic time series*, Rev. Geophys. **40**, 1–25 (2002)

12.13. N. Golyandina, V. Nekrutkin, A. Zhigljavsky, *Analysis of Time Series Structure:* SSA *and related techniques*, Monographs on Statistics and Applied Probability, **90**, (Chapman & Hall/CRC, Boca Raton, FL, 2001), p. 305
12.14. W.W. Hsieh, *Nonlinear principal component analysis by neural networks*. Tellus, **53**(A), 599–615 (2001)
12.15. W. Hsieh, K. Hamilton, *Nonlinear singular spectrum analysis of the tropical stratospheric wind*, Q J R Meteorol. Soc. **129**, 2367–2382 (2003)
12.16. H.E. Hurst, *Long-term storage capacity of reservoirs*, Trans. Am. Soc. Civil Eng., **116**, 770–808 (1951)
12.17. M. Kirby, *Geometric Data Analysis. An Empirical Approach to Dimensionality Reduction and the Study of Patterns* (Wiley, New York, NY, 2001), p. 363
12.18. E. Koscielny-Bunde, J.W. Kantelhardt, P. Braun, A. Bunde, S. Havlin, *Long-term persistence and multifractality of river runoff records: Detrended fluctuation studies*. J. Hydrol. **332**, 120–137 (2006)
12.19. M.A. Kramer, *Nonlinear Principal Component Analysis Using Autoassociative Neural Networks*. J. Am. Inst. Chem. Eng. (AIChE), **37**, 233–243 (1991)
12.20. H. Lange, K. Bernhardt, Long-term components and regional synchronization of river runoffs, in *Hydrology: Science and Practice for the 21st Century* ed. by B. Webb et al. (British Hydrological Society, 2004), pp. 165–170. ISBN 1-903741-10-6 (Accessed Jan. 2006)
12.21. M. Mahecha, *Linear and Nonlinear Dimensionality Reduction in Spatial and Temporal Ecological Data Sets*, Diploma Thesis, University of Bayreuth, (2006)
12.22. B.B. Mandelbrot, *Multifractals and 1/f noise*, (Springer, New York, NY, 1999), p. 442
12.23. B.B. Mandelbrot, J.R. Wallis, *Robustness of the rescaled range R/S in the measurement of noncyclic long-run statistical dependence*, Water Resourc. Res., **5**, 967–988 (1969)
12.24. D. Markovic, M. Koch, *Wavelet and scaling analysis of monthly precipitation extremes in Germany in the 20th century: Interannual to interdecadal oscillations and the North Atlantic Oscillation influence*, Water Resourc. Res. **41**, W09240, (2005)
12.25. A. Montanari, R. Rosso, M.S. Taqqu, *A seasonal fractional ARIMA model applied to the Nile River monthly flows at Aswan*, Water Resourc. Res. **36**, 1249–1259 (2000)
12.26. P. Pekárova, P. Miklánek, J. Pekár, *Spatial and temporal runoff oscillation analysis of the main rivers of the world during the 19th–20th centuries*. J. Hydrol. **274**, 62–79 (2003)
12.27. S.T. Roweis, L.K. Saul, *Nonlinear dimensionality reduction by locally linear embedding*, Science, **290**, 2323–2326 (2000)
12.28. O. Samko, P. Rosin, D. Marshall, *Selection of the optimal parameter value for the Isomap algorithm*, Pattern Recognit. Lett. (2006)
12.29. A. Saxena, A. Gupta, A. Mukerjee, *Non-linear Dimensionality Reduction by Locally Linear Isomaps*, L N C S, **3316**, 1038–1043 (2004)
12.30. T. Shun, C. Duffy, *Low-frequency oscillations in precipitation, temperature, and runoff on a west facing mountain front: A hydrogeologic interpretation*, Water Resourc. Res. **35**, 191–201 (1999)
12.31. J.B. Tenenbaum, Mapping a manifold of perceptual observations, in *Advances in Neural Information Processing Systems*, Vol. 10, ed. by M. Jordan, M. Kearns, S. Solla (MIT Press, Cambridge, 1998), pp. 682–687
12.32. J.B. Tenenbaum, V. de Silva, J.C. Langford, *A Global Geometric Framework for Nonlinear Dimensionality Reduction*. Science **290**, 2319–2323 (2000)

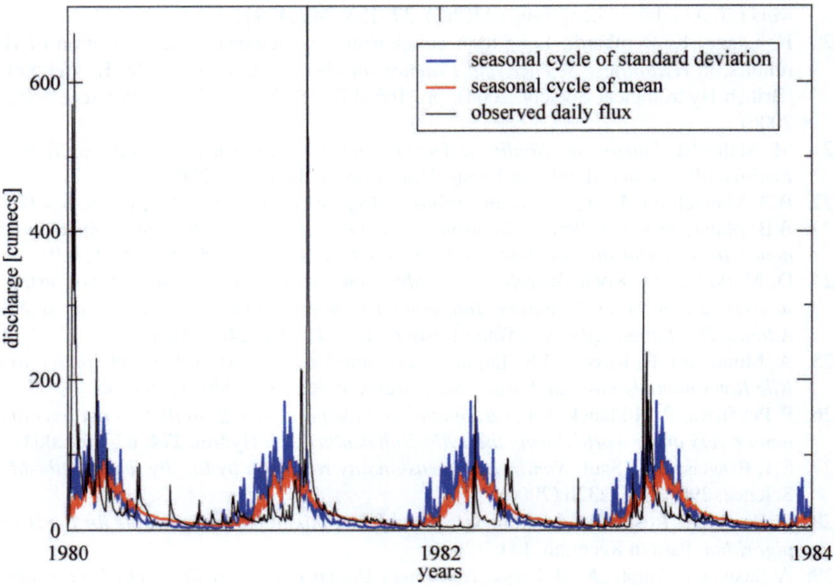

Johnstone river flux with seasonal cycles of mean and standard deviation

Chapter 13
Seasonality Effects on Nonlinear Properties of Hydrometeorological Records

Valerie N. Livina, Yosef Ashkenazy, Armin Bunde, and Shlomo Havlin

Climatic time series, in general, and hydrological time series, in particular, exhibit pronounced annual periodicity. This periodicity and its corresponding harmonics affect the nonlinear properties of the relevant time series (i.e. the long-term volatility correlations and the width of the multifractal spectrum) and thus have to be filtered out before studying fractal and volatility properties. We compare several filtering techniques and find that in order to eliminate the periodicity effects on the nonlinear properties of the hydrological time series, it is necessary to filter out the seasonal standard deviation in addition to the filtering of the seasonal mean, with conservation of linear two-point correlations. We name the proposed filtering technique "phase substitution", because it employs the Fourier phases of the series. The obtained results still indicate nonlinearity of the river data, its strength being weaker than under previously used techniques.

13.1 Introduction

To better understand the climate dynamics, the analysis of the observed climate time series is essential. Their statistical properties provide, e.g. with the average state of the climate system, the order of the natural fluctuations around this state and the climate trends [13.27].

One basic way to quantify the statistical properties of a time series is through the probability distribution function (pdf) of the data. The pdf, however, is invariant under shuffling of the time series and does not reflect the temporal statistical properties (those related to the time ordering of the time series). For instance, under shuffling of the time series, the probability distribution remains unchanged, while the autocorrelation function is affected. Still, estimation of the forms of the pdf may help, for example, to quantify the relative occurrence of extreme climate events [13.4].

V.N. Livina (✉)
School of Environmental Sciences, University of East Anglia, Norwich, NR47TJ, UK
e-mail: v.livina@uea.ac.uk

J.P. Kropp, H.-J. Schellnhuber (eds.), *In Extremis*,
DOI 10.1007/978-3-642-14863-7_13, © Springer-Verlag Berlin Heidelberg 2011

Time ordering of a climatic record may teach us about the dynamical properties of the climate system [13.11]. Pronounced periodicities, like the annual periodicities of temperature and river flow, can easily be tracked using the Fourier frequency spectrum. The harmonics of the annual periodicities may provide useful information regarding the "morphology" of the annual cycle and usually are indicators of the nonlinear underlying process (or of the nonlinearity of the time series). The periodicities are usually linked to the Fourier amplitudes; however, as is demonstrated in this chapter, some nonlinear properties (such as the periodicity in the moving variance of the time series) may be associated with the Fourier phases.[1]

In earlier works, the hydrological time series were studied either in 'raw' format (no filtering of periodicities) or after removing seasonal cycle of mean and/or standard deviation. Here, we present a comparison of the scaling properties of river records after applying various types of filtering; we also propose a new filtering technique and discuss its effects.

Figure 13.1 shows a representative hydrological record (river flux of the Johnstone river) with pronounced annual periodicity in mean and standard deviation. For this river, we show in Fig. 13.1 that after subtracting from each daily value its corresponding multi-year daily average (deseasonalising), the obtained volatility time series still has periodicities. This residual effect is due to the fact that in some seasons (e.g. summer and autumn) the river flow fluctuations are small, while in the other seasons they are very pronounced. This remaining periodicity drastically influences the measurement of the nonlinear scaling properties of river flux time series; therefore, the goal of this chapter is to compare several techniques that might cope with such periodicities. We show that it is necessary to remove the periodicities which are associated with Fourier phases, as well as those associated with the Fourier amplitudes.

The complex discrete Fourier transform of a series $x_\ell,\ \ell = 1, \ldots, N$,

$$X_\omega = \sum_{\ell=1}^{N} x_\ell \mathrm{e}^{-\mathrm{i}(2\pi\omega/N)\ell}, \quad \omega = 1, 2, \ldots, N, \tag{13.1}$$

allows us to study amplitudes $|X_\omega|$ and phases ϕ_ω of the series

$$X_\omega = |X_\omega| \mathrm{e}^{\mathrm{i}\phi_\omega}. \tag{13.2}$$

Fourier amplitudes and phases may be associated with linear and nonlinear properties of the series, respectively. By *linear* we refer to a process that appears only in the power spectrum and in the probability distribution but is independent of the Fourier phases; otherwise the process is defined as *nonlinear*. In other words, if the statistical properties, including those of large moments, remain unchanged after randomising the Fourier phases, the process is linear, while in the case where the

[1] We use the term "linear/nonlinear time series" as a synonym for the linear/nonlinear underlying process that generated the time series.

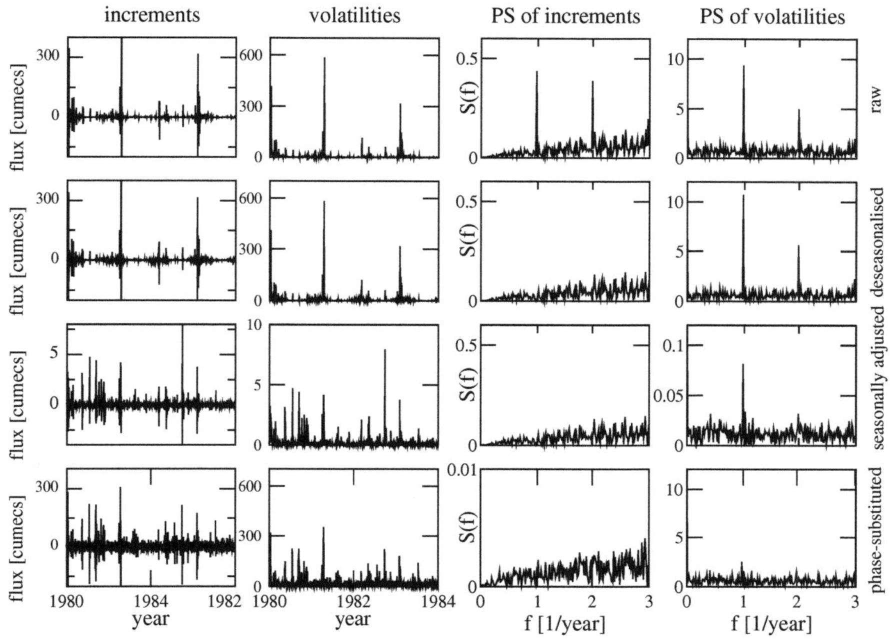

Fig. 13.1 Increments, volatilities, power spectra of increments and power spectra of volatilities for Johnstone river: raw, deseasonalised, seasonally adjusted and phase-substituted data (see detailed explanation of the filtering techniques in Sect. 13.2). The conventional deseasonalising procedure eliminates the periodicity in the power spectrum of increments – much the same as the phase substitution does – but fails to eliminate the periodicity in the power spectrum of volatilities of increments. The new phase substitution filtering method does eliminate the periodicity peak in the power spectrum of volatilities: compare the panels in the last column

statistical properties of large moments are modified after randomising the Fourier phases, the process is defined as nonlinear. By using Schreiber's test for nonlinearity [13.26], one ensures that a linear time series that was transformed by a monotonic and static nonlinear function will still be regarded as linear; for more details see [13.26]. Accordingly, Schreiber's test verifies whether the null hypothesis for linearity is rejected or not. To ensure stationarity, we concentrate mainly on the increment series and not on the original series.

Temporal correlations of climate records have been studied extensively. In the 1950s, Hurst [13.9] reported long-term correlations in river flow records, which was confirmed in further studies [13.17, 13.23]; long-term temporal correlations were also found in other climatological time series, such as temperature records [13.15, 13.16]. The surprising finding is that even time-distant events are statistically correlated, and that these long-term correlations may be quantified by a single scaling exponent H in the corresponding fluctuation function.

This scaling exponent, sometimes called Hurst exponent, is related to the power spectrum scaling exponent β by $H = (1 + \beta)/2$ and, following Schreiber's definition of linearity, quantifies the linearity of the time series. It was found that even the higher moments temporal correlation functions of hydrological time series also obey scaling laws [13.22, 13.28], where these scaling laws reflect the nonlinearity of the processes generating the time series. In the case of a linear process, all different moments may be characterised by a single exponent, and the series is *monofractal*, while a nonlinear time series has a spectrum of exponents called the *multifractal spectrum* – the stronger the nonlinearity, the broader the multifractal spectrum.

To measure the scaling exponents of the fluctuations of a climate record, the periodicities should be filtered out first; otherwise, the verification and estimation of the scaling exponents may be affected by these periodicities. When dealing with two-point correlation functions (the second moment), it is sufficient to filter out the periodicities from the power spectrum. However, when dealing with nonlinearities, or general moment scaling exponents, it is also necessary to filter out the periodicities associated with these moments. In Fig. 13.1, we show that even after eliminating the annual periodicity and its harmonics, the series still has periodicities which are not captured by the power spectrum but rather lie in the Fourier phases. Our point is that an inadequate filtering procedure does not eliminate these effects of periodicities and thus may lead to the observation of an artificial nonlinearity in the data. We show that by removing these higher moment periodicities, the nonlinearity of the time series weakens and the scaling range is extended.

Here we compare five approaches in studying hydrological time series:

Approach 1: 'Raw' time series, no filtering applied.
Approach 2: The seasonal cycle of mean values is filtered out from the original time series.
Approach 3: The seasonal cycles of mean and standard deviation are filtered out from the original time series.
Approach 4: The new technique (described below) – 'phase substitution' – is applied to the original time series.
Approach 5: The new technique – 'phase substitution' – is applied to the *increments* of the original time series.

Taking increments $x_{\ell+1} - x_\ell$ in Approach 5 aims at the nonstationarities of the hydrological time series (scaling exponent $H > 1$) and is discussed in detail in Sect. 13.4.

The chapter is organised as follows. First, we review briefly the earlier filtering techniques and discuss the problems with some of them. We then propose a technique that filters out the periodicity from both the power spectrum and the volatility series (given a time series x_ℓ, the volatility series is defined as $|x_{\ell+1} - x_\ell|$). Next, we demonstrate the use of this technique on hydrological time series and show that the nonlinearity weakens (or the multifractal spectrum width becomes narrower) when using the proposed technique, as well as when using some of the other procedures that can filter out the periodicity in the standard deviation. We then summarise the results and draw some conclusions.

13.2 Methodology

13.2.1 Detrended Fluctuation Analysis and Multifractal Detrended Fluctuation Analysis

Long-range correlated time series may be characterised by scaling laws with specific scaling exponents. It is possible to show that the power spectrum $S(\omega) \equiv |X_\omega|^2$ of long-range correlated time series scales with the frequency ω as $S(\omega) \sim \omega^{-\beta}$. However, trends that exist in climate records may cause inaccuracies in measuring this exponent, in particular, at low frequencies, and thus more advanced techniques that systematically exclude trends from the data have been developed. These techniques include wavelets [13.1, 13.16] and the detrended fluctuation analysis (DFA) [13.12, 13.24].

In recent years, the DFA method has become a widely used tool for measuring scaling exponents of natural time series. It was applied, e.g. to DNA [13.3] and heart-rate sequences [13.2, 13.5, 13.25], financial time series [13.19, 13.21] and climate records [13.8, 13.16]. In this method, one considers fluctuations of the cumulative sum ("profile") $Y_m = \sum_{\ell=1}^{m} x_\ell$ of the original series x_ℓ, $\ell = 1, \ldots, N$. We divide the time axis into nonoverlapping windows of size s. Next, within each window, we calculate the best polynomial fit and determine the variance

$$F^2(\nu, s) = 1/s \sum_{\ell=1}^{s} \left[Y((\nu-1)s+\ell) - y_\nu(\ell)\right]^2, \tag{13.3}$$

where $y_\nu(\ell)$ is the least-square polynomial fit of order k in segment ν of size s. Finally, we average $F^2(\nu, s)$ over all windows ν with fixed size s to obtain the fluctuation function $F(s)$.

When a series obeys scaling, the fluctuation function $F(s)$ follows a power law:

$$F(s) \sim s^\alpha, \tag{13.4}$$

where α is the DFA fluctuation exponent (equivalent to the Hurst exponent H). We calculate $F(s)$ for time scales between $s = 5$ and $N/4$, where N is the length of the series. We are mainly interested in the long-term persistence, i.e. in scales beyond 1 year (denoted by vertical dashed lines in Fig. 13.2.

For uncorrelated records $\alpha = 0.5$, while for long-term correlated (persistent) records, where the autocorrelation function decays with time lags as $C(s) \sim s^{-\gamma}$, $0 < \gamma < 1$, the exponents are related as $\alpha = 1 - \gamma/2$. In DFAk, trends of order $k-1$ are eliminated (kth order trends from the profile). The fluctuation exponent α is also related to the power spectrum scaling exponent $S(f) \sim 1/f^\beta$ by $\beta = 2\alpha - 1 = 1 - \gamma$.

To study the multifractal properties of the hydrological time series, we use the multifractal DFA (MF-DFA) method [13.14], which is a generalisation of the DFA method. In the MF-DFA procedure, similar to the conventional DFA, we calculate

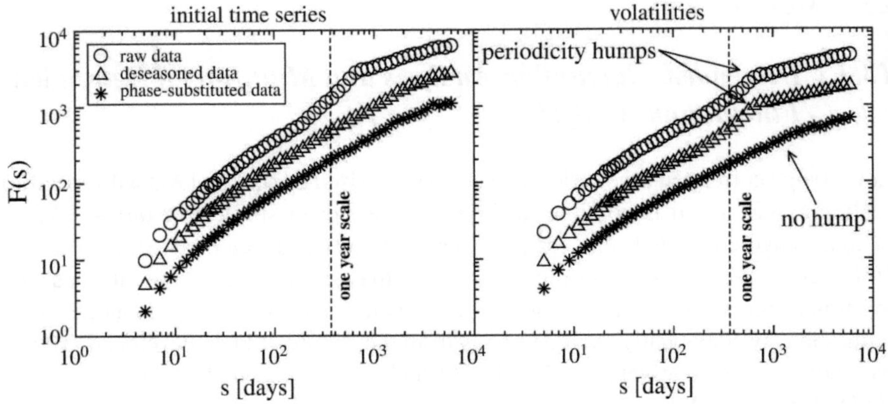

Fig. 13.2 DFA3 results for Johnstone river. *Left panel*: curves for raw (*circles*), deseasonalised (*triangles*) and phase-substituted (*stars*) data. *Right panel*: the same for volatilities. Note that conventional deseasonalising eliminates the periodicity hump in the initial data (like the phase substitution procedure), but fails to eliminate the periodicity hump in the DFA curve of the volatility series, where the phase substitution works well

the generalised fluctuation function for arbitrary moments q (for the conventional DFA, $q = 2$):

$$F_q(s) = \left\{ \frac{1}{N_s} \sum_{\nu=1}^{N_s} [F^2(\nu, s)]^{q/2} \right\}^{1/q}, \quad F_q(s) \sim s^{h(q)}, \tag{13.5}$$

where $h(q)$ is the generalised Hurst exponent. For monofractal time series, $h(q)$ does not depend on q and is constant. For positive values of q, $h(q)$ characterises the scaling behaviour of segments with large fluctuations; for negative values of q, $h(q)$ is associated with the scaling behaviour of the segments with small fluctuations. For our analysis, we use MF-DFA with third-order polynomials (MF-DFA3), thus eliminating quadratic trends in the original data.

Exponents $h(q)$ can be related to the standard scaling exponent $\tau(q)$ [13.7] by

$$\tau(q) = qh(q) - 1. \tag{13.6}$$

One often characterises the multifractality by the singularity spectrum $f(\alpha)$, which is related to $\tau(q)$ via the first-order Legendre transform [13.7]:

$$\alpha = \tau'(q), \quad f(\alpha) = q\alpha - \tau(q). \tag{13.7}$$

The width of the spectrum $f(\alpha)$ characterises the strength of multifractal effects in the data. For monofractal data, the spectrum $f(\alpha)$ collapses into a single point, and both functions $\tau(q)$, $h(q)$ are linear.

13.2.2 Volatility Correlations and Surrogate Test for Nonlinearity

We define the volatility series as $|\Delta x_\ell| = |x_{\ell+1} - x_\ell|$, $\ell = 1, 2, \ldots, N - 1$ [13.2, 13.10]. Correlations in the volatilities reflect the way the magnitudes of fluctuations are arranged; when the fluctuations are clustered into patches of large and small fluctuations, the volatility series is correlated. When the fluctuations are more homogeneous, the volatility series is less correlated.

Since the information regarding the nonlinearity of the time series is stored in the Fourier phases, it is possible to generate linearized surrogate time series by randomising the Fourier phases. However, when randomising the Fourier phases, the histogram of the original time series may be affected. A method to preserve both the histogram and the power spectrum of the original time series was introduced by Schreiber and Schmitz [13.26].

In a surrogate data test, the NULL hypothesis is that the series under consideration is linear. The surrogate data have the same pdf and almost the same power spectrum as the original series, but with random Fourier phases. If a statistical measure obtained for the original series is significantly different from that of the surrogate data, the NULL hypothesis is rejected and the series is considered to be nonlinear. Applying the surrogate test, we compare periodic volatility and volatility correlations of initial data and surrogates to measure the nonlinearity.

We apply DFA3 to study correlations in the volatility series. In our previous work on river records [13.18], we showed that in the time range between 1 week and 1 year, the volatility fluctuation exponent $\alpha \sim 0.75$ and for time scales larger than 1 year $\alpha \sim 0.65$. When applying the surrogate data test for nonlinearity, the volatility exponent decreases to $\alpha \sim 0.55$ for time scales above 1 week. The results of surrogate data test, i.e. the decrease in the volatility exponent from large value to $\sim$0.5, indicate the existence of nonlinearity in the original river flow time series (see figures in [13.18]).

Since the surrogate test should be applied to stationary time series (i.e. with DFA exponent smaller than 1), we will apply the surrogate data test on the increment series, $\Delta x_\ell = x_{\ell+1} - x_\ell$, when considering hydrological time series that are nonstationary with $\alpha > 1$ for time scales smaller than 1 year.

13.3 The Way Periodicities Affect the Estimation of Nonlinearities: Conventional Seasonal Detrending

By the conventional deseasonalising (seasonal detrending) we refer to elimination of the general climatic periodicity, i.e. subtracting the average annual cycle: $x_\ell - \langle x \rangle_{\text{day}}$, where $\langle x \rangle_{\text{day}}$ indicates the average of the particular day (or month in the case of monthly data) over the years.

More advanced conventional method of seasonal detrending is “seasonal adjustment”, in which not only seasonal cycle of mean values is subtracted but also new series is divided by the seasonal cycle of standard deviation (see, e.g. [13.6]):

$$\frac{x_\ell - \langle x \rangle_{\text{day}}}{\sqrt{\langle x^2 \rangle_{\text{day}} - \langle x \rangle_{\text{day}}^2}}. \tag{13.8}$$

In Fig. 13.2, the DFA3 curves for raw (circles) and conventionally deseasonalised data (triangles) are compared, left panel for the initial time series and right panel for the volatility series. One can see that though the conventional deseasonalising copes well with the periodicity hump in the initial data (left panel), in the curves for volatilities (right panel) the deseasonalising is not enough, and the periodicity crossover is still present and certainly influences the estimation of the scaling exponent.

It can be shown that the seasonal adjustment (13.8) alters the scaling exponent of the filtered data. Therefore, the fragile two-point correlations should be monitored in the course of processing the data. For more extended nonlinear analysis, such as the multifractal and volatility analyses, both conventional filtering techniques, deseasonalising and seasonal adjustment, are insufficient.

13.4 A New Method for Filtering Out Periodicities

In a recent study [13.17], it was shown that the river data are nonstationary at short time scales (fluctuation exponent above 1 in time scale up to several months) and are long-term correlated with $\alpha \sim 0.75$ in the asymptotic time scale. Therefore, to eliminate the nonstationarities, we consider the increment series $\Delta x_\ell = x_{\ell+1} - x_\ell$, $\ell = 1, \ldots, N-1$. By this, the fluctuation exponent is reduced by 1. This linear transformation (discrete differentiation) is harmless in the sense of correlations: once the series has been integrated back, the original fluctuation exponent is restored. The obtained increment series is stationary (fluctuation exponent is less than 1), and one can use it for further analysis (for instance, apply conventional deseasonalising) and then integrate the series back. This would provide the series with removed periodicity from the second moment curve (two-point correlations) and improve the result of the linear correlation analysis.

If the conventional deseasonalising is to be applied, we can consider either the initial series or the integrated increments. The importance of incrementing the data becomes more pronounced in the case of the seasonal adjustment which is essential for multifractal and volatility analyses. In Fig. 13.3, where the volatility analysis is performed on deseasonalised and phase substitution filtered data (the latter filtering described below), one can see that applying the new filtering onto data without taking increments still provides some retained periodicities, whereas the volatility curve for the phase substitution filtering applied to the increment series with further integration provides a flatter region in the seasonal scale of the curve.

To account for the above considerations, we process the data as follows:

(I) Possible nonstationarities in the series are eliminated by taking the increments $\Delta x_\ell = x_{\ell+1} - x_\ell$, $\ell = 1, \ldots, N$.
(II) We filter the time series by subtracting the daily (or monthly) climatological average (periodic seasonal average).

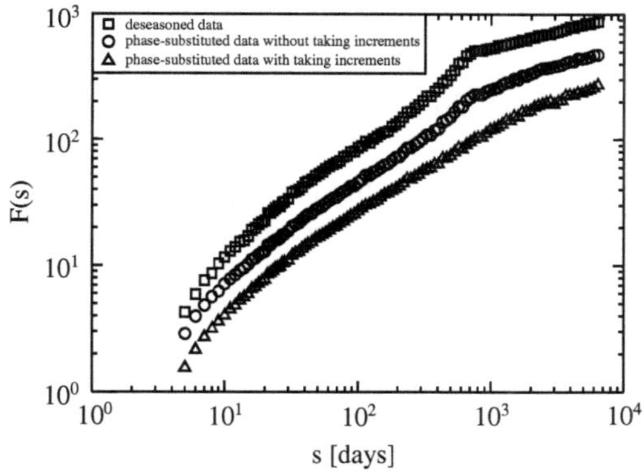

Fig. 13.3 Motivation for analysing the increments of the data: DFA3 results for volatilities of the Johnstone river. Applying phase substitution on the initial data as compared to the increments with back-integration shows that there is some periodicity which still cannot be eliminated due to the nonstationarities of the record

(III) To further eliminate the volatile periodicity of fluctuations, we divide the obtained series by the annual cycle of the climatological standard deviation; see (13.8). Combining steps II and III is seasonal adjustment.
When applying seasonal adjustment, which is a nonlinear operation, we have observed that the correlations in the increment series may weaken (possible procedure artefacts). In order to preserve both two-point correlations and nonlinear correlations (that are stored in the Fourier phases), we perform one more step.

(IV) Since the linear properties are linked to the power spectrum and the nonlinear properties are linked to the Fourier phases, we now combine the Fourier amplitudes obtained from the series in step II with the Fourier phases obtained in step III. We denote the Fourier transform of the deseasonalised data from step II as $X_\omega^{(1)} = |X_\omega^{(1)}|e^{i\phi_\omega^{(1)}}$, and the Fourier transform of the seasonally adjusted data from step III as $X_\omega^{(2)} = |X_\omega^{(2)}|e^{i\phi_\omega^{(2)}}$. The new Fourier transform is formed as $W_\omega = |X_\omega^{(1)}|e^{i\phi_\omega^{(2)}}$ and transforms back (inverse Fourier transform):

$$z_\ell = 1/N \sum_{\omega=1}^{N} W_\omega e^{2\pi i \ell \omega / N}, \tag{13.9}$$

thus providing the final filtered series z_ℓ. After that, we integrate the data back to restore the fluctuation exponent which was reduced by 1 when considering the increment series in step I. In this way, both linear and nonlinear scaling properties are preserved, and the periodicities are eliminated.

The obtained "phase-substituted" data are further studied in the next section.

13.5 Results

13.5.1 Tests of Artificial Data

To perform an extended test of the technique on simulated time series with known properties, we have generated four ensembles of artificial data, using

(i) monofractal long-term correlated data (LTC) and
(ii) multifractal long-term correlated data (MF).

The prescribed fluctuation exponent was $\alpha = 0.9$ in both cases. To each record, we added two types of asymmetric seasonal trends:

(i) with only the cycle of seasonal mean and
(ii) with the additional cycle of standard deviation. The asymmetric annual cycle used for these seasonal trends is defined as

$$C_\ell = C_{k+nT} = \begin{cases} 1 + \cos(2\pi f k), & 0 \le k < \frac{2}{3}T, \\ 1 - \cos(4\pi f k), & \frac{2}{3}T \le k < T, \end{cases} \tag{13.10}$$

where n is number of year cycles of period $T = 365$ days, $f = 0.75T$, $\ell = 1, \ldots, N-1$. Thus, C_ℓ increases 1/3 of the time period and decreases for 2/3 of T. One seasonal cycle of the function is illustrated in Fig. 13.4.

Using the asymmetric function C_ℓ, we generate four ensembles for testing the filtering techniques as follows:

Data 1. Monofractal long-term correlated process with asymmetric annual cycle of mean

$$D_\ell^1 = C_\ell + A_1 \cdot [\mathrm{LTC}]_\ell$$

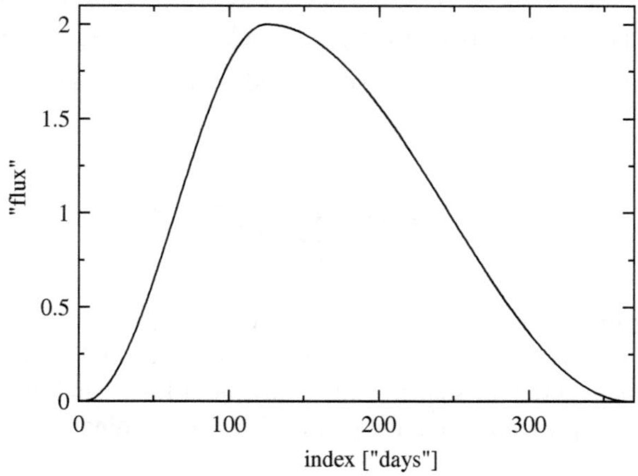

Fig. 13.4 Asymmetric seasonal function C_ℓ, 1 "year" is shown. The function increases 1/3 of the time period ($T = 365$) and decreases for the rest $2/3$ of T

Data 2. Monofractal long-term correlated process with asymmetric annual cycles of mean and standard deviation

$$D_\ell^2 = C_\ell + A_2 \cdot C_\ell \cdot [\mathrm{LTC}]_\ell + B_2 \cdot [\mathrm{LTC}]_\ell$$

Data 3. Multifractal long-term correlated process with asymmetric annual cycle of mean

$$D_\ell^3 = C_\ell + A_3 \cdot [\mathrm{MF}]_\ell$$

Data 4. Multifractal long-term correlated process with asymmetric annual cycles of mean and standard deviation

$$D_\ell^4 = C_\ell + A_4 \cdot C_\ell \cdot [\mathrm{MF}]_\ell + B_4 \cdot [\mathrm{MF}]_\ell,$$

where A_i, B_i are rescaling constants.

Each set contained 20 samples of length 65,536 data points. The algorithm for generating monofractal long-term correlated data is described in [13.20] and for multifractal long-term correlated data in [13.10]. The results are shown in Figs. 13.5 and 13.6.

In the first column of Fig. 13.5, we present samples of four processes (3-"year" data), DFA curves for raw (green curves), deseasonalised (black curves) and phase-substituted (red curves) data in second column, and DFA curves for volatilities (third column). One can see that DFA curves of deseasonalised and phase-substituted data practically coincide (second column), which means that the seasonal cycle heavily affecting the DFA curve of the raw data (green curves) is successfully removed (in both black and red curves). The DFA curves of volatilities demonstrate a pronounced difference in the cases of Data 2, 4 (where both cycles of seasonal mean and standard deviation are imposed). One can see that the phase substitution filters out this complicated periodicity better than the conventional deseasonalising.

A difference between two methods can also be observed in Fig. 13.6, where the multifractal exponents and averaged multifractal spectra are compared (deseasonalised data – black curves, phase-substituted data – red curves). By construction, only data 3, 4 are multifractal, whereas data 1, 2 are monofractal. If the conventional deseasonalising applied, the artifical multifractality is falsely detected (black spectra), whereas the spectra for phase-substituted data (red curves) are corrupted and narrower in both panels, correctly indicating monofractality. In the case of intrinsically multifractal data 3, 4, the spectra of the phase-substituted data (red curves) are smooth and wide, which means detection of multifractality – yet they are more narrow than those of deseasonalised data (black curves), where not all seasonal periodicity artefacts are properly removed.

13.5.2 Results of Volatility Analysis of Observed River Data

Having performed the test with artificial data, we apply the developed technique to the observed river flux records. The effect of the proposed phase substitution

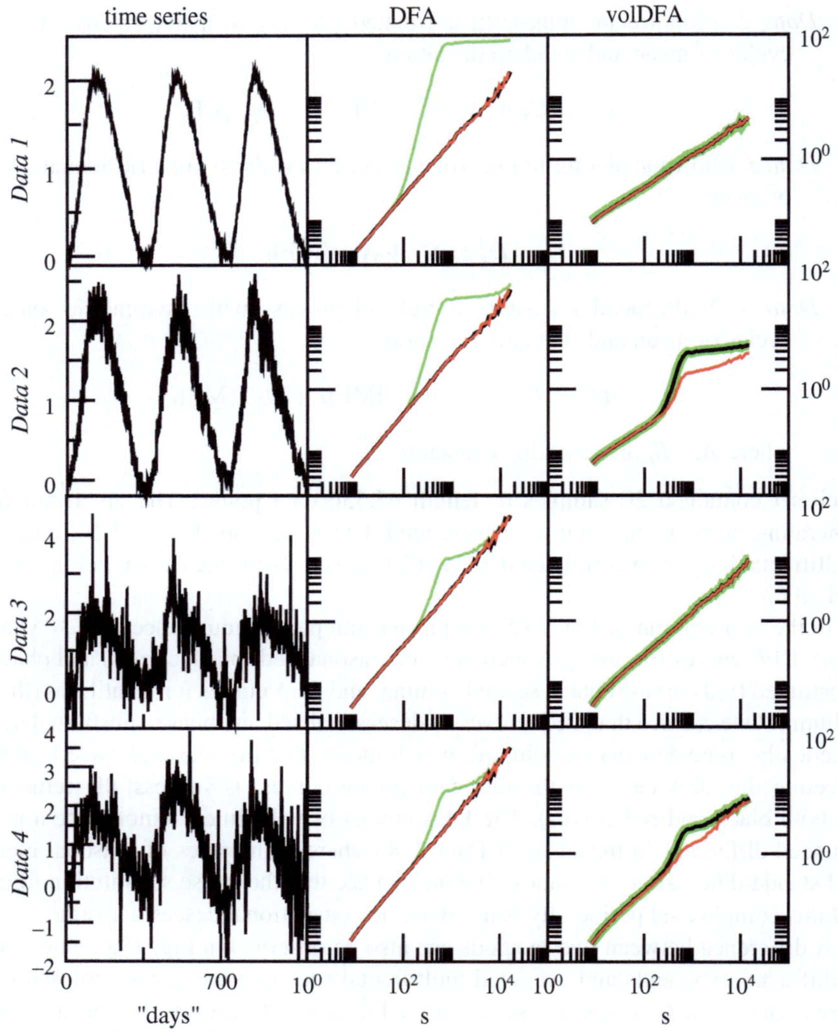

Fig. 13.5 Data and DFA results of the experiment with four ensembles. *Data 1:* monofractal long-term correlated process with asymmetric seasonal cycle of mean values. *Data 2:* monofractal long-term correlated process with asymmetric seasonal cycles of mean and standard deviation. *Data 3:* multifractal long-term correlated process with asymmetric seasonal cycle of mean values. *Data 4:* multifractal long-term correlated process with asymmetric seasonal cycles of mean and standard deviation. Three-year data samples, DFA results, and volatility DFA results (raw data – *green curves*, deseasonalised data – *black curves*, phase-substituted data – *red curves*)

filtering technique on the volatility analysis can be demonstrated by comparing second moment curves (Fig. 13.2, right panel), where the DFA3 curve of phase-substituted data has no periodicity hump, whereas the curve for the conventionally deseasonalised data does have it.

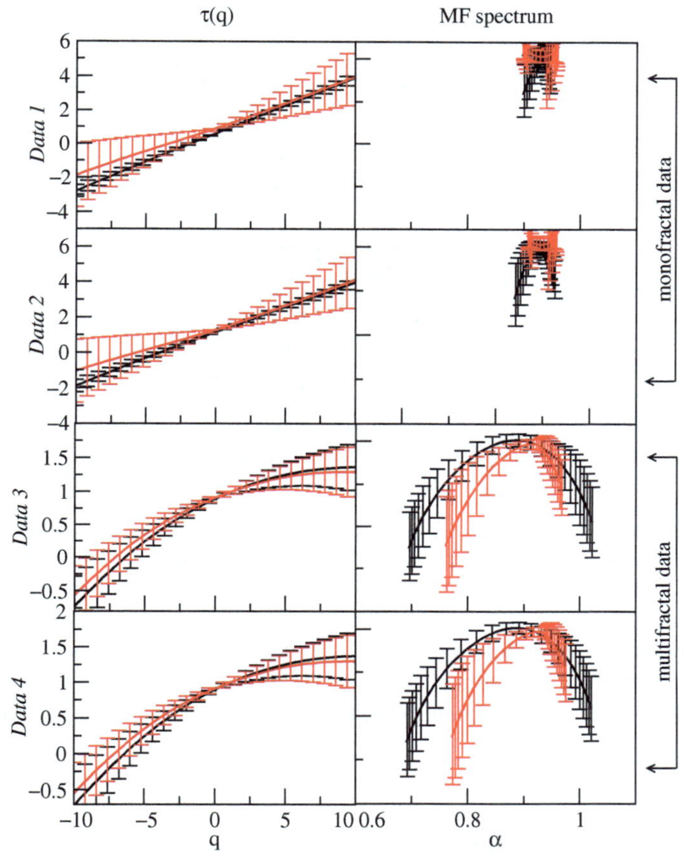

Fig. 13.6 Multifractal results of the experiment with four ensembles of artifical data: exponents $\tau(q)$ and averaged multifractal spectra (deseasonalised data – *black curves*, phase-substituted data – *red curves*)

In Fig. 13.7, we summarise the results of the volatility analysis for 30 world rivers with data preprocessed using the filtering techniques described in Sect. 13.1 (Approaches 1–5). We compare initial time series and their volatilities. For the two-point correlation analysis, the plain raw data show an exponent lower than it should be due to the seasonal periodicity. On the other hand, for the volatility series the filtering techniques that filter out the seasonal standard deviation (Approaches 3–5) show comparable results. This suggests that in order to estimate the long-term non-linear properties of the hydrological time series, one has to filter out the periodicities associated with the standard deviation.

In our earlier paper [13.18], we obtained an averaged power spectrum exponent of volatilities $\beta \sim 0.27$ for the time scales above 1 year. Since the DFA fluctuation exponent α is related to the power spectrum exponent β by $\alpha = (1 + \beta)/2$, the power spectrum results correspond to $\alpha \sim 0.63$, which is close to the present results

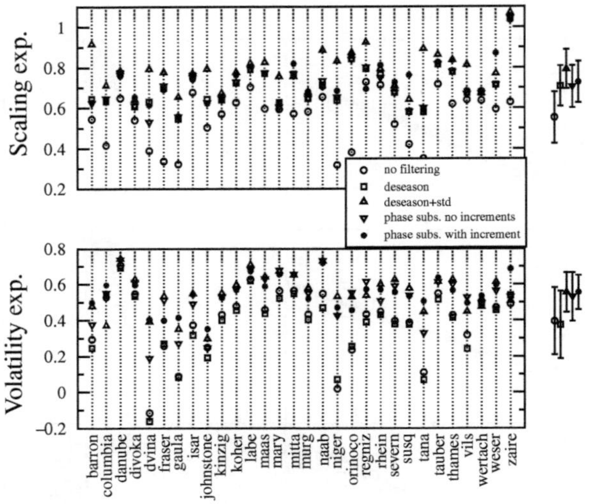

Fig. 13.7 Results of fluctuation analysis for 30 rivers using five filtering approaches. *Top panel*: exponents in the data, *lower panel*: exponents in the volatilities; estimated in the asymptotic regime

obtained by direct calculation of α in the time scale above 1 year. Therefore, the earlier suggested method of scaling analysis of volatilities by means of logarithmically binned power spectrum has been proven to be efficient and supports our present results.

13.5.3 Results of Multifractal Analysis of Observed River Data

In Fig. 13.8, we show MF-DFA partition functions of two river records processed with two filtering techniques. The curves represent discrete values of moments q from -10 to 10, from top to bottom. Crossovers in the higher moment curves deviate the results of multifractal analysis (deseasonalised data, left panels). The curves of negative moments are more affected by the periodicities than those of positive moments.

In the right-hand side panels, the curves with filled symbols correspond to DFA3 second moment, and one can see that they are identical to those in the left panels; therefore, the filtered data preserve two-point correlations of the deseasonalised data. At the same time, the higher moments' curves have been essentially smoothed (no periodicity humps in curves of negative moments).

Earlier works studied multifractal properties of hydrological time series using the conventional deseasonalising [13.13, 13.17]. To avoid the effect of crossovers in the higher moments' scaling curves (seen in Fig. 13.8, left panel), they calculated the scaling exponents for asymptotic time scales above the "bump" range. In [13.28], the authors applied no preliminary filtering to the data and studied the short-term scaling behaviour. Their results are thus expected to be influenced by the periodicity

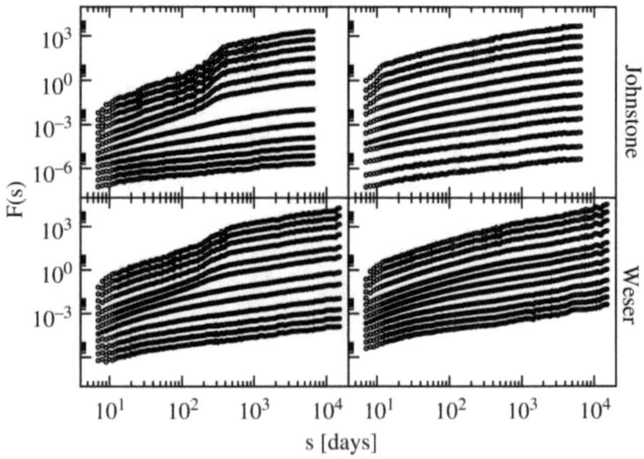

Fig. 13.8 Multifractal partition functions for Johnstone and Weser river flux, q= − 10, −8, −6, −4,−2,−1,1,2,4,6,8,10, from *top* to *bottom*. *Left panels*: deseasonalised data; *right panels*: phase-substituted data. The new filtering technique eliminates humps and crossovers that may lead to erroneous multifractal spectrum

artefacts, although their general conclusion that river flux fluctuations are multifractal is most probably valid. The method proposed here may enable extending the region of scaling that includes the annual time scale (the "bump" region), because the crossovers discussed above do not exist after applying our filtering procedure.

In Fig. 13.9, we compare multifractal spectra (left panel) and volatilities (right panel) of phase-substituted and phase-randomised data for the Johnstone river. One can see that the phase-substituted data are nonlinear and that the nonlinearities are diminished by the surrogate test. This effect supports our filtering technique, because it indicates that the phase substitution preserves the essential nonlinear features of the record.

In Fig. 13.10, we compare multifractal spectra of conventionally deseasonalised and phase-substituted data of six rivers. Results for the phase-substituted data demonstrate weakened nonlinearity: the spectra become narrower.

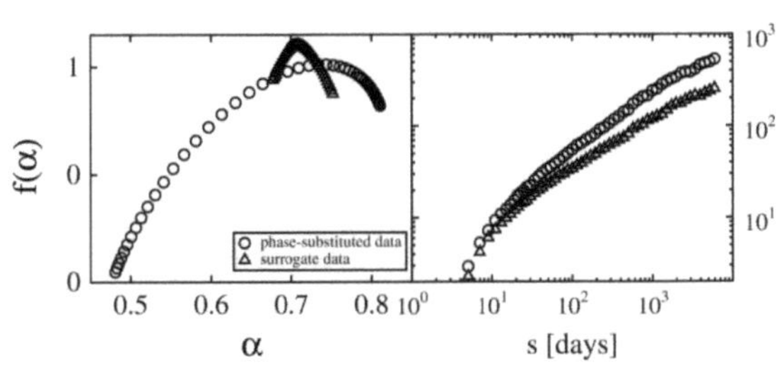

Fig. 13.9 Multifractal spectra (*left panel*) and DFA3 curves for volatilities (*right panel*) of phase-substituted (*circles*) and surrogate (*triangles*) series. The phase substitution preserves nonlinear properties of the data which can be eliminated by randomizing the phases

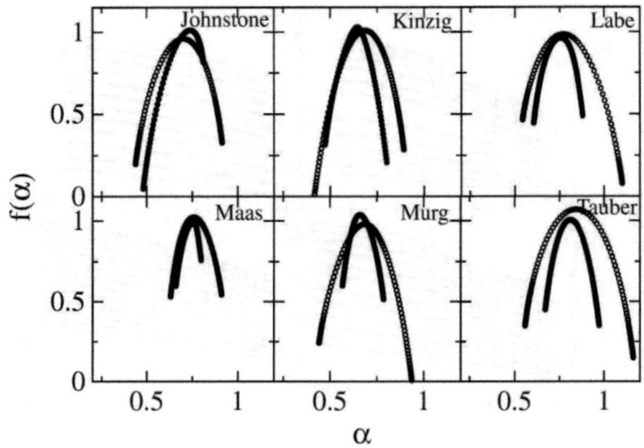

Fig. 13.10 Multifractal spectra of six rivers: deseasonalised (*open symbols*) and phase-substituted (*filled symbols*) data (the Johnstone river – Australia, 74 years; the Kinzig river – Germany, 82 years; the Labe river – Czechia, 102 years; the Maas river – Netherlands, 80 years; the Murg river – Germany, 77 years; the Tauber river – Germany, 66 years). The spectra of phase-substituted data are narrower: less influenced by the periodicity artefacts in the data

Having performed the multifractal analysis of the phase-substituted data at time scales above 1 year, we have obtained the multifractal spectra of width 0.34 ± 0.26. For the conventionally deseasonalised data, in the time scale above 1 year we obtained spectra of width 0.49 ± 0.28. Since the error bars are quite wide, we can conclude that both methods provide close results in the asymptotic scale. We hypothesise that if one would consider a scaling range that includes the crossover, the difference between the two filtering techniques would have become more significant. For that, we measure spectra width in a longer time scale range, above 100 days, where the scaling curves for conventionally deseasonalised data have the periodicity crossovers. For the phase-substituted data, in time scales above 100 days the spectra width for the 30 rivers is 0.46 ± 0.13, as compared to the results using the conventionally deseasonalised approach, 0.68 ± 0.20. We note that even though the periodicities were filtered out, the wider MF spectrum we obtain for windows scales above 100 days may still be attributed to the crossovers in the scaling curves.

13.6 Conclusion

Periodicities in climate records might influence the estimation of the two-point scaling exponent, as well as nonlinear properties of hydrological time series, such as volatility correlations and multifractal spectrum width. We compared (i) raw data, (ii) data where the seasonal average was subtracted, (iii) data where seasonal average was subtracted and the series was then divided by the seasonal standard deviation, (iv) phase-substituted data without taking increment, and (v) phase-substituted data after taking the increments. We found that in order to accurately estimate two-point correlations, it is necessary to apply at least approach (ii), while in order to accurately estimate the long-range nonlinear properties (the multifractal spectrum and volatility correlations), approaches (iii)–(v) may be applied. These approaches basically filter out the periodicity in the standard deviation.

The proposed filtering technique is merely empirical and was developed for studying hydrological time series with specific scaling properties. However, the technique may also be useful for estimation of nonlinear properties in other climatic time series with similar statistical characteristics.

Acknowledgments The authors thank the German Federal Ministry of Science for financial support (grant nos. 03F0205B9 and 01LD0016). We would like to thank J. Kantelhardt for useful discussions. YA is supported by the Yeshaya Horowitz Association in the Center for Complexity Science.

References

13.1. A. Arneodo, E. Bacry, P. Graves, J. Muzy, Characterizing long-range correlations in dna-sequences from wavelet analysis. Phys. Rev. Lett., **74**, 3293 (1995)
13.2. Y. Ashkenazy, P.-C. Ivanov, S. Havlin, C.-K. Peng, A.L. Goldberger, H.E. Stanley, Magnitude and sign correlations in heartbeat fluctuations. Phys. Rev. Lett., **86**, 1900 (2001)
13.3. S.V. Buldyrev, A.L. Goldberger, S. Havlin, R.N. Mantegna, M. Matsa, C.-K. Peng, S. Simons, H.E. Stanley, Long-range correlations properties of coding and noncoding dna sequences: Genbank analysis. Phys. Rev. E, **51**, 5084 (1995)
13.4. A. Bunde, J. Eichner, J. Kantelhardt, S. Havlin, Long-term memory: a natural mechanism for the clustering of extreme events and anomalous residual times in climate records. Phys. Rev. Lett., **94**, 048701 (2005)
13.5. A. Bunde, S. Havlin, J.W. Kantelhardt, T. Penzel, J.H. Peter, K. Voigt, Correlated and uncorrelated regions in heart-rate fluctuations during sleep. Phys. Rev. Lett., **85**, 3736 (2000)
13.6. K. Dahlstedt, H. Jensen, Fluctuation spectrum and size scaling of river flow and level. Physica A, **348**, 596 (2005)
13.7. J. Feder, *Fractals* (Plenum Press, New York, NY, 1988)
13.8. R.B. Govindan, D. Vjushin, S. Brenner, A. Bunde, S. Havlin, H.-J. Schellnhuber Global climate models violate scaling of the observed atmospheric variability. Phys. Rev. Lett., **89**, 028501 (2002)
13.9. H.E. Hurst Long–term storage capacity of reservoirs. Transact. Ann. Soc. Civil Eng., **116**, 770 (1951)
13.10. T. Kalisky, Y. Ashkenazy, S. Havlin, Volatility of linear and nonlinear time series. Phys. Rev. E, **72**, 011913 (2005)
13.11. H. Kantz, T. Schreiber, *Nonlinear Time Series Analysis* (Cambridge University Press, Cambridge, 1999)
13.12. J.W. Kantelhardt, E. Koscielny-Bunde, H.A. Rego, S. Havlin, A. Bunde, Multifractal detrended fluctuation analysis of nonstationary time series. *Physica A*, **295**, 441 (2001)
13.13. J.W. Kantelhardt, D. Rybski, S.A. Zschiegner, P. Braun, E. Koscielny-Bunde, V.N. Livina, S. Havlin, A. Bunde Multifractality of river runoff and precipitation: Comparison of fluctuation analysis and wavelet methods. Physica A, **330**(1–2), 240 (2003)
13.14. J.W. Kantelhardt, S.A. Zschiegner, E. Koscielny-Bunde, S. Havlin, A. Bunde, Multifractal detrended fluctuation analysis of nonstationary time series. Physica A, **316**(1–4), 87 (2002)
13.15. E. Koscielny-Bunde, A. Bunde, S. Havlin, Y. Goldreich, Analysis of daily temperature fluctuations. Physica A, **231**, 393 (1996)

13.16. E. Koscielny-Bunde, A. Bunde, S. Havlin, H.E. Roman, Y. Goldreich, H.-J. Schellnhuber, Indication of universal persistence law governing atmospheric variability. Phys. Rev. Lett., **81**, 729 (1998)
13.17. E. Koscielny-Bunde, J.W. Kantelhardt, P. Braun, A. Bunde, S. Havlin, Long-term persistence and multifractality of river runoff records: detrended fluctuation studies. J. Hydrol., **322**, 120–137 (2006)
13.18. V. Livina, Y. Ashkenazy, P. Braun, R. Monetti, A. Bunde, S. Havlin, Nonlinear volatility of river flux fluctuations. Phys. Rev. E, **67**, 042101 (2003)
13.19. R.N. Mantegna, H.E. Stanley, *An Introduction to Econophysics* (Cambridge University Press, Cambridge 2000)
13.20. H. Makse, S. Havlin, M. Schwartz, H.E. Stanley, Novel method for generating long-range correlations. Chaos Solitons Fractals **6**, 295–303 (1995)
13.21. K. Matia, Y. Ashkenazy, H.E. Stanley, Multifractal properties of price fluctuations of stock and commodities. Europhysics Lett., **61**, 422 (2003)
13.22. G. Pandey, S. Lovejoy, and D. Schertzer, Multifractal analysis of daily river flows including extremes for basins of five to two million square kilometres, one day to 75 years. J. Hydrol., **208**(1–2), 62 (1998)
13.23. J.D. Pelletier, D.L. Turcotte, Long-range persistence in climatological and hydrological time series: analysis, modelling and application to drought hazard assessment. J. Hydrol., **203**(1–4), 198 (1997)
13.24. C.-K. Peng, S.V. Buldyrev, S. Havlin, M. Simons, H.E. Stanley, A.L. Goldberger On the mosaic organization of dna nucleotides. Phys. Rev. E, **49**, 1685 (1994)
13.25. C.-K. Peng, S. Havlin, H.E. Stanley, A.L. Goldberger Quantification of scaling exponents and crossover phenomena in nonstationary heartbeat time series. Chaos, **5**, 82 (1995)
13.26. T. Schreiber, A. Schmitz, Surrogate time series. Physica D, **142**(3–4), 346 (2000)
13.27. H.V. Storch, F.W. Zwiers, *Statistical Analysis in Climate Research* (Cambridge University Press, Cambridge 2002)
13.28. Y. Tessier, S. Lovejoy, P. Hubert, D. Schertzer, S. Pecknold, Multifractal analysis and modeling of rainfall and river flows and scaling, causal transfer functions. J. Geophys. Res., **101**(D21), 26427 (1996)

Photograph of the summer 2002 Danube flood near Vác, Hungary, on August 18, 2002, taken from an onboard camera at the ISS. Image by courtesy of NASA Visible Earth, http://veimages.gsfc.nasa.gov/3618/ISS005-E-10000_1000.jpg

Chapter 14
Spatial Correlations of River Runoffs in a Catchment

Reik Donner

Hydrological processes are characterised by spatio-temporal patterns with certain correlations in both time and space. Thus, time series of related quantities recorded at different locations show relevant correlations if these locations are influenced by the same patterns. The actual strength and temporal as well as spatial extension of these correlations depend crucially on the considered observable (temperature, runoff, precipitation, etc.). We analyse whether the corresponding interrelationships change significantly in the presence of changing environmental conditions. For this purpose, we systematically study a variety of measures which quantify the statistical dependence between the components of bi- and multivariate hydrological records. As a particular example, we consider runoff time series from an ensemble of gauges in the Upper Main catchment area in southern Germany. The qualitative behaviour of spatial correlations and their changes during extreme weather events are intensively discussed.

14.1 Introduction

The Earth's climate is a high-dimensional complex system which is subjected to different global and local forcings and nonlinear internal feedback mechanisms which act on very different temporal as well as spatial scales. Therefore, its behaviour is highly chaotic and characterised by an extreme sensitivity with respect to relatively small changes of certain environmental parameters. Such changes are known to be able to lead to sudden transitions in the dynamics of the entire system, with the North Atlantic thermo-haline circulation as the probably most studied example [14.1, 14.2]. Time series recording the variability of climatological observables are therefore typically highly irregular and have rather high noise levels. This holds in particular for the case of hydrological and meteorological data obtained from

R. Donner (✉)
Potsdam Institut für Klimafolgenforschung, 14412 Potsdam, Germany
e-mail: reik.donner@pik-potsdam.de

J.P. Kropp, H.-J. Schellnhuber (eds.), *In Extremis*,
DOI 10.1007/978-3-642-14863-7_14,

direct measurements since the start of the instrumental period or reconstructions of earlier time intervals. Moreover, the variability of meteorological quantities in both, observations and climate models, is often characterised by properties like non-Gaussian probability distribution functions or long-term persistence which have been subjected to a long-lasting debate [14.3–14.33].

Atmospheric patterns are characterised by scales in both time and space on which meteorological observables like temperature, air pressure, or humidity vary only weakly. Measuring the temporal evolution of such quantities at different locations influenced by the same pattern, it is therefore likely that the corresponding time series are more or less strongly correlated, with a maximum correlation occuring at a time lag that is determined by the spatial distance between the sites and the typical drift velocity of the pattern. Due to the dynamic evolution of the observed patterns during their spatial motion, the correlations between hydro-meteorological records decay with an increasing distance between the considered locations. This statement holds in general for very different spatial and temporal scales.

On a global scale, the interrelationships between 500 hPa pressure levels obtained from reanalysis data have been utilised to derive a network-like structure [14.34]. Similar features are likely to be found in simulations of climate models as well, however, the behaviour of such models is known to differ from reanalysis data not only in terms of absolute variabilities and correlations, but also with respect to their nonlinear features like the local predictability [14.35]. Recently, efforts have been reported to transfer the idea of climatological networks to other scales and observables, in particular, by considering correlation and phase synchronisation [14.36] properties of temperature and precipitation records from different German stations [14.37].

On continental scales (i.e. several hundreds to thousands of kilometers), simple linear cross-correlation functions may (depending on the particular geographic situation) not necessarily be an optimal measure for describing the interrelationships and exactly detecting the delay corresponding to the maximum dependence between meteorological time series. For temperature and precipitation records [14.38], it has been suggested that the concept of phase synchronisation may be better suited for this purpose. Recent findings indicate, however, that this result cannot be generalised [14.39]. Instead, for long-term daily temperature time series from two distinct Western and Central European observatories, clear evidence for strong linear and nonlinear correlations has been reported, whereas no significant long-term phase synchronisation could be found. This finding does not exclude the existence of time intervals of phase coherence as have been recently reported for the All-Indian Rainfall (AIR) and NINO3 indices [14.40] which manifest the known interrelationships between variations of the Indian monsoon system and the El Niño phenomenon [14.41–14.44].

Concerning regional correlations of hydro-meteorological records, there are numerous studies analysing the behaviour of, in particular, precipitation observations in different areas of local to regional scale, e.g. [14.45–14.53]. Recent studies with a very high resolution in time [14.54] and space [14.55] have shown that the performance of linear correlation analysis may also be resolution dependent.

In addition, efforts have been reported to apply nonlinear concepts like mutual information [14.56], correlation dimension, and nonlinear prediction error [14.57]; however, these analyses have yet been restricted to a separate consideration of the records from different stations.

The presented study focusses on the spatial pattern of regional runoff correlations in a river catchment. Instead of considering linear and nonlinear interrelationships between only two distinguished locations, we address the general question of ensemble correlations between a number of records on a more fundamental quantitative level. The consideration of a larger set of stations avoids possible artificial results due to an eventually problematic choice of locations. Recently, this idea has motivated the development of the concept of regional flood probabilities [14.58]. In this contribution, we investigate whether the amplitude of spatial correlations depends on the actual state of the system. For example, during extraordinary climatic events, the amplitude and spatial distribution of correlations between hydro-meteorological observables at different locations in the same region may change significantly. In this case, the strength of correlations can be used as an order parameter for the characterisation of the respective environmental conditions.

The general advantage of our approach is that multivariate data sets, formed by simultaneous observations from a larger ensemble of stations located in the same region, contain more detailed information about the spatio-temporal variability of the environmental conditions than single records from only one or two stations. As a particular example, we analyse river runoff records from different gauges in the Upper Main catchment area in southern Germany (see Sect. 14.2). River runoffs are strongly influenced by the precipitation in the corresponding catchment area; however, this dependence is neither linear nor bi-directional as the entire catchment serves as a complicated nonlinear filter transferring the inflow (mainly precipitation) into the outflow (river runoff) [14.59, 14.60]. The parameters of this filter do strongly depend on the particular geography of the considered area, involving parameters like elevation, soil conditions, land use, and vegetation. In addition, time-delayed effects like the accumulation of moisture in soils have to be taken into account here.

We start our analysis with an investigation of the mutual linear and nonlinear correlations between the different records. To appropriately quantify ensemble correlations between the respective records, we characterise the complexity of interrelationships in terms of the relative number of relevant (e.g. orthogonal) patterns obtained after a suitable statistical decomposition of the data set. As the state of the atmosphere is often determined by only some basic modes which are clearly distinguishable from noise [14.61, 14.62], this basic idea can be found in numerous applications to both climate models and observations. However, the approach of quantifying this state during a given time interval by a single statistical parameter is rather novel and has up to now only been used to study multivariate palaeoclimatic proxy data [14.63–14.65]. After a detailed description of the corresponding method, we discuss its potential to describe the behaviour of spatial correlations around an extreme event.

14.2 Description of the Data

We study river runoff records from the Upper Main catchment area in Northern Bavaria, Germany, which have been collected by the Bayrisches Landesamt für Umwelt. The area under investigation has a total size of about 4 440 km^2 and ranges from the inflow of the Regnitz near Bamberg in the south-western part of the catchment up to the hills of the Fichtelgebirge forming the north-eastern watershed. The corresponding maximum extension is about 90 km in east-west and 70 km in north-south direction.

The river runoff data used in this study have been obtained at 13 gauges displayed in Fig. 14.1 and additionally listed in Table 14.1. The corresponding locations are distributed along the Upper Main river and its inflows up to the gauge of Kemmern near Bamberg.[1] Time series with a daily resolution have been made available for

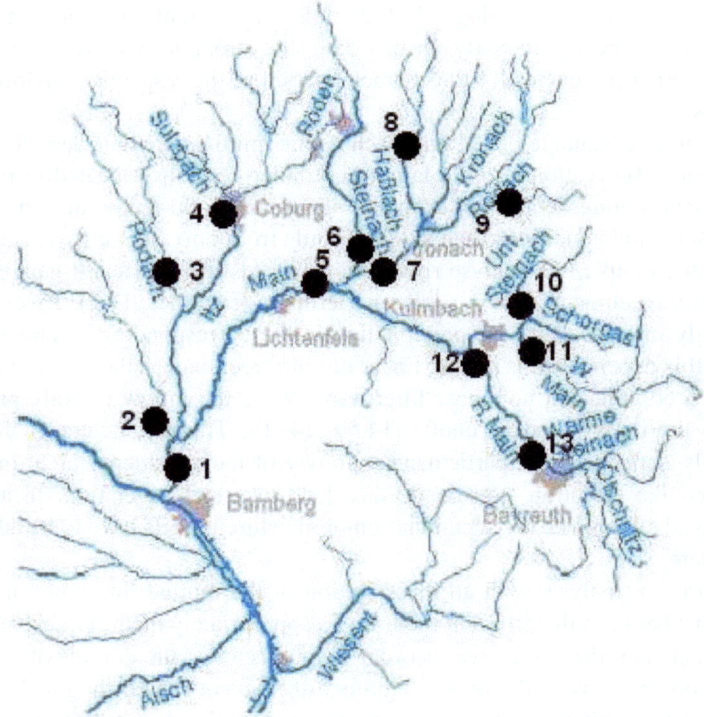

Fig. 14.1 Location of the 13 gauges in the Upper Main catchment area from which records have been available. *Numbers* indicate the stations as described in the text. The underlying map of the river network has been adapted from Hochwassernachrichtendienst Bayern (Bayrisches Landesamt für Umwelt, http://www.hnd.bayern.de)

[1] The Bavarian Landesamt für Umwelt holds data on currently 47 gauges from the catchment (see http://www.hnd.bayern.de), however, we have no continuous records from the remaining 34 stations.

Table 14.1 Gauges from the Upper Main catchment area used in our study. The first numbers correspond to the stations in Fig. 14.1. For the gauge at Steinberg, data are missing between October 31, 1967 and October 31, 1968. The catchment areas have been taken from http://www.hnd.bayern.de

No.	Gauge	River	Catchment (km^2)
1	Kemmern	Main	4 223.84
2	Leucherhof	Baunach	380.52
3	Heinersdorf	Rodach	376.38
4	Coburg	Itz	346.34
5	Schwürbitz	Main	2 418.68
6	Horb	Steinach	256.63
7	Unterlangenstadt	Rodach	713.87
8	Neukenroth	Haßlach	140.96
9	Steinberg	Kronach	94.30
10	Wallenfels	Wilde (Wild) Rodach	96.45
11	Untersteinach	Untere (Lower) Steinach	126.60
12	Ködnitz	Weißer (White) Main	313.42
13	Bayreuth	Roter (Red) Main	340.28

the time interval from January 1961 to December 1990 without significant gaps, with the exception of the runoff time series from one particular gauge where one complete year is missing. Instead of using the original data, we normalised all time series with respect to the sizes of the catchment areas of the corresponding gauges. In addition to the daily records, we also consider a subset of 11 gauges where time series with a temporal resolution of 1 h are available for the time interval from January 1989 to December 1998. This second runoff data set contains the same gauges like the first one, with the exception of the stations at Schwürbitz (gauge no. 5) and Steinberg (gauge no. 9).

The precipitation in the considered region shows a highly diversified pattern which is strongly correlated with the local orography. Its annual mean values range from about 700 mm per year in the north-western part (with typical heights of 200–500 m) up to 1 500 mm per year and more in the source regions of White Main and Steinach in the Fichtelgebirge (up to 1 040 m high) in the eastern part of the catchment [14.66, 14.67]. The main river of the region is the Upper Main with a total length of 71 km from the connection point between Red and White Main to the Regnitz inflow. According to the Hochwassernachrichtendienst Bayern, the typical time of a non-deformed flood wave to cover this distance is between 13 and 26 h. An additional time of 6–10 h has to be taken into consideration to cover the distances from the gauges closest to the main source regions, Untersteinach (gauge no. 11 in Fig. 14.1) and Bayreuth (gauge no. 13), to the connection point of Red and White Main, which are roughly about 15 and 33 km, respectively.

The probably most frequently considered hydrological station in the Upper Main area is the gauge at Kemmern near the Regnitz inflow, which is the last station along the upper part of the river and thus yields aggregated information about the hydrological conditions in the whole catchment. The extreme events recorded at this gauge are thus of particular importance. Concerning the data analysed in this study, the corresponding time interval includes the largest runoff value ever recorded during the instrumental period, with a daily average of 855 m^3/s and a peak discharge of 1 000 m^3/s (or 17.83 mm/day after normalisation with respect to the catchment area) observed on 25 December 1967. (For comparison: the minimum daily average runoff recorded was only 3.7 m^3/s, the mean value between 1961 and 1999 was about 45 m^3/s [14.66].) These few numbers illustrate the large variability of precipitation and river runoff in the area on very different time scales.

14.3 Correlation Functions and Related Quantities

To appropriately quantify the interdependences between the respective observations, one may consider different linear as well as nonlinear measures. Some particular examples are discussed in the following:

14.3.1 Pearson's Linear Correlation

The simplest measure for correlations is Pearson's correlation coefficient [14.68] defined for two time series $X(t)$ and $Y(t)$ with zero mean values as

$$r = \frac{\langle X(t)Y(t)\rangle}{\sqrt{\langle X^2(t)\rangle}\sqrt{\langle Y^2(t)\rangle}}, \tag{14.1}$$

where $\langle\cdot\rangle$ denotes the averages taken over the entire record, $t = 1, \dots, T$, which are an estimate of the expectation values of the corresponding stochastically fluctuating quantities. This coefficient can be easily generalised to the linear cross-correlation function by considering a fixed time lag τ between the compared time series as

$$C_{XY}(\tau) = \frac{\langle X(t)Y(t+\tau)\rangle}{\sqrt{\langle X^2(t)\rangle}\sqrt{\langle Y^2(t)\rangle}} \quad (C_{XY}(0) = r). \tag{14.2}$$

In the case of records with a finite length T, Pearson's linear cross-correlation is a bias-free estimate if both time series are normally distributed. In the asymptotic case of very long records ($T \to \infty$), this prerequisite is practically negligible as the values of the linear correlation function converge to those given by rank-order correlations [14.69, 14.70] described next.

14.3.2 Non-parametric (Rank-Order) Correlations

Non-parametric correlation functions do not depend on the particular values of the considered observables, but only on their ranks. Consequently, the original distribution function of the data does not influence the values of such functions, which is an important advantage compared to Pearson's linear correlation function.

The most prominent rank-order correlation coefficient is Spearman's rho [14.71] which may be computed as

$$\rho_S = 1 - \frac{6\sum_{t=1}^{T}(X_r(t) - Y_r(t))^2}{T(T^2-1)}, \tag{14.3}$$

where $X_r(t)$ and $Y_r(t)$ give the ranks of the values $X(t)$ and $Y(t)$ in the respective time series.

As an alternative, one may consider the concordance between two rank-ordered time series as a non-parametric correlation coefficient. For this purpose, one compares all pairs of observations at two different times t and t'. A pair is called concordant (discordant) if $(Y_r(t') - Y_r(t))/(X_r(t') - X_r(t)) > 0\ (< 0)$. Pairs with either $X_r(t) = X_r(t')$ (extra-y pair) or $Y_r(t) = Y_r(t')$ (extra-x pair) have to be considered separately, whereas those with both $X_r(t) = X_r(t')$ and $Y_r(t) = Y_r(t')$ are not taken into account.[2] As a measure of concordance, Kendall's tau [14.72] is defined as

$$\tau_K = \frac{N_c - N_d}{\sqrt{N_c + N_d + N_x}\sqrt{N_c + N_d + N_y}}, \tag{14.4}$$

where N_c, N_d, N_x, and N_y are the number of concordant, discordant, extra-x, and extra-y pairs, respectively. The computation of Kendall's tau requires relatively large computational efforts due to the calculation time of $\mathcal{O}(n^2)$; however, during the last years, fast algorithms have been developed for which the calculation time is dramatically reduced up to $\mathcal{O}(n \log n)$ [14.73].

Lagging both time series X and Y with respect to each other by a variable number τ of time steps, ρ_S and τ_K can be used to define appropriate non-parametric correlation functions $R_{XY}(\tau)$ and $T_{XY}(\tau)$, respectively.

14.3.3 (Cross-) Mutual Information

Mutual information [14.74] is a measure of statistical dependence that can be considered as a nonlinear generalisation of the linear covariance function. To compute this quantity, one has to consider an appropriate discretisation of the time series X and Y into symbols $\{x_i\}$ and $\{y_j\}$, respectively. The probabilities of these symbols, p_i and p_j, as well as the joint probabilities $p_{ij}(\tau)$ that x_i and y_j occur simultaneously if the time series Y is lagged by τ time steps are empirically approximated by their frequencies of occurrence in the observational records. With these probabilities, the (cross-) mutual information is defined as

$$I_{XY}(\tau) = \sum_{ij} p_{ij}(\tau) \log \frac{p_{ij}(\tau)}{p_i p_j}. \tag{14.5}$$

[2] As the rank ordering conserves all order relationships between different observations in the original time series, it is appropriate to directly use X and Y instead of X_r and Y_r.

$I_{XY}(\tau)$ is by definition restricted to non-negative values. To standardise the cross-mutual information to values in the unit interval, different approaches have been suggested (for an overview, see [14.75]). We prefer to normalise $I_{XY}(\tau)$ in a similar way as the linear covariance function, i.e.,

$$I^*_{XY}(\tau) = \frac{I_{XY}(\tau)}{\sqrt{I_X(0)}\sqrt{I_Y(0)}}. \tag{14.6}$$

Here I_X and I_Y are the univariate mutual information of the time series X and Y, respectively. The concept of mutual information may be further generalised in both the univariate and the bivariate case by replacing the Shannon entropy terms in (14.5) by a generalised (Rényi) entropy of an appropriate order [14.76]. However, the resulting generalised (cross-) mutual information functions may be more difficult to interpret and have a smaller variance [14.39].

One has to mention the conceptual difference between the (parametric or non-parametric) correlation functions on the one hand and mutual information on the other hand: Measures belonging to the first group are able to detect correlations as well as anti-correlations, which are both certain kinds of statistical dependence. Consequently, their range is usually normalised to $[-1, 1]$. In contrast to this, the mutual information detects only whether there is *any* kind of statistical dependence, which means that in its normalised version, values range between 0 (no dependence) and 1 (identical time series).

14.3.4 Recurrence Quantification Analysis (RQA)

A complementary approach to access nonlinear characteristics from given observational data is based on recurrence plots [14.77]. Originally, the concept of recurrence plots has been designed as a tool to visualise the correlation pattern within a single time series by comparing the observed values at *any* observation time t with those at *any* other time t' $(t, t' = 1, \ldots, T)$. A simple graphical representation is obtained by comparing the difference to a prescribed threshold value ϵ and encoding this difference in dependence on both times t and t' according to the order relation with respect to ϵ. Hence, recurrence plots preserve the essential topological information contained in the considered time series [14.78].

Mathematically, the corresponding recurrence matrix of a time series X is formulated using the Heavyside function as

$$R_X(t, t') = \Theta(\epsilon - ||X(t) - X(t')||). \tag{14.7}$$

(note that this matrix depends on ϵ which has to be chosen appropriately, i.e. in a way that for a total number of T observations, the normalised recurrence rate $RR = \frac{1}{T^2}\sum_{t,t'=1}^{T} R_X(t, t')$ is sufficiently different from both 0 and 1).

Recurrence plots may be used to compute several nonlinear characteristics which are based on statistics of either the diagonal or the horizontal structures. This approach includes measures for the laminarity, determinism, and predictability of the observed data and is called recurrence quantification analysis (RQA) [14.79–14.81]. In addition, the consideration of recurrence plots allows to estimate several dynamic invariants [14.82], including the second-order Renyi entropy K_2 and the mutual information already discussed above.

To define a measure for nonlinear correlations within univariate time series, one may consider the recurrence rate for a fixed time lag $\tau = t' - t$ and define the τ-recurrence rate (or generalised (auto-) correlation function [14.39]) as

$$P_X(\tau) = \begin{cases} \frac{1}{T-\tau} \sum_{t=1}^{T-\tau} R_X(t, t+\tau) & \text{for} \quad \tau \geq 0 \\ \frac{1}{T+\tau} \sum_{t=\tau}^{T} R_X(t+\tau, t) & \text{for} \quad \tau < 0 \end{cases}, \tag{14.8}$$

i.e. the τ-recurrence rate is the probability of two observations lagged by a time τ to be less separated than a prescribed threshold value ϵ.

For comparing two time series X and Y which represent the same physical quantity (for example, two runoff time series) and have the same basic statistical properties (in particular, the same length T and the same means and variances), one defines the cross-recurrence matrix [14.83]

$$CR_{XY}(t, t') = \Theta(\epsilon - ||X(t) - Y(t')||). \tag{14.9}$$

and the cross- τ-recurrence rate (or generalised cross-correlation function) $CP_{XY}(\tau)$ by replacing the univariate recurrence matrix $R_X(t, t')$ by the cross-recurrence matrix $CR_{XY}(t, t')$ in (14.8). Note that this approach should not be applied if X and Y represent different physical quantities [14.84].

(Cross-) τ-recurrence rates are more directly related to the standard (cross-) correlation functions than, for example, mutual information. However, $CP_{XY}(\tau)$ and related measures are not able to detect anti-correlations, but are restricted to the range between 0 (no correlation) and 1 (perfect correlation). One possible way to overcome this conceptual difficulty would be to additionally compute the cross-recurrence matrix between the time series X and $-Y$ and use it to define a *symmetric* (cross-) recurrence function by setting

$$CP^*_{XY}(\tau) = CP_{XY}(\tau) - CP_{X,-Y}(\tau), \tag{14.10}$$

i.e. computing the difference between the generalised correlation function $CP_{XY}(\tau)$ and the generalised *anticorrelation* function $CP_{X,-Y}(\tau)$.

14.4 Mutual Correlations Between Different Stations

In the following, we investigate the spatial structure of mutual interdependencies between the runoffs obtained at different gauges in the Upper Main catchment. As we consider time intervals of only 20–30 years, long-term trends associated with global climate change are likely to have only minor importance (i.e. are sufficiently small compared to the natural variability) and influence all records in the same way such that we assume that the necessary stationarity conditions are approximately fulfilled. To approach comparable time series, all observational records are standardised by removing the annual mean value of the respective observable for any calendar day (obtained from the entire record) and transforming the residual to zero mean and unit variance.

At first, we consider the daily records observed at all 13 gauges. Due to the missing data at the gauge at Steinberg in 1967 and 1968, we focus on the time interval between 1 January, 1969 and 31 December, 1988, where the data set is homogeneous and has no gaps. This restriction allows a better comparability of the different time series, in particular, with respect to effects of the finite number of observations (in the considered case, $T = 7\,305$ days).

In Fig. 14.2, the different correlation coefficients (Pearson's r, Spearman's rho ρ_S, and Kendall's tau τ_K) described in the previous section are compared. Whereas the latter two are non-parametric and depend only on the rank-order of the data, the standard Pearson correlation is unbiased only for Gaussian data. In contrast to this, runoff time series typically show a clearly skewed distribution, which is often approximated by log-normal, gamma, or other suitable model functions. To test the actual relevance of the probability distribution function for the resulting Pearson correlation with $T = 7\,305$ observations, we have additionally logarithmised all runoff time series to approach more symmetric distribution functions and calculated r again for the transformed data.

Comparing the correlation, mutual information, and cross-recurrence coefficients, one observes that although there are shifts in the absolute values between the respective measures, the general correlation pattern is conserved. In particular, all correlations are clearly significant. As the four considered correlation coefficients show the same behaviour, the actual distribution function of the data is only of minor importance with respect to the entire time series. However, it may become relevant if only short time intervals are considered separately, which will be discussed later.

In order to evaluate whether the above results depend qualitatively on the temporal resolution of the data, we have computed all correlation coefficients and related measures for the second runoff data set, which covers the time interval between 1989 and 1998 with a temporal resolution of 1 h. Except for the gauges at Schwürbitz (5) and Steinberg (9) where data are missing, the time series directly continue the daily records studied above. According to the results shown in Fig. 14.3, the general pattern of spatial correlations is equivalent to that of the daily data covering the preceding 20 years. This indicates that there are no significant changes of the runoff dynamics (for example, due to changing climatic conditions or anthropogenic modifications in the catchment). Hence, the temporal resolution of the runoff time series

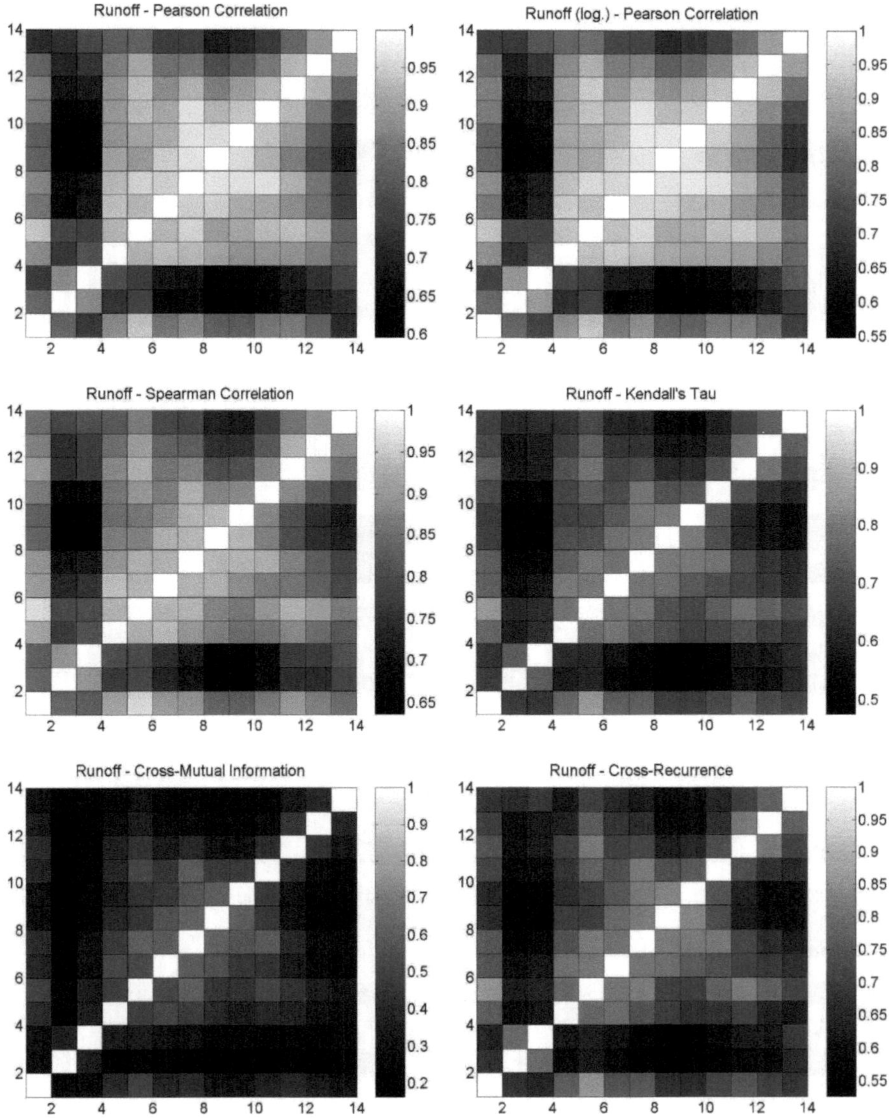

Fig. 14.2 Correlation coefficients and related measures between all daily runoff time series computed for the time interval 1969–1988. *First row*: Pearson's linear correlation coefficient r for the original (*left*) and log-transformed (*right*) data. *Second row*: Non-parametric correlation coefficients Spearman's rho ρ_S (*left*) and Kendall's Tau τ_K (*right*). *Third row*: Cross-mutual information $I^*_{XY}(0)$ (*left*) and cross-recurrence coefficient $CP_{XY}(0)$ (computed with $\epsilon = 0.25$). *Axis* labels correspond to the station numbers listed above

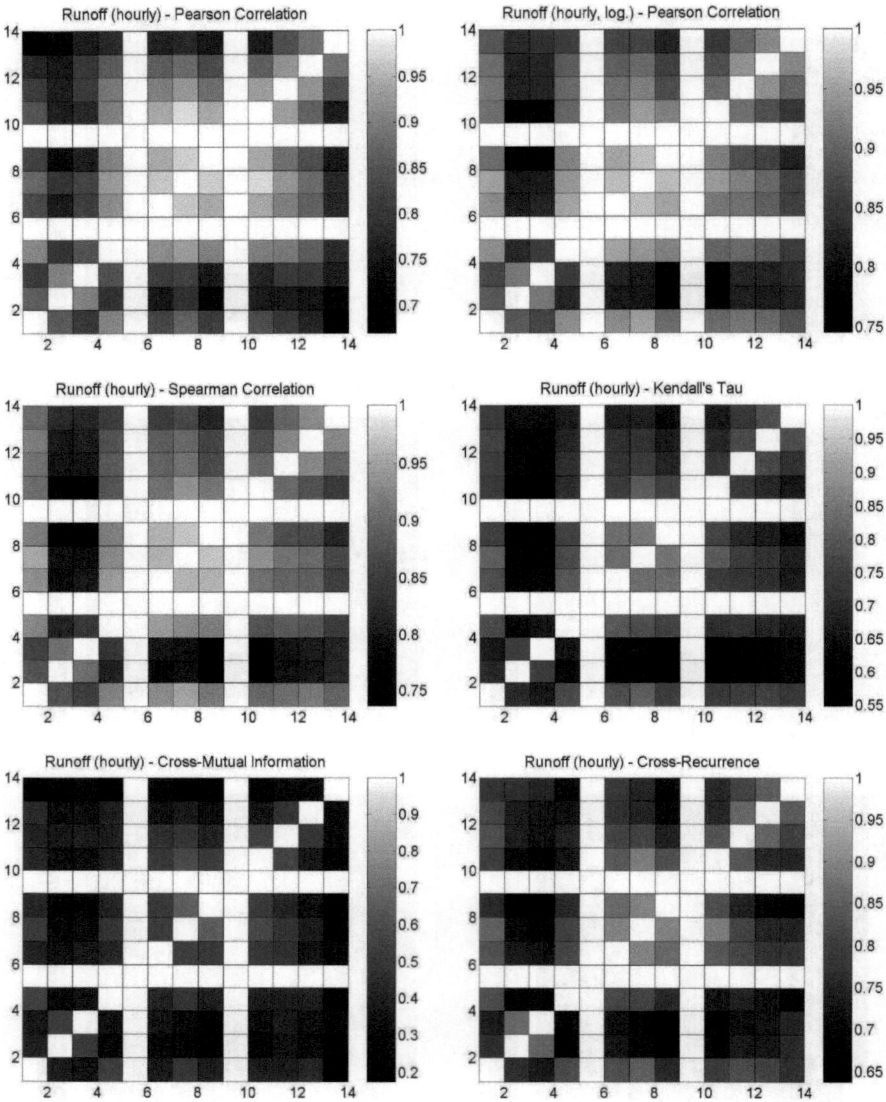

Fig. 14.3 Correlation coefficients and related measures as in Fig. 14.2 for all runoff time series with a temporal resolution of 1 h (1989–1998). The missing records from the gauges no. 5 and 9 are represented by white boxes along *horizontal* and *vertical lines*

is of minor importance for the qualitative detection of spatial runoff correlations on longer time scales.

Comparing the different measures systematically, some features are found which call for further discussion: First, one observes that although the total number of observations ($T = 87\,648$) is significantly higher than in the case of the daily data, the correlation coefficients r between the original time series on the one hand and their values for the log-transformed data as well as Spearman's rho ρ_S and Kendall's tau τ_K on the other hand still show remarkable deviations. This observation indicates that although all time series are rather long, the distribution of the data still influences the linear correlation analysis. In particular, the distribution of the original runoffs shows large algebraic tails, which are significantly reduced in the logarithmised time series.

Second, cross-mutual information and cross-recurrence coefficients show features which slightly differ from those of the correlation functions. In the case of the cross-mutual information, this is related to the fact that correlations and statistical dependences are two clearly different concepts. In addition, for the computed values of the mutual information, the choice of the partitions may influence the results. While we have restricted ourselves to partitions into equally sized classes, a data-adaptive partitioning is favourable [14.74]. Considering the cross-recurrence coefficient, we have chosen the same value of ϵ for all computations, whereas a fixation of the recurrence rate RR introduced in Sect. 14.3 instead would allow a better comparability of all records.

Power-law tails of runoff time series are accompanied by an omnipresent long-term memory. From this point of view, one possible explanation for the different behaviour of the considered correlation functions (in particular, the larger sensitivity of the Pearson correlation with respect to the distribution of the data compared to the daily records) is that the auto-correlations within hydrological records usually decay on a scale of several days. This scale is more often contained in the 20-year daily record than in the better resolved hourly record which covers only 10 years. The importance of long-term auto-correlations is also a possible reason for the strong similarity of the results obtained for daily and hourly time series. To further investigate their importance for the spatial dependencies, we have analysed whether the interrelationships between different records depend on the considered time scale. As complex signals from natural sources can be decomposed into components with different frequencies, the amplitudes of which may change with time, wavelet analysis [14.85] is an appropriate tool for this purpose. We have explicitly studied the variations on scales of 20 days, which roughly corresponds to the (theoretical) prediction limit of atmospheric processes [14.86–14.88], and 10 days, which is an upper bound for the prediction limit of current weather-forecasting algorithms.

The resulting time series (which can be considered as being smoothed over the respective temporal scale) show correlations which are qualitatively identical to those of the original time series. However, there are indeed small differences in dependence on the considered time scale. For example, Fig. 14.4 shows that the rather low correlations between the gauge at Bayreuth (13) and the other records mainly manifest on smaller time scales, whereas they become somewhat stronger when considering the 20-days variability. This finding indicates that not the distribution is mainly responsible for the corresponding pattern, but rather the long-term memory of the respective record which may depend on the particular location.

According to the rather slow evolution of hydrological processes, we have further examined whether there is a significant delay between the dynamics recorded at the

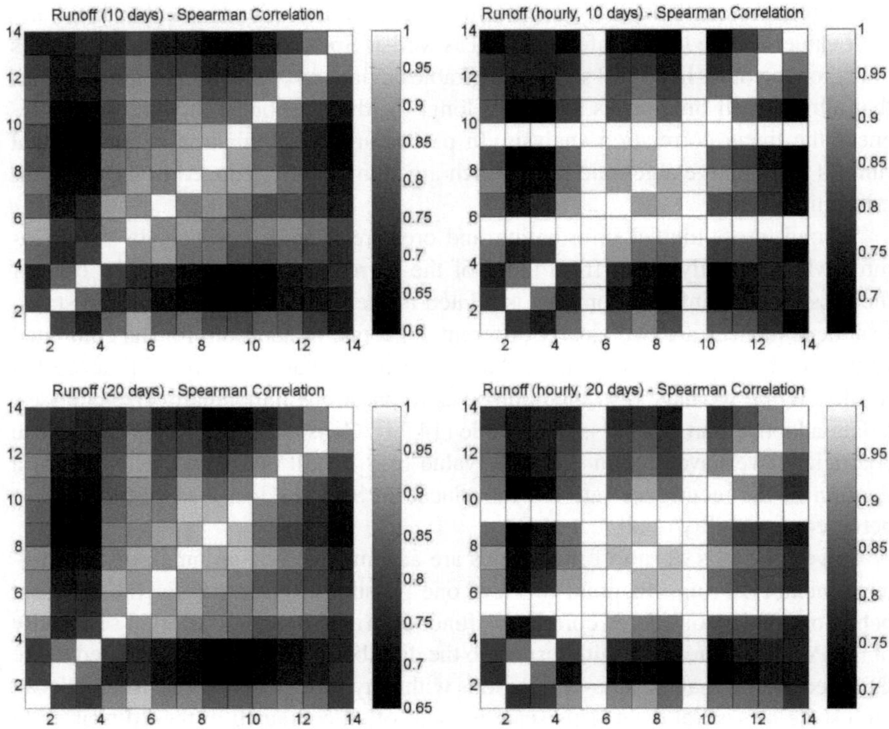

Fig. 14.4 Spearman correlations between the daily (*left panels*) and hourly (*right panels*) runoff records from the Upper Main catchment filtered on a scale of 10 (*upper row*) and 20 days (*lower row*), respectively. The corresponding Pearson correlations give completely equivalent patterns

different gauges in the catchment. As a reference for the investigation of spatial correlations, we have chosen the gauge at Kemmern (1) which is the final gauge downstream in the Upper Main catchment area and integrates the input recorded at all other gauges used in this study. If the spatial distance is larger than the flow velocity divided by the sampling interval of the time series, maximum correlations are recorded at appropriate time lags τ. Whereas in the case of daily records, a constant time lag of 1 day is found for the maximum correlations of different gauges with the Kemmern record, the hourly data resolve different time lags between rougly 1 and 2 days (see Fig. 14.5). These delays show a close, intuitive relationship to the distance between the considered gauges. To evaluate whether these correlations are stable in time, we have computed the Pearson and Spearman correlations, r and ρ_S, as well as the linear cross-correlation function $C_{XY}(\tau)$ for sliding windows of 100 days. One particular example is shown in Fig. 14.6. One clearly observes that the correlations are not stationary. However, there is no significant relationship between the average runoff and the temporal extension of cross-correlations. We will come back to this result later.

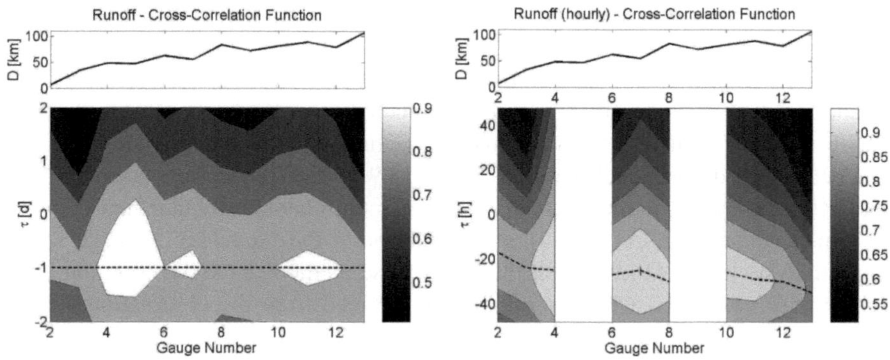

Fig. 14.5 Iso-lines of the linear cross-correlation function between the daily (*left*) and hourly (*right*) river runoff records from the gauges 2–13 and the corresponding record obtained from the gauge at Kemmern (1). The *dashed lines* indicate the delays for which the highest correlations have been found. In addition, the distance of the respective locations (in river kilometers) to Kemmern is displayed in the *upper panels*

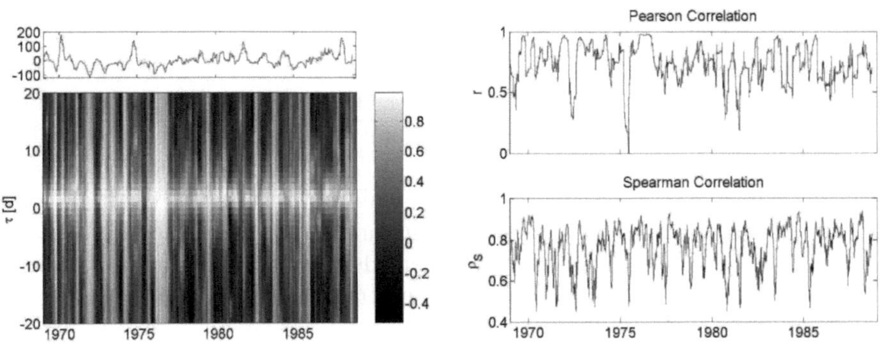

Fig. 14.6 *Left figure*: Runoff, integrated over 100 days (*upper panel*) and cross-correlation function for the daily records from Kemmern (1) and Bayreuth (13) associated with the corresponding time intervals. *Right figure*: Pearson and Spearman correlation coefficients for the corresponding data subsets as a function of time

14.5 Ensemble Correlations

Due to the complex interrelationships between precipitation and runoff in a catchment, there is no direct way to translate meteorological information into a prediction of hydrological conditions, which is a necessary task for flood forecasting [14.89]. As a possible alternative, one may exclusively consider the runoff dynamics from different gauges. This idea is closely related to the question of how to quantify ensemble correlations, i.e., the general state of all gauges in a small-scale catchment. In this section, we describe a potential framework for solving this task and discuss its eventual predictive skills.

The characterisation of ensemble correlations by a single statistical parameter requires an appropriate statistical decomposition of the corresponding multivariate time series. In principle, this decomposition can be performed by a variety of different approaches, including purely linear methods like Karhunen–Loéve decomposition (KLD) (which is often referred to as principal component analysis (PCA) [14.90]), multi-dimensional scaling (MDS) [14.91], or, referring to a separate consideration of patterns in the frequency domain, multi-channel singular spectrum (or system) analysis (MSSA) [14.92], a combination of the "standard" singular spectrum analysis (SSA) [14.93] with PCA. All these methods have the common concept that some matrix, which is suitably constructed from the observational data, is subjected to a singular value decomposition (SVD), i.e. it is decomposed into its eigenvalues and the corresponding eigenvectors. In the case of KLD, one makes use of the correlation (or scatter) matrix $S = A^{\mathrm{T}} A$ of the observed data set A (the component time series of which have to be transformed to zero means). For MDS, a transformed matrix of the squared linear inter-point distances is used, whereas MSSA is based on a Toeplitz-type lag-covariance matrix obtained from every univariate components time series.

Whereas the SVD step of all these methods may be easily and computationally efficiently performed, there are also different nonlinear generalisations. One possible way to obtain such generalisations is replacing the Euclidean metric by one defined by the local neighbourhood, for example, in terms of isometric feature mapping (ISOMAP) [14.94] or locally linear embedding (LLE) [14.95]. An alternative is realising the decomposition in terms of neural networks, including methods like nonlinear principal component analysis (NLPCA) [14.96] or independent component analysis (ICA) [14.97]. However, these nonlinear variants require a much larger amount of data for computation, while the linear methods can be applied to rather short time series as well. In addition, the methods based on neural networks do not lead to well-defined component variances such that the approach described in the following is not applicable in such cases.

14.5.1 KLD-Based Dimension Estimates

For a temporally localised characterisation of the components of multivariate data sets, we propose to focus on the linear methods only. For simplicity, we consider here KLD as a particular example, as the components derived by this method have probably the most intuitive interpretation. Again, we standardise all time series to zero means and unit variances, which has the advantage that different records become better comparable. Moreover, the entries of the symmetric equal-time correlation matrix S are in this case given by the normalised Pearson correlation coefficients r. In the following, we consider the non-negative eigenvalues σ_i^2 of S, which are rather easily computed. Without loss of generalisation, we assume that these eigenvalues are decreasingly ordered and normalised as $\sum_{i=1}^{N} \sigma_i^2 = 1$ (where N is the total number of component time series), i.e. the total variance of the multivariate data set is rescaled to 1. For the characterisation of spatio-temporal chaos, Zoldi and Greenside have introduced the concept of KLD dimension [14.98].[3] Following a recently proposed redefinition [14.63], we consider this measure as the minimum number of eigenvalues required to capture some predefined fraction f of the total variance, i.e.

$$D_{\mathrm{KLD}}(f) = \min \left\{ p \in \{0, \ldots, N\} \,\middle|\, \sum_{i=1}^{p} \sigma_i^2 \geq f \right\}. \tag{14.11}$$

As for spatio-temporal data obtained from certain model systems, D_{KLD} has been found to increase linearly with increasing N, it is appropriate to study a normalised measure, the KLD dimension density

$$\delta_{\mathrm{KLD}} = D_{\mathrm{KLD}}/N, \tag{14.12}$$

instead of D_{KLD} itself. The KLD dimension has mainly been used to characterise the dynamics of spatially extended model systems in extensive [14.98] and spiral-defect chaos [14.99] as well as that of reaction–diffusion systems [14.100]. Recently, an application to spatio-temporal observations from electrochemical turbulence was reported [14.101].

In applications to general multivariate data sets, the concept of KLD dimension has some serious disadvantages. In particular, its range is restricted to only N values, which makes it hard to consider δ_{KLD} for detecting small changes in the correlation structure of the data. In addition, for a finite length T of the considered time series and, in general, due to its dependence on the variance fraction f, the value of δ_{KLD} may deviate from the actual dimension density of a given system, which has been demonstrated for reasonable model systems [14.64, 14.65]. Thus, a strictly quantitative interpretation of δ_{KLD} is problematic, and we propose to consider this measure only as a qualitative dimension estimate.

To overcome the discrete nature of the KLD dimension density, we consider the scaling of the residual variances $V_r(p/N) = 1 - \sum_{i=1}^{p} \sigma_i^2$ with p. For rather different kinds of model systems as well as geoscientific observational data (see Fig. 14.5), this scaling is dominated by an exponential decay [14.63, 14.65][4]

$$V_r(p/N) \sim 10^{\frac{p}{N}/\delta_{\mathrm{LVD}}}. \tag{14.13}$$

In the case of a small number of component time series N, it may be problematic to fit a corresponding model to the remaining variances defined for only N different

3 Here, the term dimension is not used in the context of fractal theory, but rather topologically as the dimension of a linear subspace containing the required fraction of dynamical "information".

4 Instead of using powers of e, we define the exponential scaling law with respect to powers of 10. This convention has the advantage that 90% of variance are typically considered as a reference point for fixing the number of significant linear independent components, see [14.102] or [14.98]. In our formalism, this corresponds to a remaining variance of $V_r(p/N) = 0.1$, which is approached for $p/N = \delta_{\mathrm{LVD}}$.

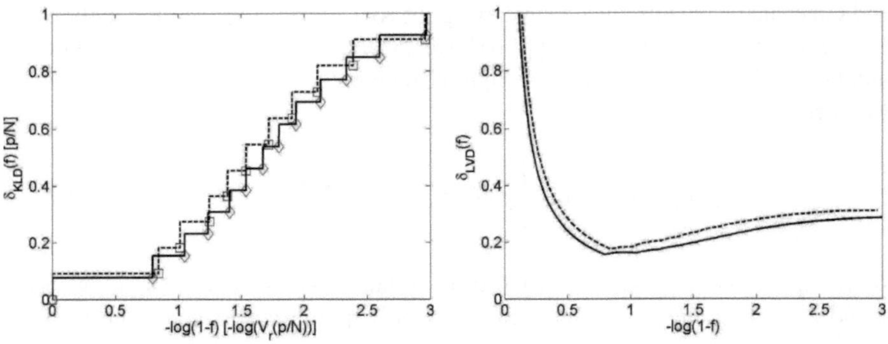

Fig. 14.7 KLD dimension density $\delta_{\mathrm{KLD}}(f)$ (*left panel*) and LVD dimension density $\delta_{\mathrm{LVD}}(f)$ (*right panel*) as a function of the cutoff variance fraction f for the daily (*solid lines*) and hourly runoff (*dashed lines*) data. In the *left panel*, the logarithm of the residual variances $V_r(p/N)$ is additionally shown as a function of the maximum component order p (daily data: ◇, hourly data: □). For the daily time series, all values have been computed for the time interval from 1 January, 1969 to 31 December, 1988. The hourly records cover the period between 1 January, 1989, and 31 December, 1998. Note that the number of values of both data sets must not be compared directly, as the number of records N differs between both data sets which influences the computed values

values of the maximum component order $p < N$. Therefore, it is more appropriate to consider δ_{KLD} which has (as a function of f) a scaling which is closely related to that of the residual variances [14.64] (see Fig. 14.7), i.e.

$$\delta_{\mathrm{KLD}}(\phi) = -\delta_{\mathrm{LVD}}(f)\log(1-\phi) \quad \text{for} \quad \phi \in [0, f]. \tag{14.14}$$

The coefficient $\delta_{\mathrm{LVD}}(f)$, which is referred to as the linear variance decay (LVD) dimension density, thus describes the exponential decay scale of the residual variance fraction with the number of components and is continuously defined for all $f \in]0,1[$. Its efficient computation is described in some more detail in [14.64, 14.65]. For palaeoclimatic multi-proxy data, it has been demonstrated that δ_{LVD} allows to detect even small changes in the correlation structure of short multivariate time series. However, δ_{LVD} can again only be considered as a qualitative indicator as its value depends on the particular choice of f and the length of the considered time intervals, M, in a system-dependent way [14.63–14.65].

14.5.2 Case Study I: The Christmas 1967 Flood

As an illustrative example for the potential application of KLD-based dimension density estimates in hydrology, we have calculated both δ_{KLD} and δ_{LVD} for short time intervals before, during, and after the major Main flood on christmas 1967. As for this time interval, no data from the gauge at Steinberg (9) are available, only 12 gauges are considered.

Figure 14.8 shows the results for both dimension estimates calculated for time intervals of $M = 14$ days. On the one hand, this window length has been chosen

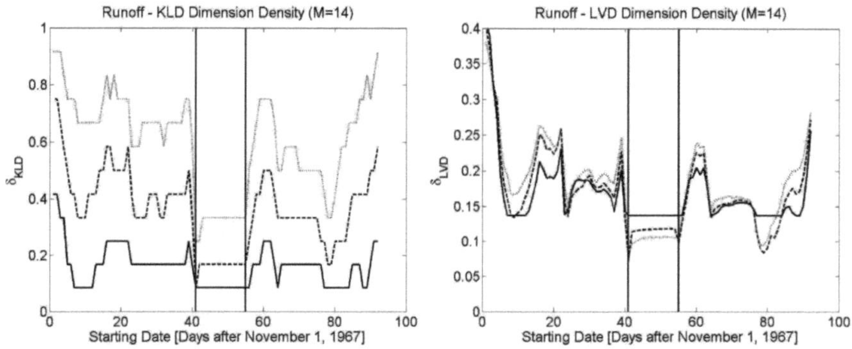

Fig. 14.8 KLD (*left panel*) and LVD dimension density (*right panel*) for time intervals of $M = 14$ days around the Main flood at 25 December, 1967. *Different line types* correspond to different cutoff levels $f = 0.9$ (*solid*), 0.99 (*dashed*), and 0.999 (*dotted*). The *vertical lines* indicate the time intervals where data from 25 December, 1967 are explicitly taken into account

such that it is not much larger than the typical duration of a flood event and of the order of the typical time scale of long-term auto-correlations. On the other hand, the corresponding time intervals must not include too few observations in order to give statistically significant results. Note that, due to the consideration of several simultaneous records, the typical number of observation points in time which is necessary for computing meaningful values is smaller than in the case of bivariate correlations and related measures.

Considering Fig. 14.8 in some detail, one observes that the values of δ_{LVD} are much more stable if the considered fraction of explained variance, f, is varied. In addition, it is found that for the runoff records, both dimension estimates show a significant decay if data from 25 December, 1967 are explicitly considered. This fact is easily understood: at this day, the recorded runoffs have been extraordinarily high at all gauges compared to the usual values such that all time series show a very similar behaviour, which increases the correlations between the different records. As a consequence, the number of components necessary for explaining a given fraction f of total variance decreases, which is recorded by the considered dimension estimates.

We have already briefly addressed the issue of statistical significance. As Fig. 14.8 illustrates, the qualitative features of the variability of δ_{KLD} and δ_{LVD} around the considered flood event do not depend on the particular choice of f, which gives a first indication of the robustness of our approach. As a second argument, Fig. 14.9 shows that the results discussed above do not change qualitatively when the width M of the considered time window is changed over a reasonable range. However, the figure also illustrates that the considered data window must not be chosen too large, as otherwise the contrast between the values of δ_{LVD} during normal and exceptional conditions decreases strongly.

To further prove the significance of the temporal variation of δ_{LVD}, the uncertainty of this measure may be explicitly estimated. For this purpose, δ_{LVD} was computed for ensembles of slightly perturbed data constructed from the original data set

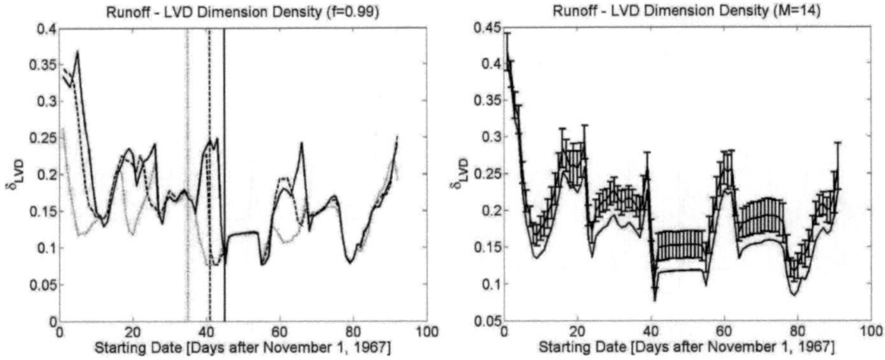

Fig. 14.9 LVD dimension density δ_{LVD} of the daily river runoff records for the period including the Main flood at 25 December, 1967. *Left panel*: Value of δ_{LVD} for different numbers of data considered for computation: $M = 10$ (*solid*), 14 (*dashed*), and 20 (*dotted*). The *vertical lines* indicate the start of the respective windows where data from December 25 contribute. *Right panel*: 95% confidence levels for δ_{LVD} with respect to a window width of $M = 14$ days. All computations have been carried out with $f = 0.99$

by substituting the data of a single, but randomly chosen time slice by a Gaussian random vector. To study the influence of such distortions on δ_{LVD} systematically, a suitably large ensemble of 1 000 of these perturbed data has been considered. In Fig. 14.9, the corresponding results are shown for windows containing $M = 14$ daily observations. While the expectation of δ_{LVD} is shifted towards higher values with respect to the original data (i.e. the surrogate data are "more stochastic" by construction), the qualitative behaviour remains unchanged. The corresponding confidence levels indicate clearly that the variations in the dimension are significant even for such small numbers of data points, which demonstrates the qualitative robustness of the considered approach.

14.5.3 Non-parametric Ensemble Correlations

As it can be inferred from Fig. 14.8, the dimension estimates of the runoff time series show remarkable minima in the case of an extreme flood. However, not every pronounced minimum corresponds to a dangerous flood event as well. Considering the strength of the decay of δ_{KLD} and δ_{LVD} (i.e. the absolute values, the difference before and after the occurrence of the "event" and its duration), there is no significant difference found compared to other situations in which minimal values occur without a severe flood. As in addition, the decay starts rather shortly before the actual flood occurs, the predictive skills of the considered dimension estimates with respect to flood forecasting are limited. However, their behaviour gives complementary information about the evolution of the hydrological state of the catchment.

To evaluate the possibility of an improvement of our method, we recall that the Pearson correlations (forming the considered correlation matrix S) depend on the distribution of observed values. In particular, in the case of short time series with strongly non-Gaussian values (as in the present analysis of ensemble correlations during a flood), the corresponding values may be unreliable. However, the use of Spearman instead of Pearson correlations offers a potential way to overcome this conceptual problem as Spearman's rho ρ_S does *not* depend on the particular distribution of the observed data. In the following, we consider NS with $\mathrm{NS}_{ij} = \rho_S(X_i, X_j)$ to be the symmetric matrix of Spearman correlations. Following the KLD dimension approach, we define non-parametric KLD and LVD dimension densities δ_{NKLD} and δ_{NLVD} by replacing S by NS and proceed further as described above.

Figure 14.10 shows that for daily runoff time series, the application of the non-parametric dimension estimates gives a signature which is comparable to that of the standard approach, but with a better statistical foundation. In particular, the decay of δ_{NLVD} during the considered flood event is of a similar significance as that of δ_{LVD}. Hence, if for a suitably chosen set of runoff records the value of δ_{NLVD} falls below a certain level (which depends on the particular choice of f), this may serve as an additional indicator for floods. The reason for this is that in advance of a flood, the river discharges increase rather continuously at all relevant gauges, which means that the corresponding rank time series become very similar for all stations. In this case, the rank-order correlation matrix NS has only entries which are close to 1, which leads to a highly dominating first eigenvalue and a corresponding very low value of the dimension densities. As a limiting case corresponding to rather extreme environmental conditions, the covariance matrix degenerates, i.e. all elements of NS become 1, which means that there is only one non-zero eigenvalue, and the corresponding dimension density thus approaches zero.

Similar to the mutual correlations, we would like to briefly address the issue of long-term stability of the ensemble correlations. Figure 14.11 shows the results for the daily records, covering a time interval of 20 years. As one can recognise, there are time intervals with low dimensions, i.e. large ensemble correlations. However, a comparison with Fig. 14.6 shows that these intervals do not directly correlate

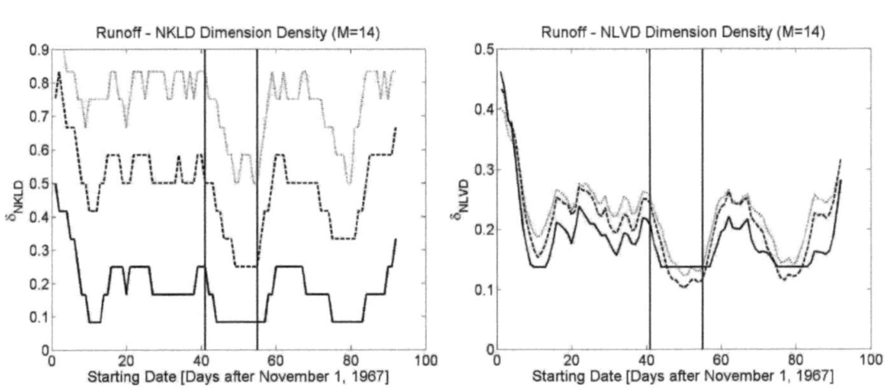

Fig. 14.10 Non-parametric KLD (*left panel*) and LVD dimension density (*right panel*) for time intervals of 14 days around the Main flood at 25 December, 1967. *Different line types* correspond to different cutoff levels $f = 0.9$ (*solid*), 0.99 (*dashed*), and 0.999 (*dotted*). The *vertical lines* indicate the time intervals where data from 25 December, 1967 are explicitly taken into account

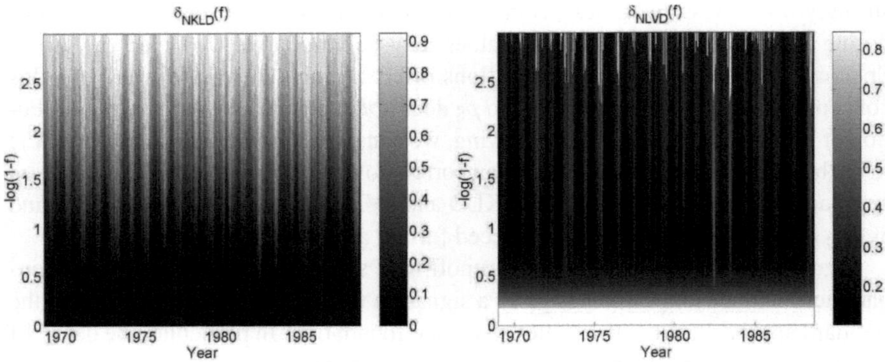

Fig. 14.11 Non-parametric KLD (*left*) and LVD (*right*) dimension densities for the daily runoff records computed for windows of $M = 100$ days. In the *right panel, white areas* correspond to regions where δ_{LVD} and δ_{NLVD} could not be computed or gave values > 1. The standard dimension estimates δ_{KLD} and δ_{LVD} gave qualitatively equivalent results

with periods of long-term cross-correlations between particularly chosen records or with intervals including extreme low or high runoffs. This observation does not contradict the existence of corresponding interdependencies on short time scales as it is influenced by the choice of the window width M which is largely compared to the typical duration of hydrological "events". The results remain qualitatively unchanged if the wavelet coefficients on a scale of 10 or 20 days are considered instead of the original record, i.e. do not depend on the resolution of the considered data. In general, one has to conclude that the general temporal pattern of ensemble correlations does not change very much over the considered time interval of 20 years.

14.5.4 Case Study II: The January 1995 Flood

In order to improve the performance and potential predictive skills of the multivariate dimension estimates, it is recommended to study time series with a significantly higher temporal resolution. As a possible modification, one may additionally consider an adjustment of the time scale of all records according to their respective delays (see Sect. 14.4). The latter transformation assures that the signatures of events which simultaneously affect a larger scale (for example, spatially extended strong precipitation) are transferred to the time at which they act cooperatively at the final gauge of the catchment or another especially exposed location.

As an example, we have computed δ_{NLVD} for 1-day ($M = 24\,\text{h}$) sliding time windows of hourly records around another severe flood at 27 January, 1995, for which a maximum runoff of $568\,\text{m}^3/\text{s}$ has been recorded at Kemmern. The results shown in Fig. 14.12 demonstrate that a consideration of the non-adjusted time series has the advantage that the fast decay of the dimension estimates occurs with a certain

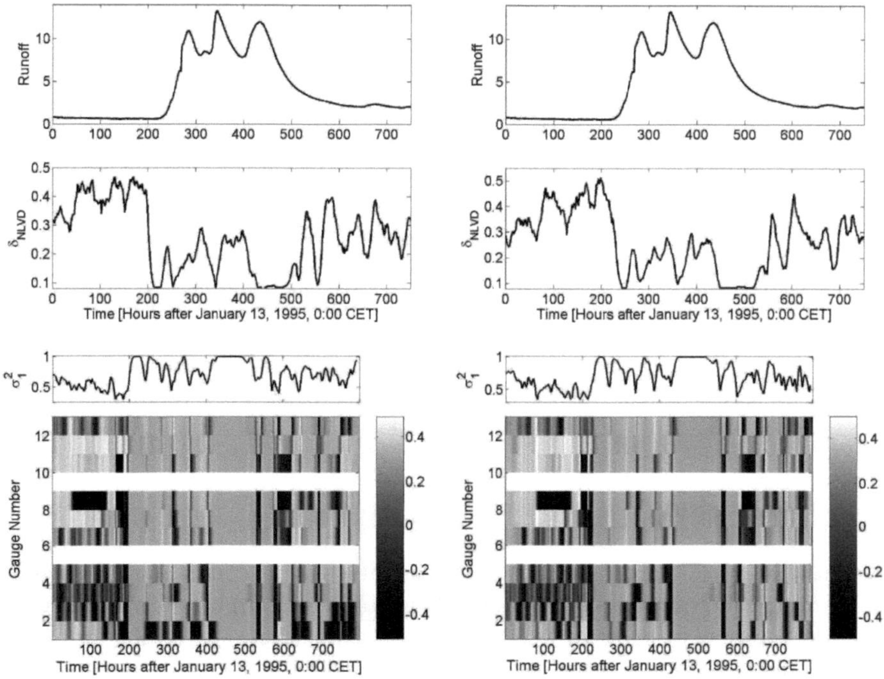

Fig. 14.12 *Upper figures*: Hourly runoff for the gauge at Kemmern (1) and non-parametric LVD dimension density δ_{NLVD} for windows of $M = 24\,\mathrm{h}$. *Lower figures*: Leading eigenvalue σ_1^2 and corresponding eigenvector of the rank-order correlation matrix NS. All values have been computed without (*left panels*) and with (*right panels*) adjusting the records according to the optimum delay to the observations at Kemmern

warning time for the behaviour at locations influenced later by the flood. The corresponding critical level of δ_{NLVD} is reached relatively fast, because it is computed on the basis of the current-time runoffs which are simultaneously recorded at all gauges. In contrast to this, for the adjusted records, the critical values of the dimension estimates are reached slower and at a later time, however, there is a stronger and direct temporal relationship between the changes in δ_{NLVD} and the runoff at the reference station. For prediction purposes, it might be useful to use a combined analysis of non-adjusted and adjusted time series to estimate the severity of a flood in advance.

To illustrate the meaning of low-dimension densities, Fig. 14.12 additionally shows the temporal evolution of the major eigenvalue σ_1^2 of the Spearman correlation matrix NS as well as that of the associated eigenvector. Around the strongest phases of the flood, the non-parametric LVD dimension density reaches a pronounced minimum, which is accompanied by major eigenvalues of only slightly below 1 indicating a completely synchronous evolution of the runoffs at all gauges. The corresponding eigenvectors change very fast and show in the most extreme cases a pattern where all records contribute equally. In general, an analysis of the eigenvalues and eigenvectors gives additional, in particular, spatial information about the state of the system, whereas dimension estimates have the advantage of being only one integrated parameter characterising the present conditions.

It has to be mentioned that the KLD-based dimension estimates detect only a common increase or decrease of runoffs, without any information about the corresponding amplitudes or the speed of these changes. Hence, they should be considered only as additional parameters for prediction purposes. However, if the number of considered gauges could be increased substantially, it is likely that the influence of geographical conditions is so large that a common behaviour of all records occurs only under very extreme conditions, i.e. is a potential indicator for an extreme event. In general, the meaning of the dimension estimates should be supported by other hydro-meteorological indicators, like strong or long-lasting precipitation or a strong increase of air temperatures in the presence of a significant snow volume.

14.6 Conclusion

We have used different approaches to quantify spatial correlations of hydro-meteorological observables in a river catchment. Mutual correlations may be quantified by a variety of linear as well as nonlinear approaches. Our results for runoff records from the Upper Main catchment area in southern Germany support earlier findings about the decay of correlations with an increasing distance between the considered locations. This holds for linear as well as nonlinear correlation measures.

For the considered time series, we have shown that correlation analysis yields significant time delays of the variability at different gauges if the spatial distance is larger than the scale defined by the typical flow velocity and the sampling interval of the data. Similar results have been found for temperature time series from stations which are separated on larger (continental) scales (see, e.g., [14.38, 14.39]).

As hydro-meteorological time series may strongly differ on rather small spatial scales due to the topography of the considered area, we have found that an analysis of ensemble correlations between the records from a whole set of nearby stations leads to more robust results. We have demonstrated that dimension estimates based on Karhunen–Loève decomposition (KLD) are well suited for quantifying such ensemble correlations. In particular, the linear variance decay (LVD) dimension density allows to detect successive changes of the strength of linear interrelationships between the components of a multivariate data set. An improvement of this approach is achieved by replacing the matrix of linear correlation coefficients with that of rank-order correlations in the KLD.

As an example, we have studied the behaviour of our measures during two severe floods in the considered area. We have demonstrated that our dimension estimates may be applied to very short data subsets. In particular, the presence of long-term auto-correlations in the runoff records is favourable for statistically significant results, as otherwise, a daily resolution of the data might cause our method to become insufficient to detect signatures of processes like floods which emerge on a time scale of only a few days.

KLD-based dimension estimates record changes of the hydrological conditions in a catchment during floods by an abrupt decay to rather small values, which is explained by a common behaviour of the discharges at all considered gauges. The significance and predictive skill of the corresponding results may be increased by improving the temporal resolution of the records and additionally considering temporally adjusted records which explicitly compensate delays in the underlying spatial correlation pattern. However, severe floods are no necessary condition for a common behaviour of all records; hence, the values and differences between the dimension estimates for different time intervals can hardly be used as a unique indicator for this kind of extreme events and should therefore be accompanied by other parameters.

Acknowledgments This work has been financially supported by the Volkswagen foundation (project no. I/78218) and the project "Extreme Events - Causes and Consequences (E2C2)" granted by the European Community (NEST contract no. 012975-E2C2). The data used in this study have been collected by the Bavarian Landesamt für Umwelt and provided by the Potsdam Institute for Climate Impact Research. Discussions with J. Kropp, M. Kallache, H. Rust, A. Witt, and J. Kurths are gratefully acknowledged. For the calculation of the cross-mutual information, we made use of an implementation based on the TISEAN package.

References

14.1. W.S. Broecker, Science **278**, 1582–1588 (1997)
14.2. P.U. Clark, N.G. Pisias, T.F. Stocker, A.J. Weaver, Nature **415**, 863–869 (2002)
14.3. E. Koscielny-Bunde, A. Bunde, S. Havlin, Y. Goldreich, Physica A **231**, 393–396 (1996)
14.4. E. Koscielny-Bunde, H.E. Roman, A. Bunde, S. Havlin, H.-J. Schellnhuber, Philos. Mag. B **77**, 1331–1340 (1998)
14.5. E. Koscielny-Bunde, A. Bunde, S. Havlin, H.E. Roman, Y. Goldreich, H.-J. Schellnhuber, Phys. Rev. Lett. **81**, 729–732 (1998)
14.6. A.A. Tsonis, P.J. Roebber, J.B. Elsner, Geophys. Res. Lett. **25**, 2821–2823 (1998)
14.7. A.A. Tsonis, P.J. Roebber, J.B. Elsner, J. Climate **12**, 1534–1541 (1999)
14.8. J. Pelletier D. Turcotte, Adv. Geophys. **40**, 91–166 (1999)
14.9. P. Talkner R.O. Weber, Phys. Rev. E **62**, 150–160 (2000)
14.10. R.B. Govindan, D. Vjushin, S. Brenner, A. Bunde, S. Havlin, H.-J. Schellnhuber, Physica A **294**, 239–248 (2001)
14.11. A. Bunde, S. Havlin, E. Koscielny-Bunde, H.-J. Schellnhuber, Physica A **302**, 255–267 (2001)
14.12. A. Király, I.M. Jánosi, Phys. Rev. E **65**, 051102 (2002)
14.13. R.B. Govindan, D. Vjushin, A. Bunde, S. Brenner, S. Havlin, H.-J. Schellnhuber, Phys. Rev. Lett. **89**, 028501 (2002)
14.14. D. Vjushin, R.B. Govindan, S. Brenner, A. Bunde, S. Havlin, H.-J. Schellnhuber, J. Phys. Condens. Matt. **14**, 2275–2282 (2002)
14.15. A. Bunde S. Havlin, Physica A **314**, 15–24 (2002)
14.16. A. Bunde, J. Eichner, R. Govindan, S. Havlin, E. Koscielny-Bunde, D. Rybski, D. Vjushin, in *Nonextensive Entropy - Interdisciplinary Applications*, ed. by M. Gell-Mann, C. Tsallis (Oxford University Press, New York, NY, 2003), pp. 207–220

14.17. R.A. Monetti, S. Havlin, A. Bunde, Physica A **320**, 581–589 (2003)
14.18. K. Fraedrich, R. Blender, Phys. Rev. Lett. **90**, 108501 (2003)
14.19. A. Bunde, J.F. Eichner, S. Havlin, E. Koscielny-Bunde, H.-J. Schellnhuber, D. Vyushin, Phys. Rev. Lett. **92**, 039801 (2004)
14.20. K. Fraedrich, R. Blender, Phys. Rev. Lett. **92**, 039802 (2004)
14.21. R. Blender, K. Fraedrich, Geophys. Res. Lett. **30**, 1769 (2003)
14.22. J.W. Kantelhardt, D. Rybski, S.A. Zschiegner, P. Braun, E. Koscielny-Bunde, V. Livina, S. Havlin, A. Bunde, Physica A **330**, 240–245 (2003)
14.23. J.F. Eichner, E. Koscielny-Bunde, A. Bunde, S. Havlin, H.-J. Schellnhuber, Phys. Rev. E **68**, 046133 (2003)
14.24. M. Pattantyús-Ábrahám, A. Király, I.M. Jánosi, Phys. Rev. E **69**, 021110 (2004)
14.25. A. Bunde, J.F. Eichner, S. Havlin, J.W. Kantelhardt, Physica A **342**, 308–314 (2004)
14.26. M.L. Kurnaz, Fractals **12**, 365–373 (2004)
14.27. D. Maraun, H.W. Rust, J. Timmer, Nonlin. Proc. Geophys. **11**, 495–503 (2004)
14.28. D. Vyushin, I. Zhidkov, S. Havlin, A. Bunde, S. Brenner, Geophys. Res. Lett. **31**, L10206 (2004)
14.29. R. Blender, K. Fraedrich, Geophys. Res. Lett. **31**, L22213 (2004)
14.30. A. Bunde, J.F. Eichner, J.W. Kantelhardt, S. Havlin, Phys. Rev. Lett. **94**, 048701 (2005)
14.31. A. Beretta, H.E. Roman, F. Raicich, F. Crisciani, Physica A **347**, 695–703 (2005)
14.32. E. Koscielny-Bunde, J.W. Kantelhardt, P. Braun, A. Bunde, S. Havlin, J. Hydrol. **322**, 120–137 (2006)
14.33. J.W. Kantelhardt, E. Koscielny-Bunde, D. Rybski, P. Braun, A. Bunde, S. Havlin, J. Geophys. Res. D **111**, 01106 (2006)
14.34. A.A. Tsonis, P.J. Roebber, Physica A **333**, 497–504 (2004)
14.35. W. von Bloh, M.C. Romano, M. Thiel, Nonlin. Proc. Geophys. **12**, 471–479 (2005)
14.36. A. Pikovsky, M. Rosenblum, J. Kurths, *Synchronization - A Universal Concept in Nonlinear Sciences* (Cambridge University Press, Cambridge, 2003)
14.37. D. Rybski, S. Havlin, A. Bunde, Geophys. Res. Abstr. **8**, 04918 (2006)
14.38. D. Rybski, S. Havlin, A. Bunde, Physica A **320**, 601–610 (2003)
14.39. R. Donner, Proc. NDES 2006, 26–29 (2006)
14.40. D. Maraun, J. Kurths, Geophys. Res. Lett. **32**, L15709 (2005)
14.41. P.J. Webster, S. Yang, Q. J. R. Meteorol. Soc. **118**, 887–926 (1992)
14.42. C. Torrence, P.J. Webster, J. Climate **12**, 2679–2690 (1999)
14.43. K.K. Kumar, B. Rajagopalan, M.A. Cane, Science **284**, 2156–2159 (1999)
14.44. S. Sarkar, R.P. Singh, M. Kafatos, Geophys. Res. Lett. **31**, L13209 (2004)
14.45. F.A. Huff, W.L. Shipp, J. Appl. Meteorol. **8**, 542–550 (1969)
14.46. F.A. Huff, Wat. Resour. Res. **6**, 254–260 (1970)
14.47. I. Zawadzki, J. Appl. Meteorol. **12**, 459–472 (1973)
14.48. R. Berndtsson, Water Res. Res. **24**, 1511–1517 (1988)
14.49. G. Sumner, C. Ramis, J.A. Guijarro, Theor. Appl. Climatol. **51**, 199–221 (1995)
14.50. M.F. Hutchinson, Agric. Forest Meteor. **73**, 237–264 (1995)
14.51. C. Uvo, R. Berndtsson, J. Geophys. Res. D **101**, 4221–4233 (1996)
14.52. E. Habib, W.F. Krajewski, G.J. Ciach, J. Hydromet. **2**, 621–629 (2001)
14.53. M. Mohapatra, U.C. Mohanty, S. Behera, Int. J. Climatol. **23**, 1867–1887 (2003)
14.54. A. Gaume, B. Sivakumar, M. Kolasinski, L. Hazoumé, J. Hydrol. **328**, 56–64 (2006)
14.55. M. Gebremichael, W. Krajewski, J. Appl. Meteor. **43**, 1180–1199 (2004)
14.56. A. Molini, P. La Barbera, L.G. Lanza, J. Hydrol. **322**, 89–104 (2006)
14.57. B. Sivakumar, J. Hydroinf. **7**, 175–184 (2005)
14.58. B.M. Troutman, M.R. Karlinger, Water Resour. Res. **39**, 1095 (2003)
14.59. M.C. Romano, M. Thiel, U. Schwarz, J. Kurths, H. Lange, M. Hauhs, Int. J. Bifurcation Chaos **11**, 2567–2578 (2001)
14.60. J.O. Skøien, G. Blöschl, Hydrol. Earth Syst. Sci. **10**, 645–662 (2006)
14.61. M.S. Santhanam, P.K. Patra, Phys. Rev. E **64**, 016102 (2001)

14.62. P. Termonia, Meteor. Atmos. Phys. **78**, 143–156 (2001)
14.63. R. Donner, A. Witt, Int. J. Bifurcation Chaos. **17**, 3685–3689 (2007)
14.64. R. Donner, A. Witt, Nonlin. Proc. Geophys. **13**, 485–497 (2006)
14.65. R. Donner, PhD thesis, University of Potsdam (2007)
14.66. D. Schwandt, PhD thesis, University of Potsdam (2003) (in german).
14.67. E. Barth, T. Molnar, H. Overland, KLIWA-Berichte **4**, 103–120 (2004) (in german).
14.68. K. Pearson, Phil. Trans. R. Soc. A **187**, 253–318 (1896)
14.69. W.C. Conover, *Practical Nonparametric Statistics*, 2nd ed. (Wiley, New York, NY, 1980)
14.70. W.H. Press, S.A. Teukolsky, W.T. Vetterling, B.P. Flannery, *Numerical Recipes in C - The Art of Scientific Computing* 2nd ed. (Cambridge University Press, Cambridge, 1992)
14.71. C. Spearman, Am. J. Psychol. **15**, 72–101 (1904)
14.72. M.G. Kendall, Biometrika **30**, 81–93 (1938)
14.73. D. Christensen, Comp. Stat. **20**, 51–62 (2005)
14.74. A.M. Fraser, H.L. Swinney, Phys. Rev. A **33**, 1134–1140 (1986)
14.75. I. Kojadinovic, Proc. ASMDA 2005, VII.16 (2005)
14.76. B. Pompe, J. Stat. Phys. **73**, 587–610 (1993)
14.77. J.-P. Eckmann, S.O. Kamphorst, D. Ruelle, D., Europhys. Lett. **4**, 973–977 (1987)
14.78. M. Thiel, M.C. Romano, J. Kurths, Phys. Lett. A **330**, 343–349 (2004)
14.79. J.P. Zbilut, C.L. Webber Jr., Phys. Lett. A **171**, 199–203 (1992)
14.80. C.L. Webber Jr., J.P. Zbilut, J. Appl. Physiol. **76**, 965–973 (1994)
14.81. N. Marwan, N. Wessel, U. Meyerfeldt, A. Schirdewan, J. Kurths, Phys. Rev. E **66**, 026702 (2002)
14.82. M. Thiel, M.C. Romano, P.L. Read, J. Kurths, Chaos **14**, 234–243 (2004)
14.83. J.P. Zbilut, A. Giuliani, C.L. Webber Jr., C.L., Phys. Lett. A **246**, 122–128 (1998)
14.84. M.C. Romano, M. Thiel, J. Kurths, W. von Bloh, Phys. Lett. A **330**, 214–223 (2004)
14.85. M. Holschneider, *Wavelets - An Analysis Tool* (Oxford University Press, Oxford, 1995).
14.86. E.N. Lorenz, *The Nature and Theory of the General Circulation of the Atmosphere* (World Meteorological Organization, Geneva, 1967)
14.87. K. Hasselmann, in *The Science of Disasters - Climate Disruptions, Heart Attacks, and Market Crashes*, edited by A. Bunde, J. Kropp, H.-J. Schellnhuber (Springer, Berlin, 2002), pp. 141–169
14.88. R. Kleeman, Physica D. **230**, 65–71 (2007)
14.89. A. Piotrowski, J.J. Napiórkowski, P.M. Rowiński, Nonlin. Proc. Geophys. **13**, 443–448 (2006)
14.90. R.W. Preisendorfer, *Principal Component Analysis in Meteorology and Oceanography* (Elsevier, Amsterdam, 1988)
14.91. T.F. Cox, M.A.A. Cox, *Multidimensional Scaling*, 2nd edn (Chapman and Hall, London, 2000)
14.92. G. Plaut, R. Vautard, J. Atmos. Sc. **51**, 210–236 (1994)
14.93. D.S. Broomhead, G.P. King, Physica D **20**, 217–236 (1986)
14.94. J. Tenenbaum, V. de Silva, J.C. Langford, Science **290**, 2319–2323 (2000)
14.95. S. Roweis, L. Saul, Science **290**, 2323–2326 (2000)
14.96. M.A. Kramer, Amer. Inst. Chem. Engin. Journal **37**, 233–243 (1991)
14.97. A. Hyvärinen, J. Karhunen, E. Oja, *Independent Component Analysis* (Wiley, New York, NY, 2001)
14.98. S.M. Zoldi, H.M. Greenside, Phys. Rev. Lett. **78**, 1687–1690 (1997)
14.99. S.M. Zoldi, J. Liu, K.M.S. Bajaj, H.S. Greenside, G. Ahlers, Phys. Rev. E **58**, 6903–6906 (1998)
14.100. M. Meixner, S.M. Zoldi, S. Bose, E. Schöll, Phys. Rev. E **61**, 1382–1385 (2000)
14.101. H. Varela, C. Beta, A. Bonnefort, K. Krischer, Phys. Rev. Lett. **94**, 174104 (2005)
14.102. S. Ciliberto, B. Nicolaenko, Europhys. Lett. **14**, 303–308 (1991)

Subject Index

Entries in this index are generally sorted with page number as they appear in the text. Page numbers that are marked in **bold face** indicate that the entry appears in a title or subheading. Page numbers that are marked in *italics* indicate that the entry appears in the caption of a table or figure.

J.P. Kropp, H.-J. Schellnhuber (eds.), *In Extremis*,
DOI 10.1007/978-3-642-14863-7,

MIX
Papier aus verantwortungsvollen Quellen
Paper from responsible sources
FSC® C105338
FSC
www.fsc.org

If you have any concerns about our products,
you can contact us on
ProductSafety@springernature.com

In case Publisher is established outside the EU,
the EU authorized representative is:
Springer Nature Customer Service Center GmbH
Europaplatz 3, 69115 Heidelberg, Germany

Printed by Libri Plureos GmbH
in Hamburg, Germany